T0295286

Modern Astrodynamics

Modern Astrodynamics

Editor: Jensen Reed

New York

Published by NY Research Press
118-35 Queens Blvd., Suite 400,
Forest Hills, NY 11375, USA
www.nyresearchpress.com

Modern Astrodynamics
Edited by Jensen Reed

International Standard Book Number: 978-1-63238-850-6 (Hardback)

Cataloging-in-Publication Data

Modern astrodynamics / edited by Jensen Reed.
p. cm.
Includes bibliographical references and index.
ISBN 978-1-63238-850-6
1. Astrodynamics. 2. Astronautics. 3. Dynamics. 4. Astrophysics. I. Reed, Jensen.
TL1050 .M63 2022
629.41--dc23

Contents

Preface

Astrodynamics is the term used to describe the application of ballistics and celestial mechanics to the problems concerned with the motion of rockets and other spacecraft. Ballistics is a mechanical field that deals with the science of designing and accelerating the projectiles to achieve the desired performance. Celestial mechanics is an astronomical branch that focuses on the motion of objects in outer space. Their motion is calculated on the basis of Newton's laws of motion and laws of universal gravitation. Numerous complex calculations related to escape velocities, trajectories and orbits take place within this field. While understanding the long-term perspectives of astrodynamics, this book makes an effort in highlighting its impact as a modern tool for the growth of the discipline. It unravels the recent studies in the field of astrodynamics. This book serves as a resource guide for both students and experts and contributes to the growth of the discipline.

This book is a result of research of several months to collate the most relevant data in the field.

When I was approached with the idea of this book and the proposal to edit it, I was overwhelmed. It gave me an opportunity to reach out to all those who share a common interest with me in this field. I had 3 main parameters for editing this text:

1. Accuracy – The data and information provided in this book should be up-to-date and valuable to the readers.

2. Structure – The data must be presented in a structured format for easy understanding and better grasping of the readers.

3. Universal Approach – This book not only targets students but also experts and innovators in the field, thus my aim was to present topics which are of use to all.

Thus, it took me a couple of months to finish the editing of this book.

I would like to make a special mention of my publisher who considered me worthy of this opportunity and also supported me throughout the editing process. I would also like to thank the editing team at the back-end who extended their help whenever required.

Editor

1

Optimal Control Scheme of the Tethered System for Orbital Transfer under a Constant Thrust

Liang Sun, Guowei Zhao, Hai Huang, and Ming Chen

Beihang University, Beijing 100191, China

Correspondence should be addressed to Guowei Zhao; zhaoguowei@buaa.edu.cn

Academic Editor: Mahmut Reyhanoglu

The tethered system with a long tether has its unique advantages in space environment exploration. With the development of requirement, the orbital transfer of the tethered system under a constant thrust and its related optimal control become significant and challenging. Three different optimal control schemes of the tethered system are proposed, including the tension control, the thrust vector control, and the mixed control. In the tension control, in order to ensure the smoothness of pendular motion of the tethered system, different cost functions are adopted and compared. In the thrust vector control, the constraint of thrust direction angle is fully considered. In the mixed control, equivalent conditions to other control schemes are investigated. The advantages and disadvantages of three optimal control schemes are compared and analyzed, which provides a reference for research on the optimal control problem of the tethered system under a constant thrust.

1. Introduction

In practical application, the tethered system has a broad application prospect in space environment exploration and space resource development [1]. This system is generally formed by a base satellite and a detector joined by a long tether. With the development of requirement, the global spatial distribution of physical parameters needs to be explored, which cannot be obtained only by deploying and retrieving the detector. Thus, the tethered system is facing the problem of orbital transfer under a thrust. In practical engineering, conventional orbital transfers can be utilized, such as the Hohmann orbital transfer and the Bielliptical orbital transfer [2]. During conventional orbital transfers, a certain windlass mechanism needs to be installed on the base satellite for damping out the pendular motion of the tethered system by adjusting the length of a tether, which is essentially consistent with the application of the traditional tethered system on the ellipse orbit. Whereas, this method actually has several limitations in applications. For example, it is not suitable for the tethered system with a short tether due to a sudden impact of impulse thrust; besides, if the windlass mechanism is broken, the pendular motion of the tethered system cannot be inhibited. Thus, a continuous thrust is more appropriate for the orbital transfer of the tethered system, and it can also be implemented to stabilize the attitude of the tethered system. Meanwhile, with the progress of small thrust (microthrust) technology, a small, continuous, and constant thrust becomes a practical and effective method, which possesses merits, such as low control requirements, high safety, and repeatability. Accordingly, proper control must be applied during orbital transfer subject to boundary conditions, cost functions, and path constraints. Therefore, the study on the optimal control of the tethered system under a constant thrust has an important theoretical and practical significance.

In recent five years, the requirement of space debris disposal becomes more urgent and important; thus, research on the tethered system with a thrust is mainly focused on the area of active debris removal (ADR). Aslanov et al. studied the dynamics of large debris connected to a space tug by a tether and also the chaotic motions of tethered satellites with low thrust [3–6]. Jasper and

Schaub explored an active debris removal method, which utilizes fuel reserves on a recently launched upper stage to rendezvous with and tether to debris [7]. Benvenuto et al. made dynamics analysis and an GNC design of flexible systems for space debris active removal [8]. Linskens and Mooij focused on the preliminary design of a guidance and control system to achieve space-debris removal as well as on the influence of different tether parameters on mission performance [9]. Sun et al. derived analytical solutions of the librational angles; besides, a method of hierarchical sliding mode tension control and a method of thrust control are presented to solve the control problems of the tethered system under a thrust, respectively [10–12]. Huang et al. studied the coupling dynamics modelling of a tethered space robot, explored a space tethered towing method which utilizes thrust to fulfill transfer, and bounded tension to stabilize tether heading; besides, Huang et al. studied the dynamics analysis and a controller design for maneuvering a tethered space net robot and also discussed the design, measurement, control, and experiment of a dexterous tethered space robot [13–16].

In these papers, a continuous thrust is normally implemented during active debris removal; meanwhile, a certain windlass mechanism can be adopted for damping out the pendular motion of the tethered system in orbit maneuvering. Nevertheless, boundary conditions, cost function, and path constraints of the tethered system are not fully considered in the design of a controller. Research on the optimal control of the tethered system under a thrust is limited. Cho and McClamroch proposed an optimal orbit transfer method in which a continuous thrust on the base satellite is controlled to achieve orbital transfer between two circular orbits [17]. In the research, the value of thrust is variable and the length of a tether is fixed; besides, linearized equations are adopted for the control design. Zhong and Zhu proposed a timescale separate optimal control of tethered space-tug systems for space debris removal, in which a Hohmann transfer orbit is implemented [2]. In practical engineering, optimal control methods adopting different actuators present different properties. Based on the previous research, the article is mainly focused on the optimal control of the tethered system under a small, continuous, and constant thrust. In the article, different optimal control schemes are proposed and studied; meanwhile, the advantages and disadvantages of methods are compared and analyzed. Moreover, the effect of cost function and path constraint on control performance is deeply analyzed.

2. Mathematical Model

Normally, the base satellite in practical engineering is within the range of 800~1200 kg. The detectors applied for space environment exploration include a three-dimensional imager, a γ-ray spectrometer, an X-ray spectrometer, and a microwave radiometer; hence, the mass of the detector is usually within the range of 50~150 kg. The Kevlar rope is widely used in the tethered system. The length of a tether is within the range of 1~5 km, and the linear density of a tether $\rho = 1\,\text{g/m}$. With respect to the mass of a detector, the mass of a tether is very small and then is treated as a rigid rod ignoring mass, elastic strain, and damping (a dumbbell model).

It should be noticed that there are several other kinds of tether dynamical models such as the lumped mass model and the beam model, in which the mass, elastic strain, and damping of a tether are fully considered; thus, the deformation of a tether can be described and it also leads to a high-order vibration, which is superimposed on the pendular motion of the tethered system. However, this kind of vibration is limited and does not affect the stability and control of the tethered system. Besides, the dumbbell model is helpful to improve the calculation efficiency of a controller. Thus, a dumbbell model is adopted to derive dynamics equations involved in the control problem (Figure 1).

2.1. Attitude Dynamics. The tethered system is formed by two-end masses joined by a rigid rod of length of l. The mass of the base satellite is M, and the mass of a detector is m. In addition, the base satellite actively drags the detector under a small, continuous, and constant thrust in an orbital plane.

Let $Ex'y'z'$ be the geocentric inertial frame. Its origin E is the mass center of the Earth. The axis Ex' points to the first Aries point, the axis Ey' is in the equatorial plane, and the coordinate system is right-hand oriented. Subsequently, let $Oxyz$ (Figure 1) be the orbital frame with origin at the mass center of the tethered system O. The Oz axis points to the mass center of the Earth E, the Ox axis is normal to the Oz axis in the orbital plane and points along the direction of an increasing polar angle α, and the coordinate system is right-hand oriented. Afterwards, let $Ox_oy_oz_o$ be the original frame with origin at the mass center O. The Oz_o axis points to the detector, and the noninertial frame is determined by one rotation from the orbital frame $Oxyz$. Following the rotations of angle θ about the Oy axis and angle ϕ about Ox_o axis, θ and ϕ are defined as in-plane and out-of-plane angles, respectively. Compared with the in-plane angle θ, the initial out-of-plane angle ϕ is very small; moreover, the coupling of in-plane angle θ and out-of-plane angle ϕ is not obvious during orbital transfer in an orbital plane; thus, for simplicity, the out-of-plane motion is not considered in the study.

Since the method of derivation is similar to that in [8–10], the process of derivation is omitted. According to the Euler-Lagrange equation, the attitude dynamics equations are derived as

$$\begin{aligned} \ddot{l} - l\left(\dot{\alpha} - \dot{\theta}\right)^2 + \frac{\mu l}{r^3}\left(1 - 3\cos^2\theta\right) &= \frac{F\sin(\gamma-\theta)}{M} - \frac{T(m+M)}{mM}, \\ l\left(\ddot{\alpha} = \ddot{\theta}\right) + 2\dot{l}\left(\dot{\alpha} - \dot{\theta}\right) - 3\frac{\mu l}{r^3}\sin\theta\cos\theta &= \frac{F\cos(\gamma-\theta)}{M}, \end{aligned} \tag{1}$$

where $\mu = 3.986 \times 10^5\,\text{km}^3/\text{s}^2$ is the gravitational constant of the Earth. r and α are the geocentric distance and the orbital polar angle of the mass center O, respectively. The thrust direction angle γ is the intersection angle from F_α to F. F is the value of thrust F, and F_α is the transversal component of F. Besides, T is the tether tension.

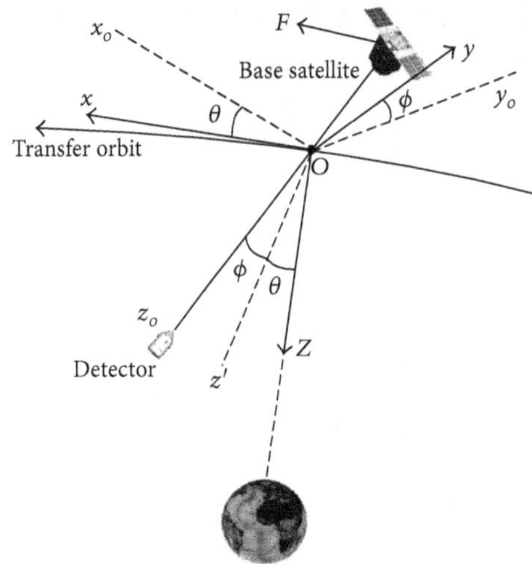

FIGURE 1: Description of a dumbbell-modeled tethered system.

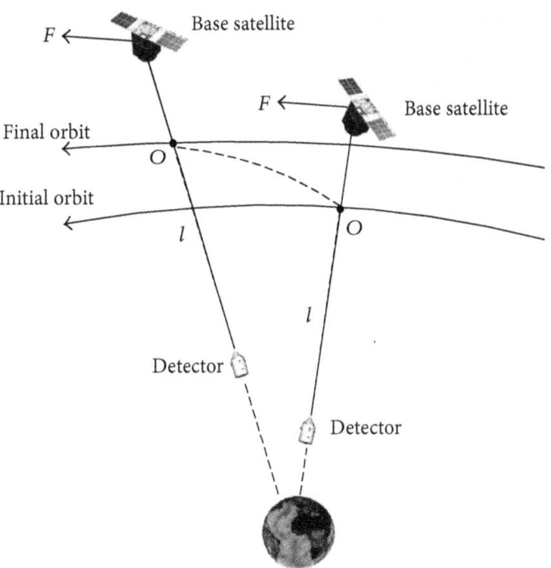

FIGURE 2: Description of an optimal control problem.

2.2. Orbit Dynamics. The thrust imposed on the base satellite always remains in the orbital plane. In practical engineering, the value of thrust F is within the range of 100 mN~10 N and the specific impulse I_s is in the range of 300 s~400 s. Besides, the transfer time usually reaches to 1~2 orbit periods; therefore, the fuel consumption approaches to 2 kg approximately during orbital transfer. With respect to the mass of the base satellite, the fuel consumption can be ignored.

Moreover, because the tethered system is applied for the space environment exploration in the low orbit, the aerodynamic forces can affect the librational motion of the tethered system, especially in a long term. However, the aerodynamic force on the tethered system is smaller than 10^{-5} N. During orbital transfer, the value of thrust F is within the range of 10^{-1} N~10 N and the gravity is greater than 10^{-1} N; hence, the aerodynamic force can be neglected with respect to the thrust and the gravity. The orbit dynamics equations of the tethered system in polar coordinate can be expressed as

$$\begin{aligned} \ddot{r} + \frac{\mu}{r^2} - \dot{\alpha}^2 r &= \frac{F \sin \gamma}{M+m}, \\ \ddot{\alpha} r + 2\dot{\alpha}\dot{r} &= \frac{F \cos \gamma}{M+m}. \end{aligned} \tag{2}$$

3. Nonlinear Optimal Control

3.1. Description of an Optimal Control Problem. During orbital transfer, the base satellite actively drags the detector under a small, continuous, and constant thrust F in the orbital plane. Initially, the attitude of the tethered system is stable without pendular motion, which means $\theta = \dot{\theta} = 0$. Then, the thrust F is imposed on the base satellite. The value of thrust is constant, but the direction of thrust can be controlled during orbital transfer; in addition, the tether tension T can also be adjusted. Finally, as the mass center O reaches to the expected circular orbit, the length of a tether comes back to the original value (Figure 2) and $\theta = \dot{\theta} = 0$.

3.2. Control Schemes. According to different actuators, optimal control schemes of tethered system under a small, continuous, constant thrust can be divided into three categories.

(1) Tension control: the transfer orbit of the tethered system is ascertained in advance, which is not related to the pendular motion of the tethered system. Besides, the thrust direction angle γ is decided by the motion of the mass center; thus, it becomes an independent problem of orbit transfer. During orbital transfer, the tether tension T is controlled to ensure the desired state of the tethered system at the final moment. Moreover, the smoothness of the pendular motion of the tethered system should be taken into consideration, which belongs to an optimal control problem of tether tension.

(2) Thrust vector control: during orbital transfer, the tether of the tethered system is with fixed length. The only control variable is the thrust direction angle γ, which is requested not only to achieve the orbital transfer between two circular orbits in the shortest possible time but also to guarantee the desired state of the tethered system at the final moment. Besides, to avoid the adverse effect of thrust on the base satellite, the constraint on thrust direction angle γ should be considered.

(3) Mixed control: except for the thrust direction angle γ, the tether tension can also be controlled. These two control variables are applied to achieve the orbital transfer and the desired state of the tethered system in the meantime.

3.3. Control Method. According to (1) and (2), the dynamic equations of the tethered system under a constant thrust in the orbital plane can be obtained as

$$\begin{aligned} \dot{x}_1 &= x_2, \\ \dot{x}_2 &= f_1(x,u), \\ \dot{x}_3 &= x_4, \\ \dot{x}_4 &= f_2(x,u), \\ \dot{x}_5 &= x_6, \\ \dot{x}_6 &= f_3(x,u), \\ \dot{x}_7 &= x_8, \\ \dot{x}_8 &= f_4(x,u), \end{aligned} \tag{3}$$

where $x_1 = l, x_2 = \dot{l}, x_3 = \theta, x_4 = \dot{\theta}, x_5 = r, x_6 = \dot{r}, x_7 = \alpha, x_8 = \dot{\alpha}$, and the state vector x is defined as $x = [x_1\ x_2\ x_3\ \ x_4\ x_5\ x_6\ x_7\ x_8]^T$; besides, $u_1 = \gamma$, $u_2 = T$, and the control vector u is defined as $u = [u_1 u_2]^T$. The nonlinear scalar functions of x and u are given by

$$\begin{aligned} f_1(x,u) &= x_1(x_4 - x_8)^2 - \mu x_1 x_5^{-3}\left(1 - 3\cos^2 x_3\right) \\ &\quad + FM^{-1}\sin(u_1 - x_3) - M^{-1}m^{-1}(M+m)u_2, \\ f_2(x,u) &= -2x_1^{-1}x_2(x_4 - x_8) - 2x_5^{-1}x_6x_8 - 3\mu x_5^{-3}\sin x_3\cos x_3 \\ &\quad + F(M+m)^{-1}x_5^{-1}\cos u_1 - Fm^{-1}x_1^{-1}\cos(u_1 - x_3), \\ f_3(x,u) &= -\mu x_5^2 + x_5 x_8^2 + F(M+m)^{-1}\cos u_1, \\ f_4(x,u) &= -2x_5^{-1}x_6x_8 + F(M+m)^{-1}x_5^{-1}\cos u_1. \end{aligned} \tag{4}$$

Equation (3) can be directly used for a controller design, as the mixed control is adopted. But if the tension control is implemented, γ will be ascertained in advance and the control variable $u = T$; besides, when the thrust vector control is applied, $\dot{l} = 0$, $\ddot{l} = 0$, and $u = \gamma$.

According to the description of the optimal control problem, the boundary conditions can be expressed as

$$\begin{aligned} x(t_0) &= \begin{bmatrix} l_0 & 0 & 0 & 0 & r_0 & 0 & \alpha_0 & \sqrt{\mu r_0^{-3}} \end{bmatrix}^T, \\ x(t_f) &= \begin{bmatrix} l_f & 0 & 0 & 0 & r_f & 0 & \alpha_f & \sqrt{\mu r_f^{-3}} \end{bmatrix}^T, \end{aligned} \tag{5}$$

where l_0 and l_f are the initial and final lengths of a tether, respectively; besides, $l_0 = l_f = l_c$ and l_c is the original length of a tether. r_0 and r_f are the geocentric distances of initial and final circular orbits, respectively. α_0 and α_f are the initial and final polar angles, respectively.

The inequality path constraint is written as

$$C(x,u,t) \le 0. \tag{6}$$

In addition, the cost function (Lagrange type) can be expressed as

$$J(x,u) = \int_0^{t_f} L(x,u,t)\,dt. \tag{7}$$

In the study, a MATLAB software, GPOPS (general pseudospectral optimization software) [18], is adopted for solving the optimization control problem. The method employed by GPOPS is the Radau pseudospectral method (RPM) where the collocation points are the Legendre-Gauss-Radau (LGR) points; besides, the hp-adaptive approach is implemented to solve the NLP (nonlinear programming) problem [18].

4. Case Study

In this section, three different optimal control schemes are studied, respectively; besides, the effects of the cost function $J(x,u)$ and the path constraint $C(x,u,t)$ on the control performance are analyzed.

4.1. Tension Control Scheme. In the tension control scheme, the thrust direction angle γ and the orbital trajectory of the mass center are ascertained in advance. Taking the orbital transfer between two circular orbits as an example, different orbital transfer methods can be adopted, such as an optimized orbital transfer or a transversal orbital transfer. In order to study the optimization of the tether tension, a transversal orbital transfer is implemented, in which the thrust direction angle $\gamma \equiv 0$ during orbital transfer.

According to the transversal orbital transfer, a small, continuous, and constant thrust is imposed on the base satellite and along the transversal direction of the mass center O; then, the polar angle α and the geocentric distance r can be expressed as [8]

$$\begin{aligned} \alpha &\approx \alpha_0 + wt - \frac{3f_\alpha}{2r_0}t^2 - \frac{4f_\alpha}{w^2 r_0}\cos wt + \frac{4f_\alpha}{w^2 r_0}, \\ r &\approx r_0 + \frac{2f_\alpha}{w}t - \frac{2f_\alpha}{w^2}\sin wt, \end{aligned} \tag{8}$$

where $\omega = \sqrt{\mu r_0^{-3}}$. As $\alpha - \alpha_0 = 2n\pi (n = 1, 2 \ldots)$, the orbital trajectory of the mass center returns to an another circular orbit; meanwhile, the transfer time t_f and the transversal thrust acceleration f_α can be derived as

$$\begin{aligned} t_f &= \frac{2n\pi}{w}, \\ f_\alpha &= \frac{w^2(r_f - r_0)}{4n\pi}. \end{aligned} \tag{9}$$

In practical engineering, the mass of the base satellite is within the range of 800~1200 kg and the mass of the detector is within the range of 50~150 kg. The tethered system is mainly utilized to explore the space environment in a low Earth orbit; normally, the orbit altitude is within the range of 400~800 km. Because the length of a tether is within a limited range of 1~5 km, the tethered system can only obtain the spatial environment data in the vicinity by deploying and retrieving the detector. In order to obtain the global spatial

TABLE 1: System parameters of the tethered system under tension control.

Parameters	Values
Number of orbit laps n	1
Geocentric distance of the initial circular orbit r_0, km	7000
Geocentric distance of the final circular orbit r_f, km	7010
Transversal thrust acceleration, m/s^2	9.25×10^{-4}
Transfer time t_f, s	5835
Mass of the base satellite M, kg	1000
Mass of the detector m, kg	50
Original length of the tether l_c, m	1000
Initial value of the in-plane angle θ_0, °	0
Final value of the in-plane angle θ_f, °	0
Initial value of the in-plane angle θ'_0, °/s	0
Final value of the in-plane angle θ'_f, °/s	0
Initial value of tether tension T_0, N	$3\mu l_0 r_0^{-3}$
Final value of tether tension T_f, N	$3\mu l_0 r_f^{-3}$

data, the orbit altitude of the tethered system is requested to be increased by at least 10 km. Without loss of generality, the geocentric distance of the initial circular orbit is set to 7000 km and the geocentric distance of the final circular orbit is set to 7010 km in the case study. In addition, the other simulation parameters of the tethered system are listed in Table 1.

In the GPOPS [18], the node number $N_k \in [6, 12]$ and the tolerance is smaller than 10^{-6}; besides, the tension tether $T > 0$. In order to ensure the smoothness of the pendular motion of the tethered system during orbital transfer, the Lagrange function $L = k_l \dot{l}^2 + k_\theta \dot{\theta}^2$ is adopted. As $k_l \neq 0$ and $k_\theta = 0$, the optimal controller is designed to minimize the variation of tether length. And if $k_l = 0$ and $k_\theta \neq 0$, the variation of the in-plane angle is requested to be the minimum. Besides, when $k_l \neq 0$ and $k_\theta \neq 0$, both of $\dot{l}^2$ and $\dot{\theta}^2$ are integrated into account in the Lagrange function. Considering the magnitudes of $\dot{l}^2$ and $\dot{\theta}^2$ and without loss of generality, the Lagrange function is selected as $L = \dot{l}^2$, $L = \dot{\theta}^2$, and $L = 1 \times 10^{-3} \dot{l}^2 + 1 \times 10^3 \dot{\theta}^2$, respectively.

As shown in Figure 3, the tethered system returns to the equilibrium position after completing orbital transfer. The curves of l and θ present approximate symmetrical distributions. Without constraints on l and θ, when the Lagrange function $L = \dot{l}^2$, the length of the tether l changes within the reasonable range of 400~1000 m and meanwhile the in-plane angle θ varies in the range of −48°~0°. Though the variation of θ is large with respect to $L = \dot{\theta}^2$, the curve of θ is smooth and the flight safety can be ensured during orbital transfer. Inversely, if the Lagrange function $L = \dot{\theta}^2$, the in-plane angle varies in a small range of −18°~0°, but the length of the tether l reaches to 3000 m and $\dot{l}$ attains to ±5 m/s; then, it is negative for the stability of the tethered system and also brings difficulty to the mechanism. Compared with $L = \dot{\theta}^2$, the Lagrange function $L = \dot{l}^2$ is conductive to the smoothness of pendular motion and the flight safety. Besides, in order to further improve the performance of optimal control, two methods can be adopted. In the first method, the constraints on l and θ can be considered, but it may lead to a sudden change of the tether, which is not easy to realize in practical engineering. In the second method, both of $\dot{l}^2$ and $\dot{\theta}^2$ are integrated into account in the Lagrange function. For instance, as the Lagrange function $L = 1 \times 10^{-3} \dot{l}^2 + 1 \times 10^3 \dot{\theta}^2$, with respect to $L = \dot{l}^2$, the variation of in-plane angle θ is reduced to the range of −35°~0° and meanwhile the curve of l changes a little. In general, this method is more suitable in engineering applications.

Remark 1. In order to improve the calculation efficiency of the controller, the dumbbell model is adopted in this paper. Whereas, in practical engineering, though the effect of tether vibration on the pendular motion is limited, several effects associated with the tether flexibility, elasticity, and the detector rotation, such as "bounce back" and "tail wagging," should be taken into account. In order to soften this problem, two methods can be adopted. One method is to introduce the lumped model instead of the dumbbell model, which can improve the control accuracy; but even so, the qualitative trends, results, and conclusions are consistent. The other method is to guarantee the good initial condition of the tether by utilizing the windlass mechanism; then, the "bounce back" phenomenon can be avoided.

Remark 2. In the tension control scheme, the tension is defined as the control input, which is provided by a windlass mechanism; whereas, there is no actuator that can directly provide the desired time-varying tension signal. In practical engineering, the tension of the tether can be measured by a force sensor and transmitted to a motor controller. The controller drives the motor to operate, and then the desired tension of the tether can be provided. It should be noticed that the error, delay, and transmission of tether tension ought to be fully considered in the feedback controller.

Remark 3. GPOPS generates the open-loop control, and the control inputs are obtained online. In order to ensure the closed-loop stability and the robustness to disturbances, the real-time merits of the presented schemes can be further exploited to design a closed-loop controller based on more advanced control design techniques such as model-predictive control.

4.2. Thrust Vector Control. In the thrust vector control scheme, the thrust direction angle γ can be adjusted, but the length of the tether is with fixed length. In order to make a comparison, the value of thrust acceleration is still selected as 9.25×10^{-4} m/s^2 and other simulation parameters are consistent with Table 1. The Lagrange function L is set as 1 and $\dot{\theta}^2$, respectively. When $L = 1$, the transfer time is requested to

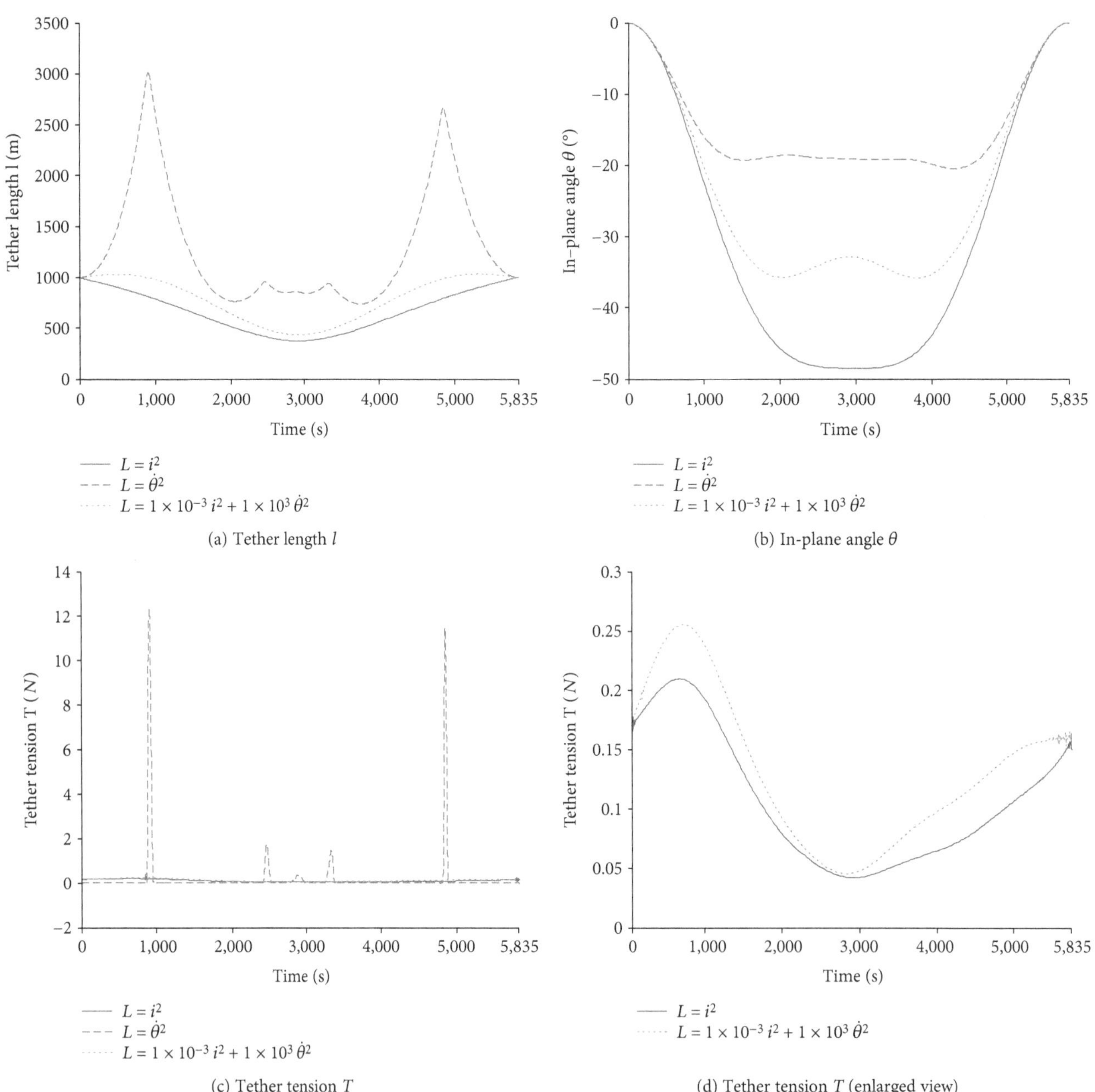

FIGURE 3: Time responses of the tethered system under tension control.

be shortest; when $L = \dot{\theta}^2$, the variation of the in-plane angle is requested to be the minimum.

As shown in Figure 4, when the Lagrange function $L = 1$, the thrust direction angle γ changes in the range of $-15^\circ \sim 55^\circ$. Obviously, as γ is utilized to control the orbital transfer and the pendular motion at the same time, the transfer time t_f between two circular orbits is 6214 s, which is longer than the transfer time of the transversal thrust transfer orbit (5835 s). Meanwhile, the in-plane angle θ varies within the range of $-30^\circ \sim 0^\circ$ and the tether tension maintains in the range of 0.02~0.12 N. Though the variation of θ is large with respect to $L = \dot{\theta}^2$, the curve of γ is smooth and the controller can be realized in practical engineering. To be opposite, if the Lagrange function $L = \dot{\theta}^2$, the in-plane angle varies in a small range of $-7^\circ \sim 0^\circ$, but the transfer time reaches to 16,419 s and γ changes dramatically, which goes against the flight safety during orbital transfer.

Thus, $L = \dot{\theta}^2$ is not suitable for the optimal control of the tethered system during orbital transfer.

Furthermore, in order to avoid the adverse effect of thrust on the base satellite, the constraint on thrust direction angle γ is considered when $L = 1$; then, the constraints of $|\gamma| \leq 40^\circ$, $|\gamma| \leq 50^\circ$, and $|\gamma| \leq 60^\circ$ are selected, respectively. As shown in Figure 5, if $|\gamma| \leq 60^\circ$, the inequality path constraint does not affect the optimized result. Whereas, if more strict constraint

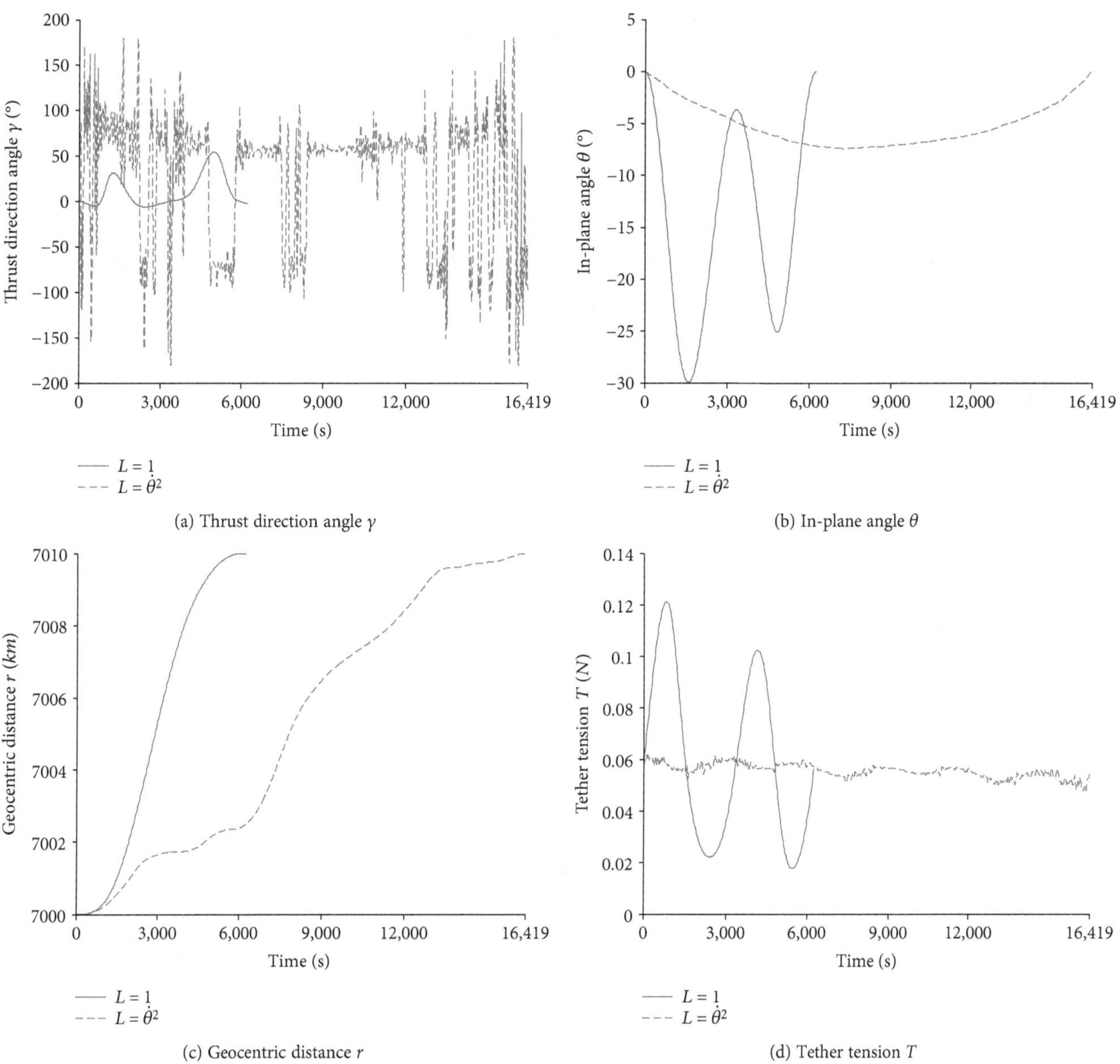

(a) Thrust direction angle γ

(b) In-plane angle θ

(c) Geocentric distance r

(d) Tether tension T

Figure 4: Time responses of the tethered system under thrust control.

of $|\gamma| \leq 40°$ is considered, the transfer time rises to 6228 s, but θ and T change a little; thus, proper constraint on γ is effective and conducive to orbital transfer. However, if the threshold value of $|\gamma|$ continues to decrease or even tends to zero, γ will switch back and forth and reasonable optimal solutions cannot be obtained. Moreover, in order to study the effect of the thrust on the optimal control, the thrust accelerations $f = 9 \times 10^{-4}\,\mathrm{m/s^2}$, $f = 9.25 \times 10^{-4}\,\mathrm{m/s^2}$, and $f = 10 \times 10^{-4}\,\mathrm{m/s^2}$ are selected, respectively (Figure 6). Accordingly, the final moments of orbital transfer are 5909 s, 6214 s, and 6358 s, respectively. Obviously, when the thrust acceleration f increases, the transfer time t_f is shortened; whereas, the variations of the in-plane angle θ and the tether tension T increase. In practical engineering, the value of thrust is decided by the specific task, but the trends of time responses present were similar.

4.3. Mixed Control. In the mixed control scheme, both γ and T are implemented to solve the problem of optimal control. Similarly, the value of thrust acceleration is selected as $9.25 \times 10^{-4}\,\mathrm{m/s^2}$ and other simulation parameters are consistent with Table 1; besides, the tension tether $T > 0$.

At first, the Lagrange function L is set to 1. As shown in Figure 7, when the transfer time is requested to be the shortest, the thrust is implemented only for orbit optimization; thus, the transfer time t_f decreases to 5821 s and the trajectory of the orbit is not related to the pendular motion of the tethered system. In other words, when

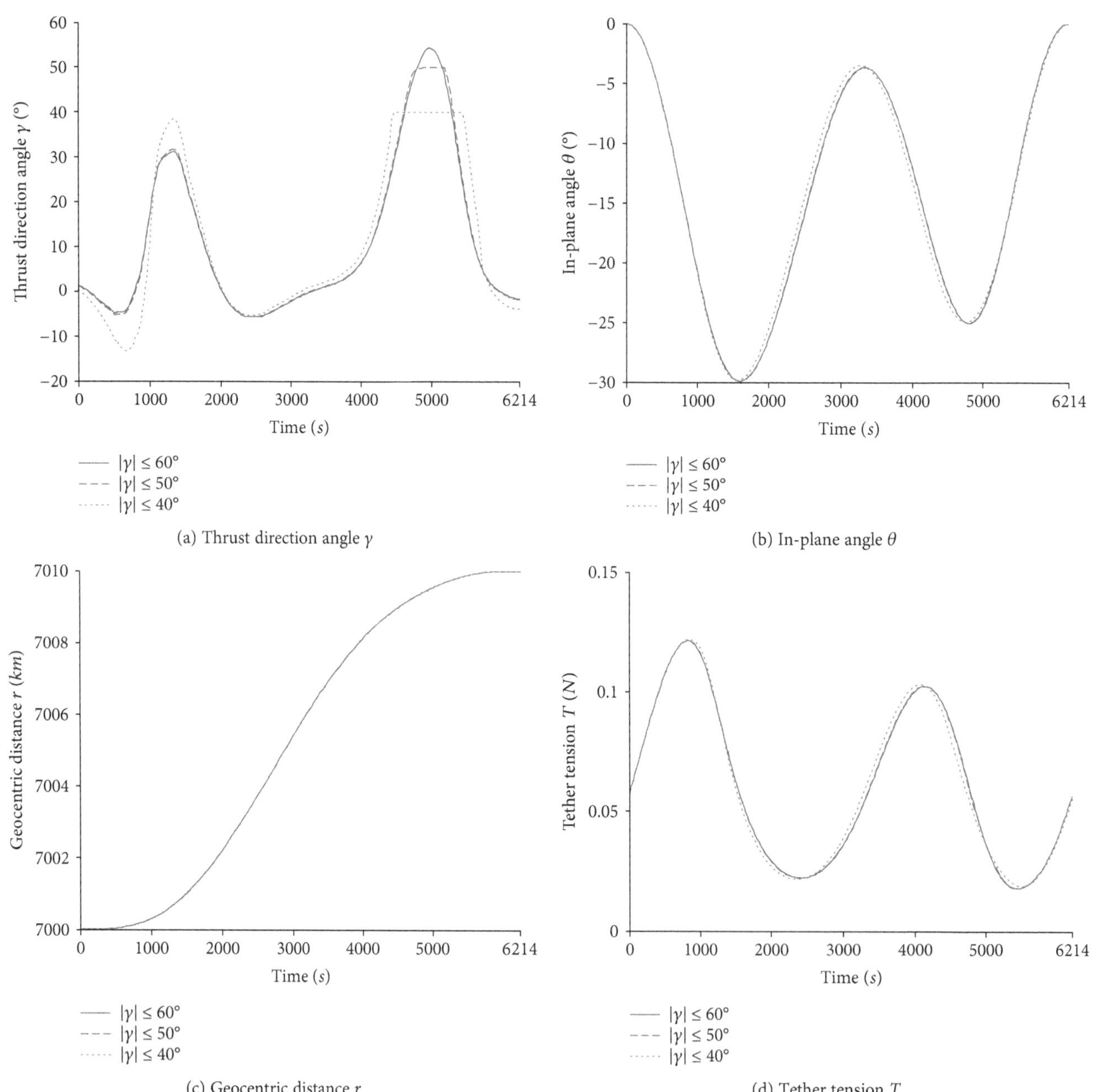

(a) Thrust direction angle γ

(b) In-plane angle θ

(c) Geocentric distance r

(d) Tether tension T

FIGURE 5: Time responses of the tethered system with different constraints on γ.

$L = 1$, the mixed control is same as the tension control in essence except for the difference that a transversal thrust transfer orbit is replaced by an optimized transfer orbit. During this optimized orbital transfer, the tethered system can be viewed as a spacecraft with no tethers.

Moreover, the Lagrange function L is set to $\dot{l}^2$, $\dot{\theta}^2$, and $1 \times 10^{-2}\dot{l}^2 + 1 \times 10^{2}\dot{\theta}^2$, respectively. As shown in Figure 8, when $L = \dot{l}^2$, the tether length is fixed at 1000 m and the transfer time attains to 6932 s; meanwhile, the in-plane angle θ varies within the range of $-21^\circ{\sim}0^\circ$ and the tether tension changes mildly. Actually, when $L = \dot{l}^2$, the mixed control is almost same as the thrust vector control but the transfer time is not requested to be the shortest. If the Lagrange function $L = \dot{\theta}^2$, the in-plane angle θ is limited to $-2^\circ{\sim}0^\circ$; however, the length of the tether reaches to 10,000 m, the transfer time approaches to 14,737 s, and the thrust direction angle γ changes seriously. Obviously, $L = \dot{l}^2$ is adverse to the orbital transfer. If $L = 1 \times 10^{-2}\dot{l}^2 + 1 \times 10^{2}\dot{\theta}^2$, the in-plane angle θ decreased to the range of $-17^\circ{\sim}0^\circ$ with respect to $L = \dot{l}^2$; in the meantime, the transfer time t_f changes to 7360 s and the variation of tether length is only 1.5 m. Overall, the mixed control has both advantages of tension control and thrust vector control; besides, the performance can be improved by adjusting the Lagrange function. Therefore, the mixed control is more practical and effective in engineering.

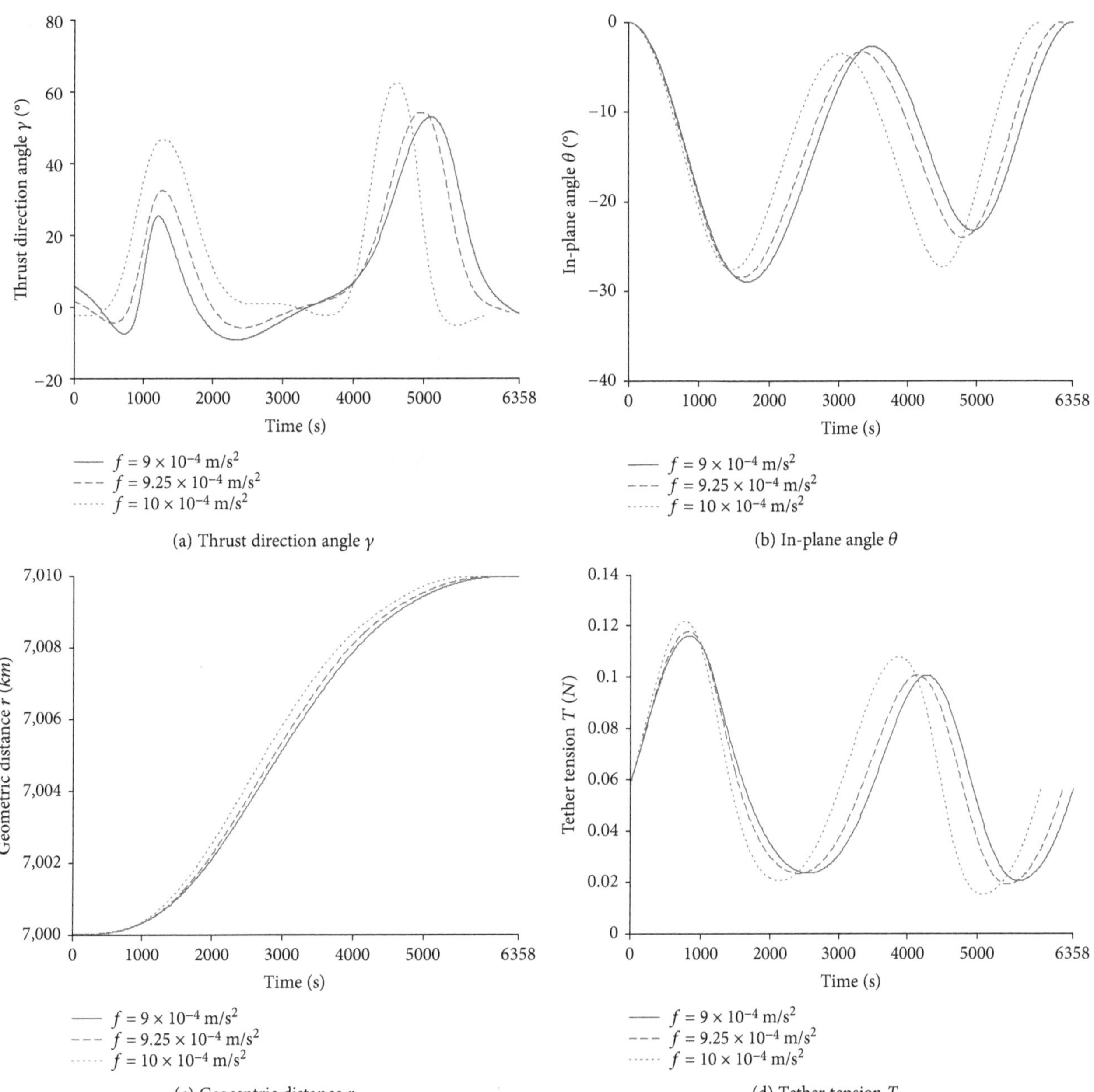

FIGURE 6: Time responses of the tethered system with different thrust accelerations.

5. Conclusion

Based on a series of analyses and simulations of optimal control of the tethered system, the following conclusions can be drawn: in practical engineering applications, all three optimal control schemes possess their own merits. In terms of the tension control, the orbit trajectory of the mass center of the tethered system is ascertained in advance without regard to the pendular motion of the tethered system, so typical transfer orbits can be implemented, such as the time-optimal transfer orbit, the transversal thrust transfer orbit, and the Hohmann transfer orbit. Furthermore, the smoothness of pendular motion can be ensured by adopting a proper cost function. In the tension control, the windlass mechanism can help avoid the slackness of the tether. Accordingly, the reliability of the windlass mechanism needs to be further verified in space applications. In the thrust vector control scheme, the only control variable is the thrust direction angle γ, which is adopted for optimizations of both orbit and attitude of the tethered system. This method is convenient due to its simple configuration of the actuator, but the fuel consumption is higher with respect to the tension control; besides, though the constraint of γ is considered, the magnitude of variation of thrust direction angle γ is still large, which will increase the control difficulty of the base satellite. As far as the mixed control is concerned, two control

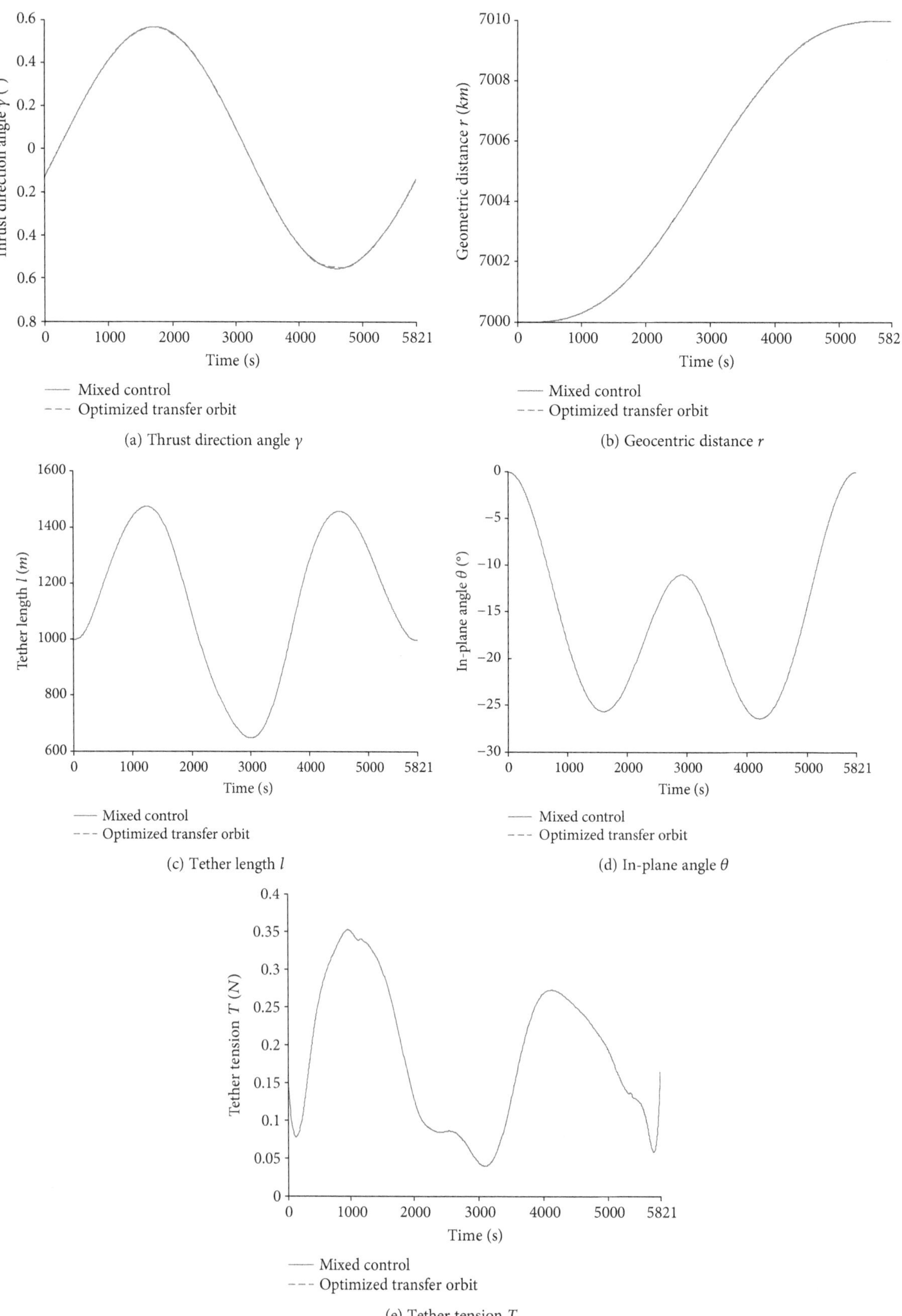

FIGURE 7: Time optimal control of the tethered system under mixed control.

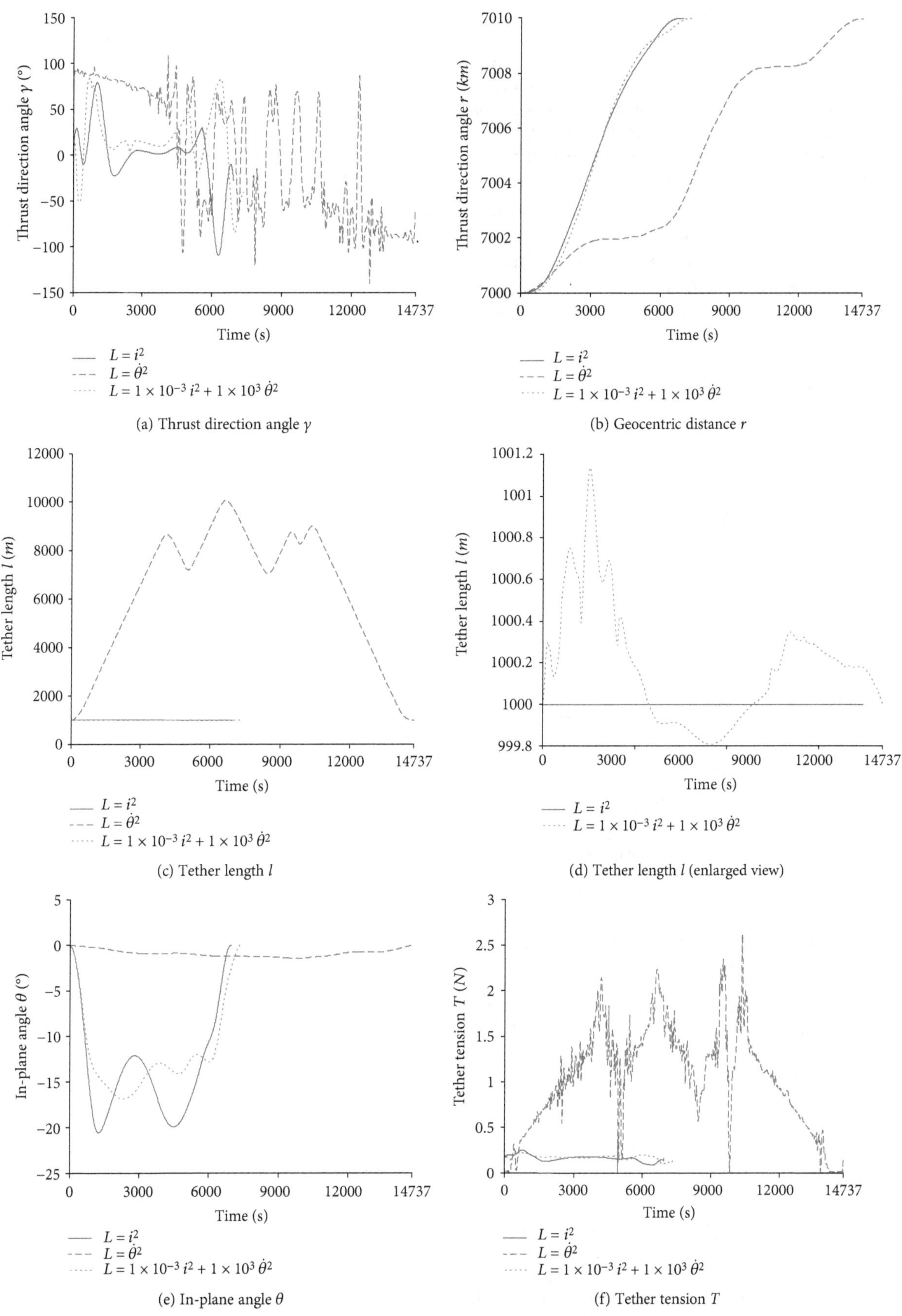

(a) Thrust direction angle γ

(b) Geocentric distance r

(c) Tether length l

(d) Tether length l (enlarged view)

(e) In-plane angle θ

(f) Tether tension T

FIGURE 8: Time responses of the tethered system under mixed control.

variables T and γ are employed; hence, this method combines the advantages of both tension control and thrust vector control on the performance and the practicability. The mixed control is best to be utilized as the main control mode; whereas, as the fuel consumption is limited or the windlass mechanism is broken, the control mode can be switched to the tension control mode or thrust vector control mode. According to the optimal performance, the Lagrange function $L = \dot{\theta}^2$ is not suitable for optimal control; inversely, $L = k_l \dot{l}^2 + k_\theta \dot{\theta}^2$ is conducive to the orbital transfer, but the values of k_l and k_θ need to be further studied in practical engineering and as well as the threshold value of γ. This paper provides a reference for the research on the optimal control problem of the tethered system under a constant thrust.

Conflicts of Interest

There is no conflict of interest in the manuscript.

Acknowledgments

This research is funded by the National Natural Science Foundation of China under Grant no. 11602008 and Grant no. 11572016, which do not lead to any conflict of interests regarding the publication of this manuscript.

References

[1] Y. Chen, R. Huang, L. He, X. Ren, and B. Zheng, "Dynamical modelling and control of space tethers: a review of space tether research," *Nonlinear Dynamics*, vol. 77, no. 4, pp. 1077–1099, 2014.

[2] R. Zhong and Z. H. Zhu, "Timescale separate optimal control of tethered space-tug systems for space-debris removal," *Journal of Guidance, Control, and Dynamics*, vol. 39, no. 11, pp. 2540–2545, 2016.

[3] V. S. Aslanov and V. V. Yudintsev, "Dynamics of large debris connected to space tug by a tether," *Journal of Guidance, Control, and Dynamics*, vol. 36, no. 6, pp. 1654–1660, 2013.

[4] V. Aslanov and V. Yudintsev, "Dynamics of large space debris removal using tethered space tug," *Acta Astronautica*, vol. 91, no. 10, pp. 149–156, 2013.

[5] V. Aslanov, A. Misra, and V. Yudintsev, "Chaotic motions of tethered satellites with low thrust," in *67th International Astronautical Congress (IAC)*, Guadalajara, Mexico, September 2016.

[6] V. Aslanov and V. Yudintsev, "Dynamics of large space debris removal using tethered space tug," *Acta Astronautica*, vol. 91, pp. 149–156, 2013.

[7] L. Jasper and H. Schaub, "Input shaped large thrust maneuver with a tethered debris object," *Acta Astronautica*, vol. 96, no. 4, pp. 128–137, 2014.

[8] R. Benvenuto, S. Salvi, and M. Lavagna, "Dynamics analysis and GNC design of flexible systems for space debris active removal," *Acta Astronautica*, vol. 110, pp. 247–265, 2015.

[9] H. T. K. Linskens and E. Mooij, "Tether dynamics analysis and guidance and control design for active space-debris removal," *Acta Astronautica*, vol. 107, pp. 150–162, 2015.

[10] G. Zhao, L. Sun, S. Tan, and H. Huang, "Librational characteristics of a dumbbell modeled tethered satellite under small, continuous, constant thrust," *Proceedings of the Institution of Mechanical Engineers, Part G: Journal of Aerospace Engineering*, vol. 227, no. 5, pp. 857–872, 2013.

[11] L. Sun, G. Zhao, and H. Huang, "Stability and control of tethered satellite with chemical propulsion in orbital plane," *Nonlinear Dynamics*, vol. 74, no. 4, pp. 1113–1131, 2013.

[12] G. Zhao, L. Sun, and H. Huang, "Thrust control of tethered satellite with a short constant tether in orbital maneuvering," *Proceedings of the Institution of Mechanical Engineers, Part G: Journal of Aerospace Engineering*, vol. 228, no. 14, pp. 2569–2586, 2014.

[13] P. Huang, Z. Hu, and Z. Meng, "Coupling dynamics modelling and optimal coordinated control of tethered space robot," *Aerospace Science and Technology*, vol. 41, pp. 36–46, 2015.

[14] B. S. Wang, Z. J. Meng, and P. F. Huang, "A towing orbit transfer method of tethered space robots," in *2015 IEEE International Conference on Robotics and Biomimetics (ROBIO)*, pp. 964–969, Zhuhai, China, December 2015.

[15] Z. Meng, B. Wang, and P. Huang, "A space tethered towing method using tension and platform thrusts," *Advances in Space Research*, vol. 59, no. 2, pp. 656–669, 2017.

[16] F. Zhang and P. Huang, "Releasing dynamics and stability control of maneuverable tethered space net," *IEEE/ASME Transactions on Mechatronics*, vol. 22, no. 2, pp. 983–993, 2017.

[17] S. Cho and N. H. McClamroch, "Optimal orbit transfer of a spacecraft with fixed length tether," *Journal of Astronautical Sciences*, vol. 51, no. 2, pp. 195–204, 2003.

[18] C. L. Darby, W. W. Hager, and A. V. Rao, "An *hp*-adaptive pseudospectral method for solving optimal control problems," vol. 32, no. 4, pp. 476–502, 2011.

2

The Influence of Orbital Element Error on Satellite Coverage Calculation

Guangming Dai, Xiaoyu Chen, Mingcheng Zuo, Lei Peng, Maocai Wang, and Zhiming Song

School of Computer Science, Hubei Key Laboratory of Intelligent Geo-Information Processing, China University of Geosciences, Lumo Road, Wuhan 430074, China

Correspondence should be addressed to Xiaoyu Chen; cxymail@126.com

Academic Editor: Enrico C. Lorenzini

This paper studies the influence of orbital element error on coverage calculation of a satellite. In order to present the influence, an analysis method based on the position uncertainty of the satellite shown by an error ellipsoid is proposed. In this error ellipsoid, positions surrounding the center of the error ellipsoid mean different positioning possibilities which present three-dimensional normal distribution. The possible subastral points of the satellite are obtained by sampling enough points on the surface of the error ellipsoid and projecting them on Earth. Then, analysis cases are implemented based on these projected subastral points. Finally, a comparison report of coverage calculation between considering and not considering the error of orbital elements is given in the case results.

1. Introduction

The calculation of satellite and constellation coverage of ground targets is a basic problem of Earth observation systems. Various kinds of satellites and constellations are available and selectable according to different requirements of mission design. As the basic purposes of constellation design, continuous coverage and maximum regional coverage are often considered. Hence, many proposed methods were devoted to how to optimize these two goals. Cote [1] offered tips on how to maximize satellite coverage while cruising in the high-latitude West Coast waters of British Columbia. Yung and Chang [2] proposed a method for maximizing satellite coverage at predetermined local times for a set of predetermined geographic locations. Draim [3] described a new four-satellite elliptic orbit constellation giving continuous line-of-sight coverage of every point on Earth's surface.

During the implementation of a mission, the visible time windows of important targets are also usually considered. For example, Lian et al. [4] proposed two static and dynamic models of observation toward Earth by agile satellite coverage to decompose area targets into small pieces and compute visible time window subarea targets. Wang et al. [5] analyzed four parameters to determine the coverage area of satellite. It also introduced and compared two current test methods of satellite area, which presents the composition of the required test equipment and test processes, and proved the effectiveness of the tests through engineering practices. Besides the numerical analysis methods, heuristic search algorithms such as particle swarm optimization [6] were also applied to the satellite coverage calculation. Huang and Dai [7] proposed an optimal design of constellation of multiple regional coverage based on NSGA-II with the aid of multiobjective optimization.

The error influence of orbital elements on the satellite positioning was mainly considered in the field GIS and radar location [8–10]. Analysis methods proposed in these papers usually employed a model that can transform the possible position of a satellite to the points in an error ellipsoid. Nelson [11] introduced and discussed some basic properties of the error ellipsoid. For example, the basic parameters of the error ellipsoid were depicted by the axial direction and the axial ratio among each axis. In the study of satellite positioning error, the error ellipsoid was used in the positioning accuracy analysis of global navigation satellite system [12]. Some papers [13–15] introduced that the axis ratio of error ellipsoid for GPS should be controlled within a certain range. Wang et al. [16] selected eight sites to study the different shapes of the error ellipsoid in the global range. Alfano [17] showed the effects of positional uncertainty on the Gaussian probability computation for orbit conjunction. A model of position uncertainty for the TIS-B System was proposed [18]. Shi et al. [19] introduced a distributed location model and algorithm with position uncertainty. Trifonov [20] introduced position uncertainty measures for a particle on the sphere.

However, for coverage calculation, the influence of the orbital element error has not been considered, which is the research purpose of this paper. Because only the error influence of initial orbital elements on coverage calculation is studied in this paper, hence, the influence of other variables, such as perturbation, are not considered here. Because of the nonlinear function between satellite position and orbital elements, assuming the error of orbital elements presents normal distribution, therefore, the uncertainty of the satellite position will also present normal distribution in the three-dimensional space. The relationship between the error of satellite position and orbital element error can be obtained according to the knowledge of error propagation and the function relationship between the satellite position and orbital elements, which is the basis of coverage uncertainty analysis.

The rest of this paper is organized as follows. In Section 2, the proposed method is introduced. In Section 3, preparations for cases are given. In Section 4, five analysis cases are implemented. Section 5 is devoted to the analysis of experimental results and summary of this paper. Section 6 gives some conclusions.

2. Coverage Uncertainty Related to Orbital Elements

The uncertainty analysis method proposed in this paper includes four steps, which are summarized as follows:

Step 1: obtain the position vector covariance matrix based on the given orbital element error matrix and transformation matrix.

Step 2: calculate the eigenvalues and eigenvectors of the position vector covariance matrix. The eigenvalue is the axial length of the error ellipsoid, and the eigenvector is the axial direction of the error ellipsoid. Then, perform the rotation and translation operations on the error ellipsoid.

Step 3: extract the edge points of the error ellipsoid projection on the surface of Earth.

Step 4: obtain the coverage analysis results to ground targets based on the extracted edge points.

In the above four steps, the implementations of steps 1 and 2 are to obtain the necessary parameters about error ellipsoid. Steps 3 and 4 are utilized to project all the points of error ellipsoid surface on Earth for the further coverage calculation. Process simulation presented in Figure 1 shows that each extracted point of the projection edge is regarded as the worst position of satellite positioning, and all the analysis work are implemented based on these points. The method employed to extract the edge points is a convex hull algorithm shown in Table 1. The analysis method metioned in step 4 is introduced in Table 2 with more details. Sections 2.1 and 2.2 describe the specific process of obtaining an error ellipsoid.

2.1. Obtain the Covariance Matrix of the Satellite Position. Considering the nonlinear relationship between the satellite position and orbital elements, the covariance matrix and propagation matrix between orbital elements and position uncertainty can only be obtained by solving partial derivative, which can be shown by

$$\underset{mm}{Dzz} = \underset{mn}{k}\ \underset{nn}{Dxx}\ \underset{nm}{k^T}, \tag{1}$$

where Dzz_{mm} is an $m * m$ square matrix. In this paper, Dzz_{mm} represents the covariance matrix of satellite position in directions X, Y, and Z. Dxx_{nn} is the error matrix of orbital elements.

When z is a nonlinear function of x, such as

$$\begin{aligned} z_1 &= f_1(x_1, x_2, \ldots, x_n), \\ z_2 &= f_2(x_1, x_2, \ldots, x_n), \\ &\vdots \\ z_m &= f_m(x_1, x_2, \ldots, x_n), \end{aligned} \tag{2}$$

the partial derivative of z by partial derivative of x needs to be obtained by the following method:

$$\begin{aligned} dz_1 &= \frac{\partial f_1}{\partial x_1} dx_1 + \frac{\partial f_1}{\partial x_2} dx_2 + \cdots + \frac{\partial f_1}{\partial x_n} dx_n, \\ dz_2 &= \frac{\partial f_2}{\partial x_1} dx_1 + \frac{\partial f_2}{\partial x_2} dx_2 + \cdots + \frac{\partial f_2}{\partial x_n} dx_n, \\ &\vdots \\ dz_m &= \frac{\partial f_t}{\partial x_1} dx_1 + \frac{\partial f_t}{\partial x_2} dx_2 + \cdots + \frac{\partial f_t}{\partial x_n} dx_n. \end{aligned} \tag{3}$$

Then, (5) can be obtained based on (3) and represented by

$$\begin{aligned} \underset{m1}{z} &= \begin{pmatrix} z_1 \\ z_2 \\ \vdots \\ z_m \end{pmatrix}, \\ d\underset{m1}{z} &= \begin{pmatrix} dz_1 \\ dz_2 \\ \vdots \\ dz_m \end{pmatrix}, \end{aligned} \tag{4}$$

$$\underset{mn}{k} = \begin{pmatrix} \frac{\partial f_1}{\partial x_1} \frac{\partial f_1}{\partial x_2} & \cdots & \frac{\partial f_1}{\partial x_n} \\ \frac{\partial f_2}{\partial x_1} \frac{\partial f_2}{\partial x_2} & \cdots & \frac{\partial f_2}{\partial x_n} \\ \vdots & & \vdots \\ \frac{\partial f_t}{\partial x_1} \frac{\partial f_t}{\partial x_2} & \cdots & \frac{\partial f_m}{\partial x_n} \end{pmatrix}. \quad (5)$$

The nonlinear function between the satellite position vector and orbital elements is shown in (6), (7), and (8), where a, e, E, i, Ω, and ω are orbital elements and P and Q are functions of Ω, ω, and i. Vector r is decomposed into three vectors in the directions x, y, and z, and in each direction, the partial derivatives of orbital elements can be obtained. For example, the partial derivative of z by that of a, e, E, i, Ω, and ω can be presented in (9). Similarly, gaining the partial derivatives of x and y by that of a, e, E, i, Ω, and ω, respectively, can finally obtain the error propagation matrix.

$$r = a(\cos E - e) \cdot P + a\sqrt{1-e^2} \sin E \cdot Q, \quad (6)$$

$$P = \begin{pmatrix} \cos\Omega \cos\omega - \sin\Omega \sin\omega \cos i \\ \sin\Omega \cos\omega + \cos\Omega \sin\omega \cos i \\ \sin\omega \sin i \end{pmatrix}, \quad (7)$$

$$Q = \begin{pmatrix} -\cos\Omega \sin\omega - \sin\Omega \cos\omega \cos i \\ -\sin\Omega \sin\omega + \cos\Omega \cos\omega \cos i \\ \cos\omega \sin i \end{pmatrix}, \quad (8)$$

$$\begin{aligned}
\frac{\partial_z}{\partial_a} &= \sin E \cos\Omega\sqrt{1-e^2} - \sin\Omega \sin i(e - \cos E), \\
\frac{\partial_z}{\partial_e} &= a \sin\Omega \sin i - \frac{ae \sin E \cos\Omega \sin i}{\sqrt{1-e^2}}, \\
\frac{\partial_z}{\partial_E} &= a \cos E \cos\Omega \sin i\sqrt{1-e^2} - a \sin E \sin\Omega \sin i, \\
\frac{\partial_z}{\partial_i} &= a \sin E \cos\Omega\sqrt{1-e^2} - a \cos i \sin\Omega(e - \cos E), \\
\frac{\partial_z}{\partial_\Omega} &= -a \cos\Omega \sin i(e - \cos E) - a \sin E \sin\Omega \sin i\sqrt{1-e^2}, \\
\frac{\partial_z}{\partial_\omega} &= 0.
\end{aligned} \quad (9)$$

2.2. Rotation and Translation of the Error Ellipsoid. Assume that the major axis direction of error ellipsoid is

$$r = (x_m, y_m, z_m). \quad (10)$$

So the rotation angle of major axis around axis z should be

$$\theta_z = \arctan\left(\frac{y_m}{x_m}\right). \quad (11)$$

Then, the rotation angle of major axis around axis y is

$$\theta_y = \arcsin\left(\frac{|z_m|}{r_m}\right). \quad (12)$$

Subsequently, all points need to be rotated around axis x. Based on the previous rotations, the third rotation angle is equal to the angle between axis y and middle axis of error ellipsoid.

Finally, if the standard position of satellite is (x, y, z), the point (x_0, y_0, z_0) needs to be translated to a new position $(x_0 + x, y_0 + y, z_0 + z)$.

2.3. Feature of the Error Ellipsoid. It is assumed that the measurement errors of six orbital elements present normal distribution. So the position error of satellite also presents three-dimensional normal distribution [5, 16]. The distribution density of possible satellite position in three-dimensional space is

$$\frac{\exp\left(-0.5\left(r^T D_{zz}{}^{-1} r\right)\right)}{(2\pi)^{1.5}\sqrt{\Delta}}, \quad (13)$$

where r is the position increment.

The denominator of (13) is a constant, while the numerator is changeable. So all points with the same distribution density can be presented by

$$r^T D_{zz}{}^{-1} r = k^2. \quad (14)$$

Equation (14) is a presentation of ellipsoid cluster, and k is a magnification factor. If r is symbolized by

$$r = \begin{pmatrix} \Delta x \\ \Delta y \\ \Delta z \end{pmatrix}, \quad (15)$$

(14) can also be shown by

$$\begin{bmatrix} \Delta x \\ \Delta y \\ \Delta z \end{bmatrix}^T \begin{bmatrix} \sigma_{xx}^2 & \sigma_{xy} & \sigma_{xz} \\ \sigma_{xy} & \sigma_{yy}^2 & \sigma_{yz} \\ \sigma_{xz} & \sigma_{yz} & \sigma_{zz}^2 \end{bmatrix} \begin{bmatrix} \Delta x \\ \Delta y \\ \Delta z \end{bmatrix} = k^2. \quad (16)$$

Because it is not convenient to show the ellipsoid in coordinate system $OXYZ$, the principal axis coordinate system $OUVW$ is employed to present it. For a real symmetric matrix D_{zz}, an orthogonal matrix T can be utilized to make

$$\begin{bmatrix} \Delta x \\ \Delta y \\ \Delta z \end{bmatrix} = T \begin{bmatrix} U \\ V \\ W \end{bmatrix}, \quad (17)$$

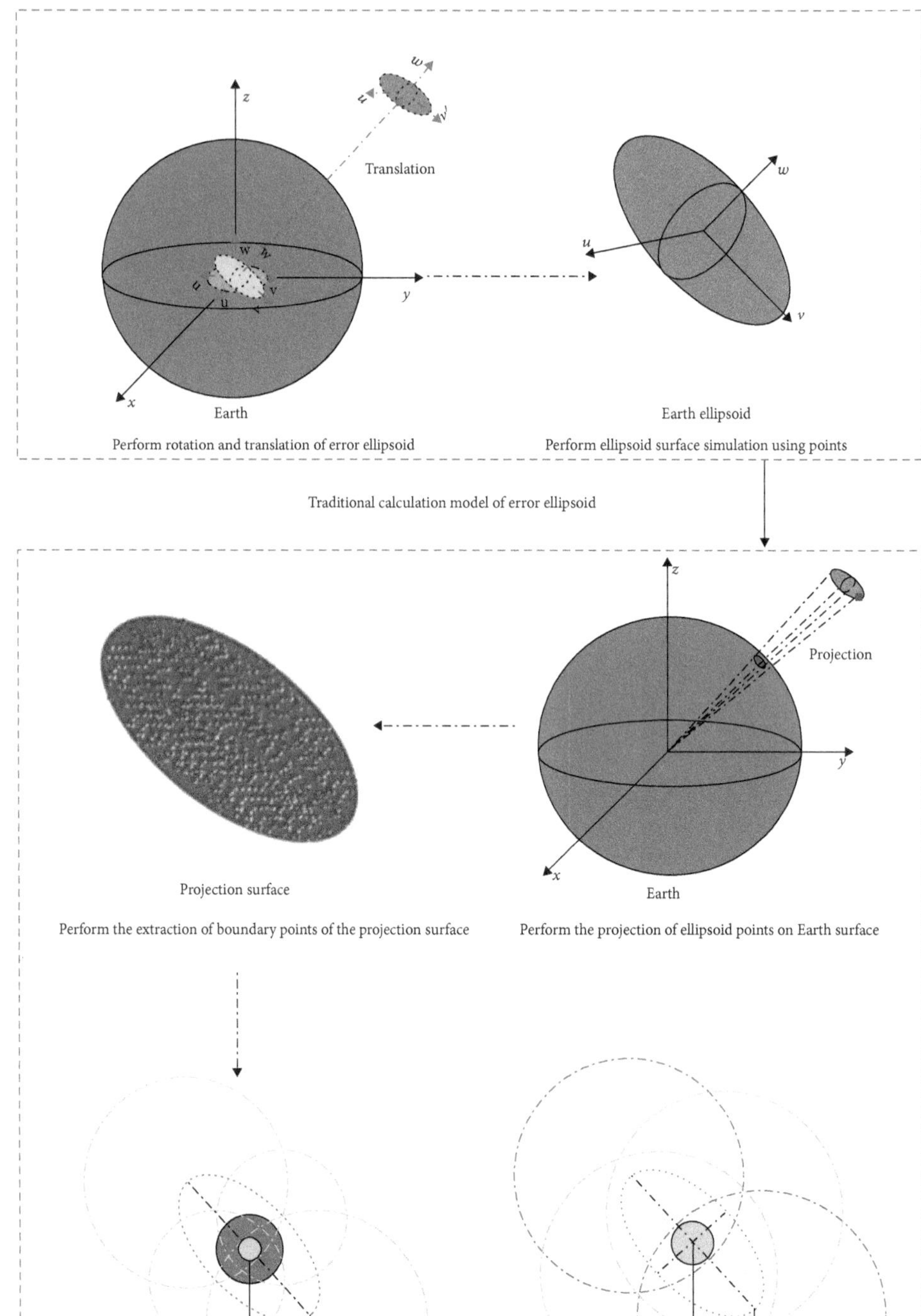

Figure 1: Simulation of uncertainty analysis process.

TABLE 1: Convex hull algorithm.

Input: Planar points set p; **Output:** convex hull $\#(p)$ of points set p; clockwise queue of $\#(p)$.
Step 1: Find point p_0 with the minimum ordinate in p. If more than one point meets this requirement, select the point with minimum horizontal ordinate.
Step 2: Order all the points according to the polar angle relative to p_0. If more than one point locates on the same direction, retain the farthest one.
Step 3: Start from the minimum polar angle, and check whether the points in this direction can be contained in $\#(p)$. If not, put this point to $\#(upper)$ and update it.
Step 4: Repeat step 3 to all the points have been checked.

TABLE 2: Method of analyzing error influence.

Step 1:	Find two points p_{l1} and p_{l2} from $\#(p)$ with the longest distance. Find two other points p_{s1} and p_{s2} which are perpendicular to the line $p_{l1}p_{l2}$.
Step 2:	If The target is a point.
Step 3:	Obtain the calculation result based on the points gained in step 1.
Step 4:	If The target cannot be covered when subastral points are p_{s1} and p_{s2}.
Step 5:	Draw the conclusion: This target cannot be covered completely.
Step 6:	Else If The target cannot be covered when subastral points are p_{s1} and p_{s2}.
Step 7:	Draw the conclusion: This target may be covered possibly.
Step 8:	Else Draw the conclusion: This target can be covered completely.
Step 9:	ELSE Obtain the calculation result based on the points gained in step 1.
Step 10:	If The target cannot be covered when subastral points are p_{s1} and p_{s2}.
Step 11:	Draw the conclusion: This target cannot be covered completely.
Step 12:	Else If The target cannot be covered when subastral points are p_{l1} and p_{l2}.
Step 13:	Obtain the least percentage that the area target can be covered.
Step 14:	Else Draw the conclusion: This target can be covered completely.

and

$$T^T DzzT = \begin{bmatrix} \lambda_1 & & \\ & \lambda_2 & \\ & & \lambda_3 \end{bmatrix}. \tag{18}$$

According to (18), (15) can be shown by

$$\begin{bmatrix} U \\ V \\ W \end{bmatrix}^T \begin{bmatrix} \frac{1}{\lambda_1} & & \\ & \frac{1}{\lambda_2} & \\ & & \frac{1}{\lambda_3} \end{bmatrix} \begin{bmatrix} U \\ V \\ W \end{bmatrix} = k^2. \tag{19}$$

Equation (19) can be expanded to

$$\frac{u^2}{\lambda_1} + \frac{V^2}{\lambda_2} + \frac{W^2}{\lambda_3} = k^2. \tag{20}$$

Assume that each axis of the error ellipsoid is symbolized by a, b, and c; hence,

$$a^2 = k^2\lambda_1,$$
$$b^2 = k^2\lambda_2, \tag{21}$$
$$c^2 = k^2\lambda_3.$$

As mentioned before, the position of the satellite is an uncertainty position distribution within error ellipsoid interior. This distribution probability can be presented by

$$p = \iiint \frac{\exp\left(-0.5\left(U^2/\sigma_1^2 + V^2/\sigma_2^2 + W^2/\sigma_3^2\right)\right)}{(2\pi)^{1.5}\sigma_1\sigma_2\sigma_3} dUdVdW, \tag{22}$$

giving a variable substitution, such as

$$u' = \frac{u}{\sqrt{2\sigma 1}},$$
$$v' = \frac{v}{\sqrt{2\sigma 2}}, \tag{23}$$
$$w' = \frac{w}{\sqrt{2\sigma 2}}.$$

The probability in the ellipsoid is converted to the probability in the ball as follows:

$$p = \frac{1}{\pi^{1.5}} \int_0^{2\pi} d\theta \int_0^{\pi} d\varphi \int_0^{k/\sqrt{2}} \exp\left(-r^2 r^2\right) \sin\varphi dr = \frac{4}{\sqrt{\pi}} \int_0^{k/\sqrt{2}} e^{-r^2} r^2 dr. \tag{24}$$

In addition, considering that

$$e^{-r^2} = 1 - r^2 + \frac{r^4}{2!} - \frac{r^6}{3!} + \cdots + (-1)^{-1} \frac{r^{2n}}{n!} + \cdots,$$
$$r \in (-\infty, +\infty), \tag{25}$$

Table 3: Data of orbital elements.

Semimajor axis (km)	Eccentricity	True anomaly (°)	Inclination of orbit (°)	Argument of perigee (°)	Right ascension of ascending node (°)
10,424	0.3	0	45	45	20

Table 4: Data of axis length.

Major axis (km)	Middle axis (km)	Minor axis (km)
20.9535	14.6836	5.4130

Table 5: Data of rotation angle.

Rotation angle of the major axis around z-axis (°)	Rotation angle of the major axis around y-axis (°)	Rotation angle of the middle axis around x-axis (°)
55.262168	−29.999022	125.621076

(24) can be expanded to

$$p = \frac{4}{\sqrt{2}\pi}\left(\frac{k^3}{6} - \frac{k^5}{20} + \frac{k^7}{112} - \frac{k^9}{864} + \cdots\right), \quad (26)$$

where the probability of the point distribution in the ellipsoid varies with the change of K as follows:

(i) When $K = 1$, the probability of the point distribution in the error ellipsoid is 19.9%.

(ii) When $K = 2$, the probability of the point distribution in the error ellipsoid is 73.9%.

(iii) When $K = 3$, the probability of the point distribution in the error ellipsoid is 95%.

3. Preparation for Coverage Case

3.1. Calculation of the Error Ellipsoid. According to the measurement accuracy in actual engineering project, the error matrix of orbital elements is set as follows:

$$Dxx = \begin{pmatrix} 3^2 & 0 & 0 & 0 & 0 & 0 \\ 0 & 0.002^2 & 0 & 0 & 0 & 0 \\ 0 & 0 & (0.05^*H)^2 & 0 & 0 & 0 \\ 0 & 0 & 0 & (0.05^*H)^2 & 0 & 0 \\ 0 & 0 & 0 & 0 & (0.05^*H)^2 & 0 \\ 0 & 0 & 0 & 0 & 0 & (0.05^*H)^2 \end{pmatrix}, \quad (27)$$

where H is a conversion factor between angle and radian.

The orbital elements of a satellite are shown in Table 3. In addition, as a requirement of parameter transformation, the conversion relationship between eccentric anomaly E and true anomaly f is needed:

$$E = 2 \cdot a\tan\left(\frac{\tan(f/2)}{((1+e)/(1-e))^{0.5}}\right). \quad (28)$$

The covariance matrix of the satellite position Dzz is obtained as follows:

$$Dzz = \begin{pmatrix} 264.2010 & 99.6929 & 30.6575 \\ 99.6929 & 251.3009 & 168.8499 \\ 30.6575 & 168.8499 & 168.4551 \end{pmatrix}. \quad (29)$$

The feature vector v of Dzz is

$$v = \begin{pmatrix} 0.4935 & 0.8516 & 0.1766 \\ 0.7117 & -0.2786 & -0.6449 \\ 0.5000 & -0.4439 & 0.7436 \end{pmatrix}. \quad (30)$$

The data of axis length and rotation angle are shown in Tables 4 and 5, respectively. The standard position coordinate of the satellite position is (3600.6, 5193.1, 3648.4). The final simulation of the error ellipsoid with different views is shown in Figure 2.

3.2. Edge Extraction of Projection on Earth. The projection of error ellipsoid on Earth is a curved surface, so in order to obtain the edge points, a mapping on plane is more convenient. 421 edge points are obtained in the case by using convex hull algorithm, with different views of simulation results shown in Figure 3.

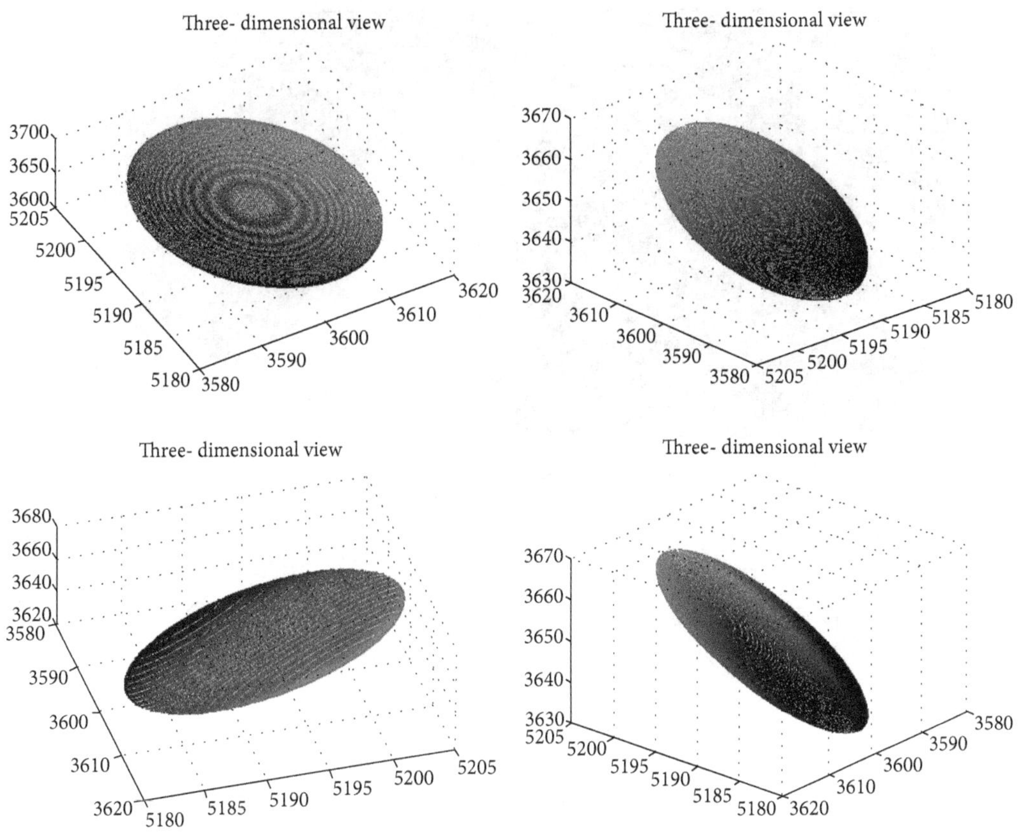

FIGURE 2: Space distribution of error ellipsoid.

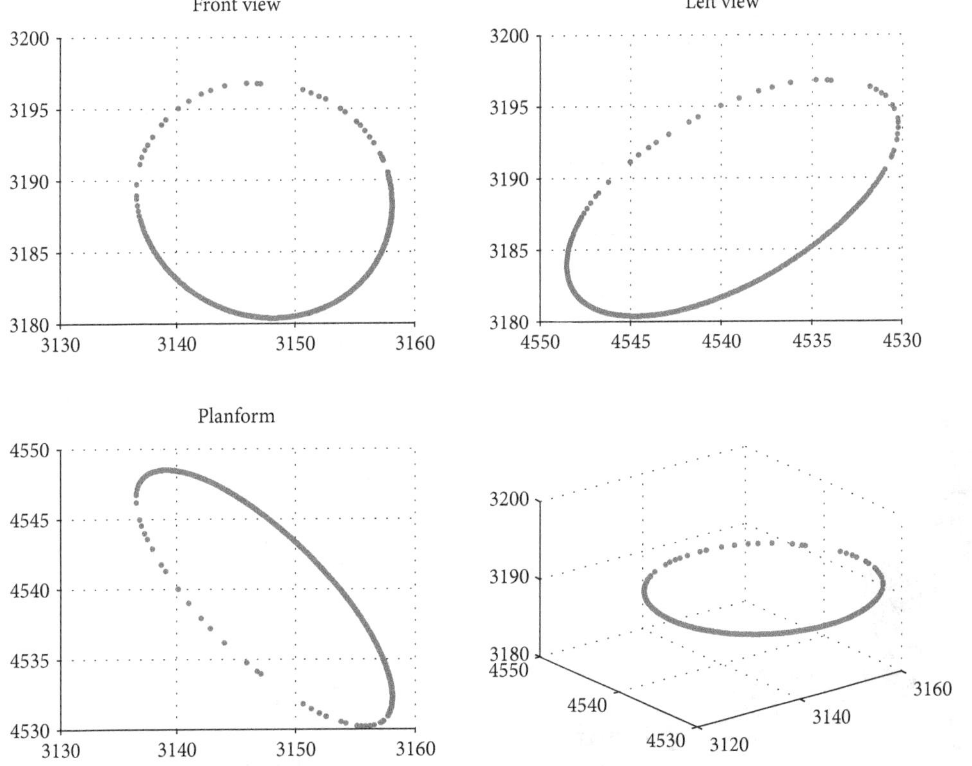

FIGURE 3: Result of edge extraction.

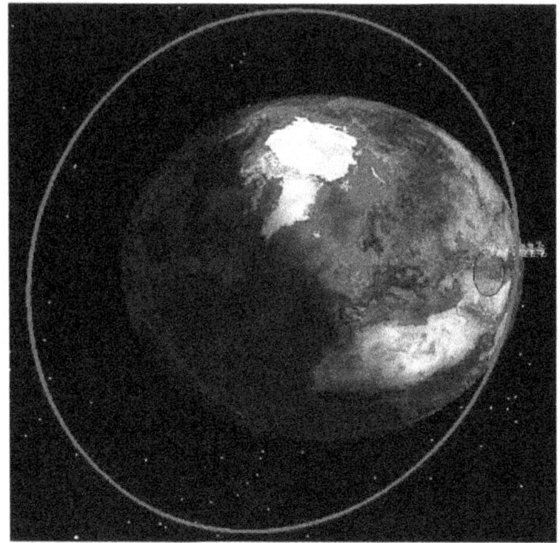
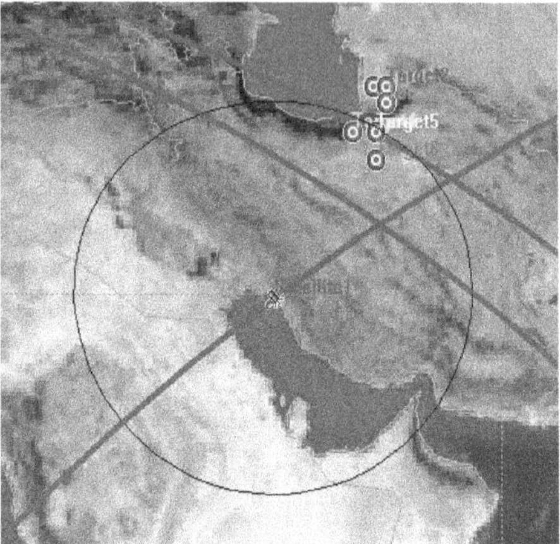

FIGURE 4: Satellite orbit and point targets.

TABLE 6: Coordinate of the six point targets.

Group	Coordinate	Latitude and longitude
1	(3700.600, 5193.100, 3648.400)	54.526°E, 37.855°N
2	(3700.600, 5293.100, 3648.400)	55.041°E, 37.848°N
3	(3700.600, 5293.100, 3748.400)	55.041°E, 37.260°N
4	(3900.600, 5293.100, 3748.400)	53.612°E, 36.112°N
5	(3900.600, 5493.100, 3748.400)	54.621°E, 36.117°N
6	(3900.600, 5493.100, 3948.400)	54.621°E, 35.068°N

TABLE 7: Results of case 1.

Group	Geocentric angle without considering error (°)	Geocentric angle with considering error (°)
1	1.356686	1.580021
2	1.595898	1.713935
3	1.940826	2.012707
4	4.564908	4.720740
5	6.070766	6.134848
6	7.341740	7.391859

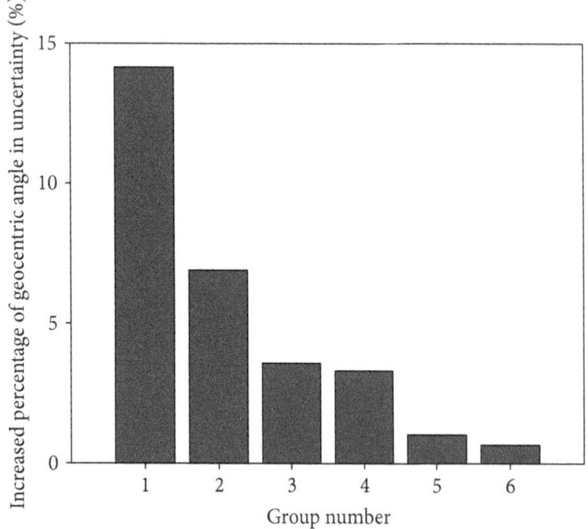

FIGURE 5: Simulation comparison of case results.

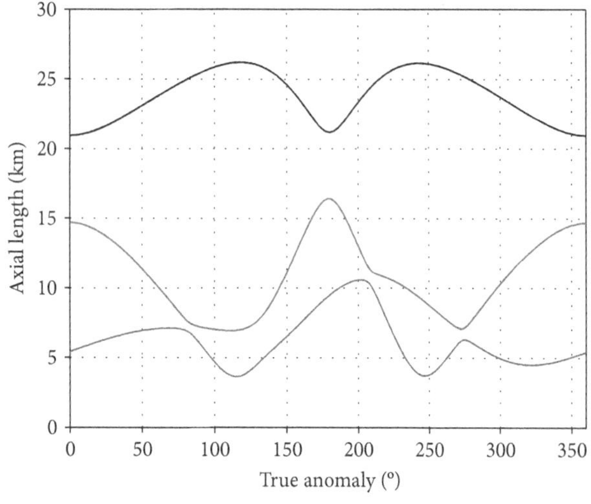

FIGURE 6: Changes of axial length.

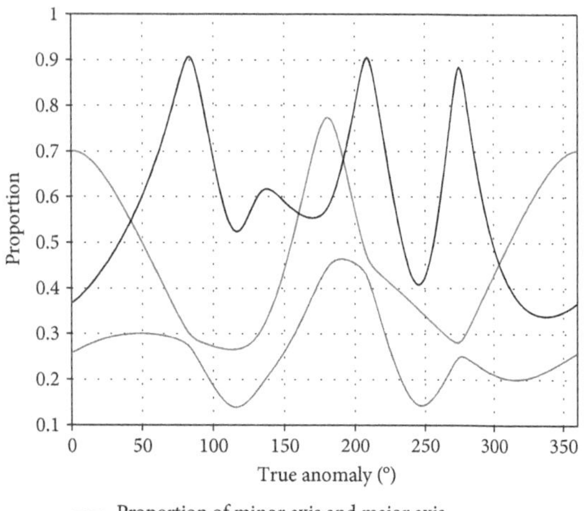

FIGURE 7: Changes of axial ratio.

Table 8: Orbital element of each satellite.

Semi major axis (km)	Eccentricity	True anomaly (°)	Inclination of orbit (°)	Argument of perigee (°)	Right ascension of ascending node (°)
10,424	0.3	0	45	45	20
10,424	0.3	0	45	45	40
10,424	0.3	0	45	45	60

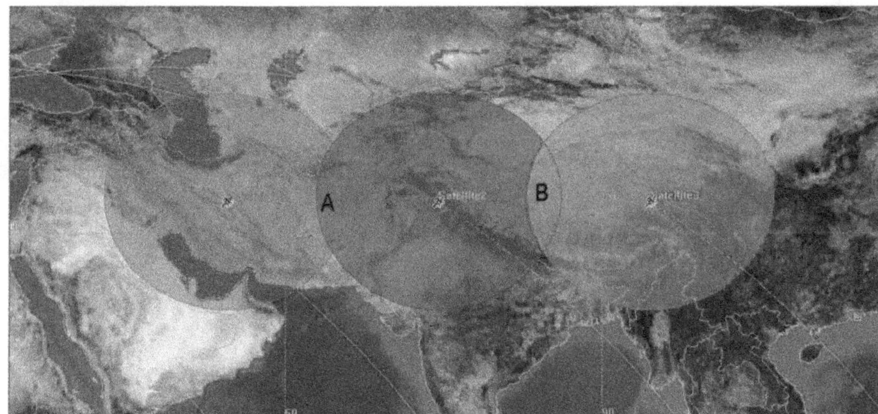

Figure 8: Positions of points A and B.

4. Numerical Case

Five cases with different configures are designed and implemented in this paper. Firstly, the uncertainty influence of single satellite positing on multiple point target coverage is given in case 1. Then, in order to deeply understand the shape change of "error ellipsoid" when the satellite moves to different positions, the changes of axial length and axial ratio are given in case 2. Lastly, the uncertainty influence of a constellation positioning on multiple point targets, constellation positioning on a single area target, and constellation positioning on two area targets are given in cases 3 to 5, respectively.

The reference system used in this paper is ECI. When the satellite is in the initial position perigee, the latitude and longitude of the satellite are 55.2°E and 30°N, respectively. The orbit and ground point targets are shown in Figure 4 with 3D and 2D views.

4.1. Influence of Single Satellite Position Uncertainty on Multiple Point Target Coverage. The purpose of case 1 is to explain the necessity of considering orbital element error in coverage calculation. Six point targets are set in this case, whose coordinates are presented in Table 6, obtaining the different minimum coverage requirements of these point targets in terms of geocentric angles considering and not considering positioning uncertainty, respectively. Case results listed in Table 7 show that the geocentric angle needs to be larger to cover all the aimed targets when considering error of orbital elements. The results shown in Figure 5 obviously present the relative percentage of needed geocentric angle when considering uncertainty.

Table 9: Minimum coverage radius of satellites.

Satellite	Point target	Minimum covering radius (km)
1	A	469.590510
2	A	1484.017928
2	B	460.448878
3	B	1459.206737

4.2. Shape Change of the "Error Ellipsoid." The purpose of case 2 is to research on the shape change of the error ellipsoid. Analyze the axial length variation while true anomaly changes from 0° to 360°. Focus on the shape change by studying the variations of axial length and axial ratio, which are shown in Figures 6 and 7, respectively. Interesting conclusions can be drawn from the case results. The most significant one is that the 3D shape of error ellipsoid is a time-dependent function, and the error ellipsoid is rather slender when true anomaly changes around 110°. While anomaly reaches around 180°, the shape is stouter.

4.3. Uncertainty Influence of a Constellation Positioning on Multiple Point Targets. The implementation of case 3 is to study the coverage uncertainty of constellation containing 3 satellites whose orbital elements are shown in Table 8. Two point targets A (3159.6162, 5472.44908, and 3648.4) and B (1097.44509, 6222.5133, and 3648.4) shown in Figure 8 are set in this case. The latitude and longitude of these two points are (60°E, 30°N) and (80°E,30°N), respectively. Case results in Table 9 show that when considering positioning uncertainty, the minimum required covering radius is larger than that of not considering uncertainty. If it is needed to cover both

TABLE 10: Coverage results of a circular area when the sensor angle is 7°.

Coverage times	Target number	Coverage times	Target number	Coverage times	Target number	Coverage times	Target number
31	2525	37	404	43	0	49	0
32	2780	38	193	44	0	50	0
33	1918	39	38	45	0	51	0
34	1398	40	0	46	0	52	0
35	1082	41	0	47	0	53	0
36	818	42	0	48	0	54	0

TABLE 11: Coverage results of a circular area when the sensor angle is 7.4°.

Coverage times	Target number	Coverage times	Target number	Coverage times	Target number	Coverage times	Target number
31	1810	37	2145	43	1392	49	821
32	5354	38	4878	44	2160	50	724
33	5371	39	3507	45	1268	51	516
34	3882	40	2590	46	1316	52	344
35	3745	41	2006	47	1331	53	215
36	4106	42	2102	48	1061	54	38,589

TABLE 12: Coverage results of a circular area when the sensor angle is 7.7°.

Coverage times	Target number	Coverage times	Target number	Coverage times	Target number	Coverage times	Target number
31	0	37	902	43	675	49	501
32	0	38	836	44	1185	50	469
33	0	39	1756	45	612	51	257
34	0	40	1441	46	1160	52	168
35	0	41	1179	47	775	53	99
36	198	42	1152	48	440	54	82,792

TABLE 13: Coverage results of a circular area when the sensor angle is 8°.

Coverage times	Target number	Coverage times	Target number	Coverage times	Target number	Coverage times	Target number
31	0	37	0	43	0	49	0
32	0	38	0	44	0	50	0
33	0	39	0	45	0	51	0
34	0	40	0	46	0	52	0
35	0	41	0	47	0	53	0
36	0	42	0	48	0	54	96,597

points A and B at the same time, the minimum radius must be 1484.017928 km.

4.4. Uncertainty Influence of Constellation Positioning on a Single Area Target. We have mentioned that 421 edge points are obtained in the case by using convex hull algorithm. In order to reduce the calculation expense, 54 edge points are chosen evenly to implement the coverage case containing 96,576 ground target points. Only when each of the 96,576 sampling points can be covered by the satellite, whose subastral point is any one of the 54 edge points, the area is deemed covered completely. Assuming that the target is a circular area, and the radius of this area is 100 km, compute the minimum sensor angle meeting the coverage requirements of this area when the angle of the satellite sensor changes from 7°. Case results are shown in Tables 10, 11, 12, and 13. In order to present this change significantly, relevant simulation results of these data are shown in Figures 9–12, respectively. As well, the percentage change of area coverage is also shown in Figure 13.

4.5. Uncertainty Influence of Constellation Positioning on Two Area Targets. Assuming that A and B are circular areas with circle radius = 100 km. The minimum coverage radius

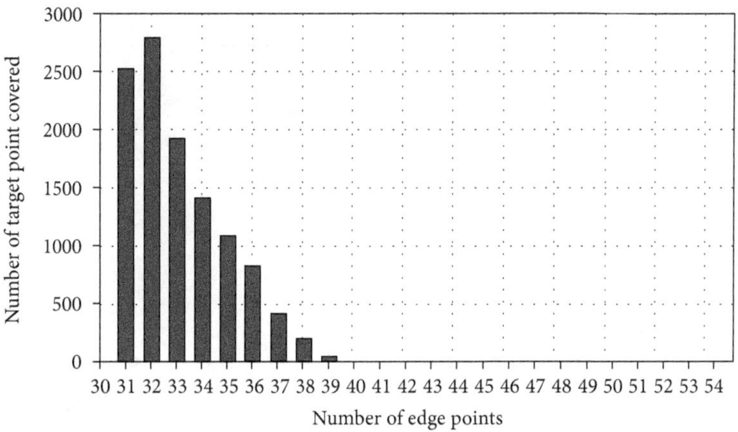

FIGURE 9: Coverage results of a circular area when the sensor angle is 7°.

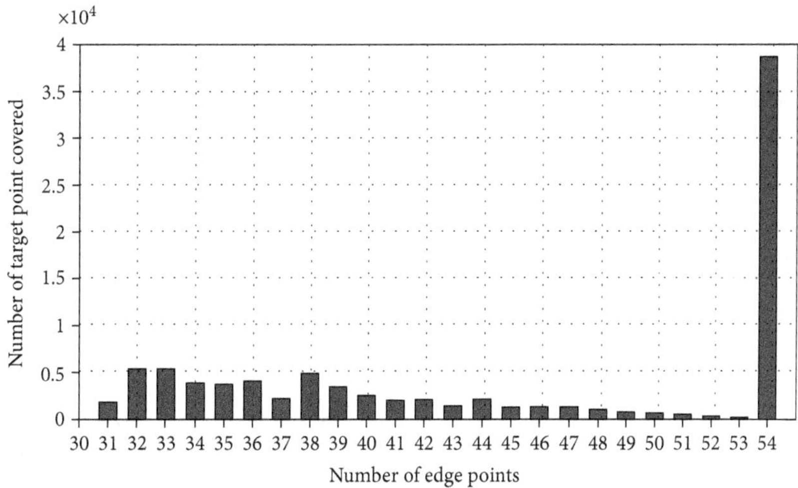

FIGURE 10: Coverage results of a circular area when the sensor angle is 7.4°.

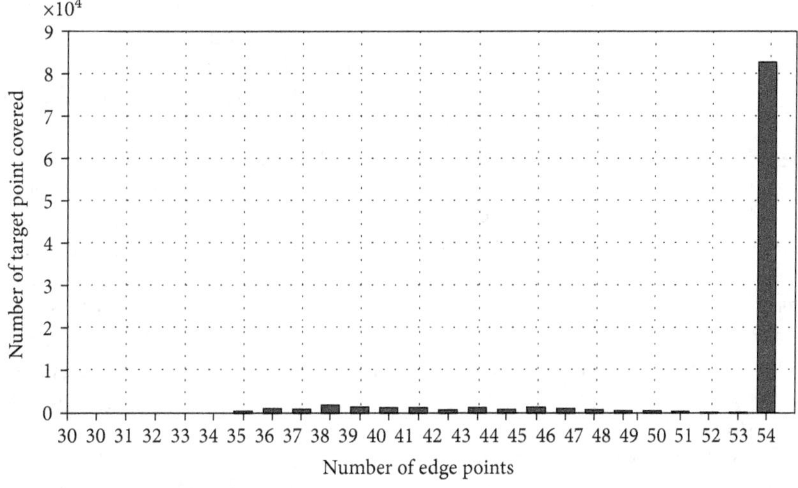

FIGURE 11: Coverage results of a circular area when the sensor angle is 7.7°.

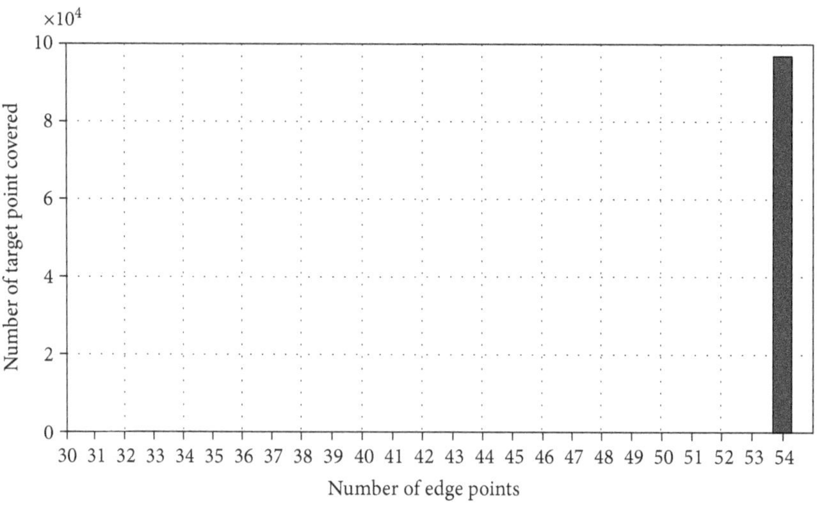

FIGURE 12: Coverage results of a circular area when the sensor angle is 8°.

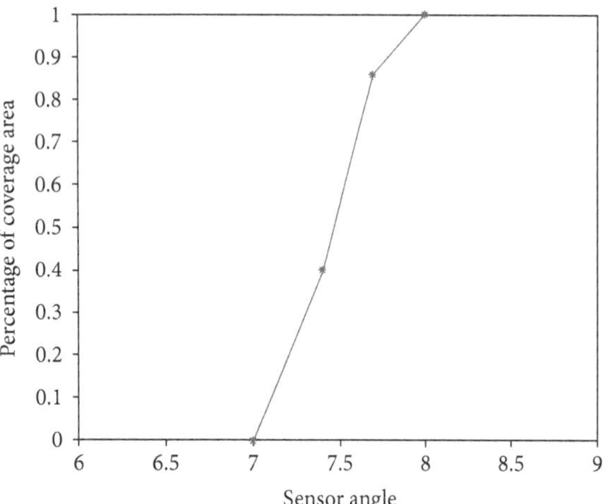

FIGURE 13: Percentage change of area coverage.

should be calculated with respect to the requirement of the complete coverage from satellite 2. From case results shown in Table 14, we know that the minimum coverage radius of satellite 2 for coverage requirement is 1684.017928 km.

5. Results and Discussion

Above cases have shown the influence of orbital element errors on coverage calculation. It is obvious that the minimum angle requirement is larger when considering error of orbital elements. From case 1, it can be seen that the influence of orbital element error on coverage is more significant when the required geocentric angle is smaller. As the required geocentric angle increases, the impact of orbital element error on coverage will be smaller and smaller. From case 2, an interesting phenomenon can be seen that when true anomaly is about 180°, and it is a turning point for the changes of axial length and axial ratio. Cases 3–5 also show the analysis results of different coverage problems based on the Monte Carlo simulation for area targets. These cases show that

TABLE 14: Minimum coverage radius for each target.

Satellite	Point target	Minimum covering radius (km)
1	A	669.590510
2	A	1684.017928
2	B	660.448878
3	B	1659.206737

under the case settings in this paper, the extra requirements of sensor angle grows by about 14.3%. The analysis implemented in this paper is based on the fixed positions of satellite or constellation. Hence, in future research, this work can be extended to analyze the real-time coverage error related to positioning uncertainty.

6. Conclusions

In this paper, we present a method of accurately calculating the constellation coverage when considering position uncertainty. The position uncertainty of satellite is shown by an "error ellipsoid," whose spatial shape is also analyzed in the paper. For the illustration of proposed method, coverage calculations for a single satellite and constellation are both considered to carry out the coverage cases. The cases show that when considering the position uncertainty related to influence of orbital error, in order to surely cover the target points, the covering radius needs to be larger than not considering position uncertainty. The work presented in this paper also has important reference value for the reliability analysis of constellation design, which can effectively deal with the influence of position uncertainty.

Conflicts of Interest

The authors declare that there is no conflict of interests regarding the publication of this paper.

Acknowledgments

This work is supported by National Natural Science Foundation of China under Grant no. 61472375, the 13th Five-Year Pre-Research Project of Civil Aerospace in China, the Joint Funds of Equipment Pre-Research and Ministry of Education of the People's Republic of China under Grant no. 6141A02022320, Fundamental Research Funds for the Central Universities under Grant nos. CUG2017G01 and CUG160207, and the Reform and Research of Graduate Education and Teaching under Grant no. YJS2018308.

References

[1] J. Cote, *Maximizing Satellite Coverage*, Pacific Yachting, 2012.

[2] K. W. Yung and D. C. Chang, *U.S. Patent No. 7,198,230*, U.S. Patent and Trademark Office, Washington, DC, USA, 2007.

[3] J. E. Draim, "A common-period four-satellite continuous global coverage constellation," *Journal of Guidance, Control, and Dynamics*, vol. 10, no. 5, pp. 492–499, 1987.

[4] Z. Lian, Y. Tan, Y. Xu, and J. Li, "Static and dynamic models of observation toward Earth by agile satellite coverage," in *Proceedings of International Workshop on Planning and Scheduling for Space*, pp. 1–6, Darmstadt, Germany: ESOC, 2011.

[5] H. B. Wang, F. Xue, and C. J. Liu, "Test method of satellite coverage area and its engineering application," *Radio Communications Technology*, vol. 3, p. 19, 2010.

[6] J. A. Ojeda Romero, *Dual Satellite Coverage using Particle Swarm Optimization Doctoral dissertation*, Virginia Tech, 2014.

[7] X. Huang and G. Dai, "An optimal design of constellation of multiple regional coverage based on NSGA-II," *International Journal of Advancements in Computing Technology*, vol. 4, no. 21, pp. 295–303, 2012.

[8] E. S. Fekpe, T. K. Windholz, and K. Beard, "Probabilistic approach for modeling and presenting error in spatial data," *Journal of Surveying Engineering*, vol. 135, no. 3, pp. 101–112, 2009.

[9] E. D. Kaplan, "Understanding GPS: principles and application," *Journal of Atmospheric and Solar-Terrestrial Physics*, vol. 59, no. 5, pp. 598-599, 2005.

[10] P. Misra and P. Enge, "*Global Positioning System: Signals, Measurements and Performance*," Ganga-Jamuna Press, 2011.

[11] W. Nelson, *Use of Circular Error Probability in Target Detection*, MITRE Corporation, Bedford MA, USA, 1988.

[12] J. C. Roh and D. W. Waters, "Position and velocity uncertainty metrics in GNSS receivers," US Patent US8525727, 2013.

[13] E. M. Mikhail and F. E. Ackermann, *Observations and Least Squares*, Harper and Row, New York, NY, USA, 1976.

[14] M. Nomura and T. Tanaka, "GPS positioning method under condition of only three acquired satellites," *SICE Journal of Control, Measurement, and System Integration*, vol. 2, no. 4, pp. 206–212, 2009.

[15] F. D. Nunes, F. M. G. Sousa, and J. M. N. Leitao, "Performance analysis of altimeter-aided GNSS receiver for indoor scenarios," in *The 7th Conference on Telecommunications*, Santa Maria da Feria, Portugal, May 2009.

[16] M. Wang, L. Ma, L. Zhang, and H. Ji, "Analysis of horizontal positioning error distribution of the Chinese area positioning system," *Astronomical Research & Technology*, vol. 9, no. 3, pp. 522–526, 2012.

[17] S. Alfano, "Relating position uncertainty to maximum conjunction probability," *Journal of the Astronautical Sciences*, vol. 53, no. 2, pp. 193–205, 2004.

[18] Z. Qingzhu, Z. Jun, Z. Yanbo, and L. Wei, "Navigation accuracy category-position models and estimate position uncertainty calculations for tis-b system," *Chinese Journal of Aeronautics*, vol. 22, no. 4, pp. 419–425, 2009.

[19] Y. L. Shi, Q. M. Cui, S. Q. Cao, and X. F. Tao, "Distributed location model and algorithm with position uncertainty," *The Journal of China Universities of Posts and Telecommunications*, vol. 21, no. 2, pp. 48–56, 2014.

[20] D. A. Trifonov, "Position uncertainty measures on the sphere," in *Proceedings of the Fifth International Conference on Geometry, Integrability and Quantization*, pp. 211–224, Institute of Biophysics and Biomedical Engineering, Bulgarian Academy of Sciences, 2004.

3

Analytical Approach for Orbital Evasion with Space Geometry Considered

Dateng Yu,[1] Hua Wang,[1] Shuai Guo,[1] and Hongyu Wang[2]

[1] *College of Aerospace Science and Engineering, National University of Defense Technology, Changsha, China*
[2] *Aerospace System Engineering Shanghai, Shanghai 200000, China*

Correspondence should be addressed to Hua Wang; wanghua@nudt.edu.cn

Academic Editor: Heidi Kuusniemi

This paper researches an optimal problem of orbital evasion with considering space geometry by using an analytical approach. Firstly, an angles-only relative navigation model is built and the definition of completely nonobservable maneuver is proposed. After algebraic analysis of relative space geometry, it is proved that the completely nonobservable maneuver is nonexistent. Based on this, the angle measurements of orbit without evasion are set as reference measurements and an analytical solution is derived to find the minimum difference between measurements and the reference measurements in a constant measuring time. Then, an object function using vector multiplication is designed and an optimization model is established so as to prove the optimality of analytical solution. At last, several numerical simulations are performed with different maneuver directions, which verify the effectiveness of the analytical method of this paper for orbital evasion problem. This method offers a new viewpoint for orbital evasion problem.

1. Introduction

Nowadays the satellites face various threats, not only orbit debris but also some noncooperative rendezvous. To increase the survivability of satellites, it is important to have some evasion strategies and perform optimally evasive maneuvers. In this paper, we focus on optimal evasion strategies for an evading satellite against a noncooperative rendezvous spacecraft. It should be mentioned that both of the relevant spacecrafts are active-spacecraft, and the debris is out of consideration in this paper.

The optimal evasion problem has been studied for many years [1–6]. In this problem, a pursuer tries to approach its target (namely, an evader) through several evasive maneuvers, at the same time the evader expects to escape from the pursuer through some optimal evasive maneuvers. Varieties of evasion strategies had been proposed. Shinar and Steinberg [7] proposed a closed form expression for a switching equation with a new navigation gain considered. Forte et al. [8] analyzed an equivalent linearization of the three-dimensional optimal avoidance problem.

The optimal evasive strategies have been applied on many aerospace problems. Kelly and Picciotto [9] proposed an optimal rendezvous evasive method by using a nonlinear optimization technology. Patera [10–12] firstly introduced an optimal evasive strategy in consideration of collision probability. Bombardelli [13] obtained an optimal maneuver method to numerically maximize the miss distance, which described the arc length separation between the maneuvering rear point and the predicted collision point. Recently, the study of evasive maneuvers has been done with different emphasis, Lee et al. [14, 15] used genetic algorithm to find a solution of minimum fuel consumption and to determine delta-V maneuvers in LEO and GEO. de Jesus and de Sousa [16] investigated the existence of symmetry in determining the initial conditions of collisions among objects. The evasion problem studies mentioned above did not consider the navigation performance influence. In addition, the approaching objects were always considered to be debris or failure vehicles in most of the orbital evasion research, which means that the evasion strategy may be useless when the object is changed to noncooperative spacecraft.

In fact, the navigation performance is actually a significant view to analyze evasion strategies. In this paper, we introduced space geometry of the two spacecraft in an orbital evasion problem to characterize the measurements as a new index. As known, the system observability can be altered by the maneuvers of evader and pursuer during angles-only navigation [17, 18]. Vallado [19] found that the diversity of relative motion had a positive nonlinear correlation with the system observability. Woffinden and Geller [20] derived an analogical correlation between system observability and maneuvers in the orbit rendezvous field. Grzymisch and Fichter [21, 22] found an optimal maneuver method for rendezvous through analyzing the observability conditions. Dateng et al. [23] used a multiobjective optimization approach to investigate orbital evasion problem with the consideration of system observability. The fundamental reason why the system observability can be altered by the maneuvers of evader and pursuer is that these maneuvers alter the relative space geometry and the measurements that the pursuer and the evader can acquire. This is the viewpoint which this paper tried to focus on.

In this work, a novel analytical evasion strategy is proposed to find optimal evasive maneuvers by considering the variation of relative space geometry and measurements. The rest of this paper is organized as follows. First, a problem overview about the orbital evasion is described. Then, an angles-only relative navigation model is established and the definition of completely nonobservable maneuver is proposed. After algebraic analysis of relative space geometry, it is proved that the completely nonobservable maneuver is nonexistent. Based on this, an analytical solution is derived to find the optimal evasive maneuver. Then, an optimization model is established and numerical solution using genetic algorithm (GA) is introduced so as to prove the optimality of analytical solution. At last, a numerical simulation is performed to verify the validity of the method. The results indicate that the analytical method proposed in this paper can reach the expected effect.

2. Problem Overview

There are usually two spacecrafts in an orbital evasion problem (a pursuer and an evader). The objective of pursuer is to capture or approach the evader, and the objective of evader is to find some evasion strategies and escape by the pursuer. This paper researches the optimal evasion strategy when the initial relative distance is about 100 km and the initial orbits are coplanar and HEO (Highly Elliptical Orbit).

At such distance, the pursuer usually only has two angle measurements, because the evader is noncooperative and the LiDAR (Light Detection and Ranging) cannot get distance measurements in such a range. Assuming that evader and pursuer both know the initial state of each other, the pursuer will acquire the future state of evader through angle measurements with the help of some filters. The angle measurements of orbit without evasion are set as reference measurements. Generally, the angle measurements will have a great change when the evader starts an evasive maneuver, meaning that pursuer can catch the maneuver through filtering immediately. When pursuer gains the evasive maneuver, it can replan its approaching to make the evasive maneuver invalid.

However, if the evasive maneuver causes minor or even no difference in angle measurements compared with the reference measurements, the accuracy of filtering will be declined or even non convergent. Thus, the accuracy of navigation has a relationship with the relative space geometry which is influenced by the evasive maneuver. In this way, the evasive maneuver can change the accuracy of navigation and the system observability.

The optimal evasive maneuver is expected to minimize the difference of angle measurements pursuer acquired between evasion and no evasion, and then the difficulty of filter tracking is increased. When the magnitude of evasive maneuver is given, the evasion problem changes to an optimal optimization problem of two control variables. In the following, we will introduce both analytical and numerical solutions to solve this problem.

3. The Relationship of Space Geometry

In this section, the relationship of relative space geometry and evasive maneuver is analyzed. Many different indexes have been used to describe the observability, such as the condition number of the observability matrix [24] and the distance error [25]. A new index that represents the level of state estimation is introduced here which is different from the previous indexes, and we call it measurement observability.

3.1. State Transfer Matrix for HEO. The relative Local Vertical Local Horizontal (LVLH) coordinate system is used as reference frame to describe orbital relative motion. Since pursuer and evader are in elliptical orbit, the Tschauner-Hempel (TH) equation [26] is introduced to describe the relative motion. The homogeneous solution is Yamanaka-Ankersen state transfer matrix [27]:

$$\begin{aligned}\Phi(t,t_0) &= \begin{bmatrix}\Phi_{rr}(t,t_0) & \Phi_{rv}(t,t_0)\\ \Phi_{vr}(t,t_0) & \Phi_{vv}(t,t_0)\end{bmatrix}\\ &= \Phi_\theta(f)\,\Phi_\theta^{-1}(f_0),\end{aligned} \tag{1}$$

where f_0 and f are true anomaly of the reference spacecraft at t_0 and t, respectively. The expressions of $\Phi_\theta(f)$ and $\Phi_\theta^{-1}(f_0)$ are as follows:

$$\Phi_\theta(f)$$

$$= \left[\begin{array}{ccc|ccc} s & c & 2-3esI & 0 & 0 & 0\\ s' & c' & -3e\left(sI+\frac{s}{k^2}\right) & 0 & 0 & 0\\ c\left(1+\frac{1}{k}\right) & -s\left(1+\frac{1}{k}\right) & -3k^2 I & 1 & 0 & 0\\ \hline -2s & e-2c & -3(1-2esI) & 0 & 0 & 0\\ 0 & 0 & 0 & 0 & c & s\\ 0 & 0 & 0 & 0 & -s & c\end{array}\right],$$

$$\Phi_\theta^{-1}(f_0) = \left[\begin{array}{ccc|ccc} -3s\dfrac{k+e^2}{k^2} & c-2e & 0 & -s\dfrac{k+1}{k} & 0 & 0 \\ -3\left(e+\dfrac{c}{k}\right) & -s & 0 & -\left(c\dfrac{k+1}{k}+e\right) & 0 & 0 \\ 3k-\lambda^2 & es & 0 & k^2 & 0 & 0 \\ \hline -3es\dfrac{k+1}{k^2} & -2+ec & \lambda^2 & -es\dfrac{k+1}{k} & 0 & 0 \\ 0 & 0 & 0 & 0 & \lambda^2 c & -\lambda^2 s \\ 0 & 0 & 0 & 0 & \lambda^2 s & \lambda^2 c \end{array}\right], \tag{2}$$

where $k = 1 + e\cos f$, $c = k\cos f$, $s = k\sin f$, $I = \int_{f_0}^{f}(1/k^2)du = (\mu^2/h^3)(t-t_0)$, $\lambda = \sqrt{1-e^2}$. e is eccentricity of the reference orbit, μ is gravitational coefficient of the earth, s' and c' are the first derivative of s and c with respect to f.

In the following, we set $\Phi_p = [\Phi_{rr} \quad \Phi_{rv}]$ and $\mathbf{B} = [\Phi_{rv} \quad \Phi_{vv}]^T = [\mathbf{B}_r \quad \mathbf{B}_v]^T$. Assume a spacecraft executes a maneuver $\mathbf{u}$ at t_0; then the relative state at t with respect to the reference orbit is

$$\mathbf{X}_t = \Phi(t, t_0)\mathbf{X}_{t_0} + \mathbf{B}\mathbf{u}. \tag{3}$$

3.2. Measurement Equations. In the actual projects, optical camera is one of the most common measuring devices and can provide two independent angle measurements (contains elevation ε and azimuth θ) at every moment. Figure 1 shows the geometric schematic of the orbital pursuit-evasion system along with the elevation and azimuth in the frame of camera, where F_{PO}, F_{MO}, and F_{EO} are the evader, pursuer, and camera coordinate systems, respectively. F_{MO} and F_{PO} are regarded as the same coordinate system in the following analysis. The definition of the orbital coordinate system is as follows: z is along the position vector which is from the center of the earth to the pursuer, x is perpendicular to z in the orbital plane and at the same side of the velocity direction, and $y = x \times z$ obeys the right-handed coordinate system. The definition of the orbital coordinate attached to the evader is omitted here, since the definition is similar.

The relative position between the evader and pursuer in measuring coordinate system is denoted as $\mathbf{r}_{PE} = [x \quad y \quad z]^T$; the relationship of $\mathbf{r}_{PE}$ and measuring parameters obtained from the optical camera is

$$\mathbf{h}(\mathbf{X}_E) = \begin{bmatrix} \varepsilon \\ \theta \end{bmatrix} = \begin{bmatrix} \arctan\left(\dfrac{z_{PE}}{x_{PE}}\right) \\ \arctan\left(\dfrac{y_{PE}}{\sqrt{x_{PE}^2 + z_{PE}^2}}\right) \end{bmatrix}, \tag{4}$$

where ε and θ are the elevation and azimuth angles, respectively.

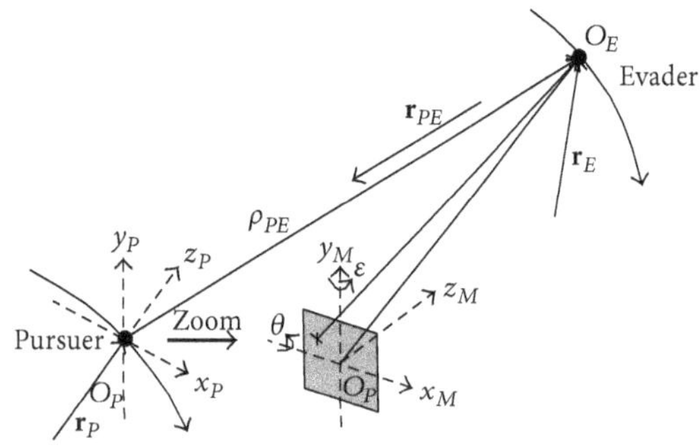

FIGURE 1: Relative observation geometry.

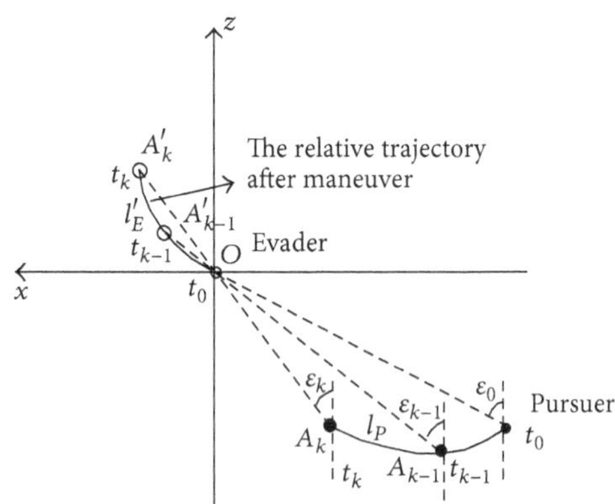

FIGURE 2: Completely unobservable maneuver.

Through linear inverse transformation, (4) is transferred to

$$\mathbf{H}(\mathbf{y})\mathbf{X}_r = 0, \tag{5}$$

where $\mathbf{X}_r$ is position component of the relative state and $\mathbf{H}(\mathbf{y})$ is as follows:

$$\mathbf{H}(\mathbf{y}) = \begin{bmatrix} \sin(\varepsilon) & 0 & -\cos(\varepsilon) \\ \sin(\theta) & -\cos(\varepsilon)\cos(\theta) & 0 \end{bmatrix}. \tag{6}$$

According to the EKF, the relative position and velocity can be estimated using the equations above during the angles-only relative navigation. It is easy to find that the maneuver of pursuer or evader will affect the relative navigation, measurement, and the observability of the system.

3.3. Space Geometry Analysis

Definition 1. Assume evader executes a nonzero evasive maneuver at t_0; if the angle measurements pursuer acquired after the evasive maneuver stays the same compared with those of no maneuver, then the evasive maneuver can be called completely unobservable maneuver.

Figure 2 shows the projection of relative motion trajectory in xz plane and describes the space geometry of completely unobservable maneuver. Assuming the evader's orbit without maneuvers as a reference orbit, then trajectories

l_P and l'_E describe the relative motion of the pursuer and of the evader when evasion maneuvers are exploited. In Figure 2, $\mathbf{A}_k\mathbf{O}$ represents the relative position vector of the pursuer, and $\mathbf{OA}'_k$ the relative position vector of the evader with an evasive maneuver, with respect to the reference orbit. Definition 1 can be illustrated as follows: the relative motion trajectory l'_E satisfies the geometry relationship described in Figure 2, namely, $\mathbf{A}_k\mathbf{O} = \lambda\mathbf{OA}'_k$, at any time after evasion ($k = 1, 2, 3, \ldots, \lambda$ is nonnegative). Thus, the angle measurements pursuer acquired after the evasive maneuver stays the same compared with those without evasion. In this situation, pursuer will never find out whether evader executes an evasive maneuver. When l'_E in Figure 2 exits from the scenario of the pursuer, the correspondent evasive maneuver $\mathbf{u}_E$ is called completely unobservable maneuver.

However, the completely unobservable maneuver is only a hypothesis about space geometry. After derivation, it can be found that the completely unobservable maneuver does not exist. We have the following.

Theorem 2. *If the initial relative distance of pursuer and evader is nonzero, the evasive maneuver evader executed cannot become completely unobservable maneuver.*

Proof. Assume a completely unobservable maneuver $\mathbf{u}_E$ exits. Set t_0 as initial time and take the initial two steps t_1 and t_2 as example. The relative position of evader after evasion and the reference orbit are as follows:

$$\begin{aligned} \mathbf{X}_{Er1} &= \mathbf{B}_r\mathbf{u}_E, \\ \mathbf{X}_{Er2} &= \mathbf{\Phi}_p\mathbf{B}\mathbf{u}_E. \end{aligned} \tag{7}$$

If the relative state of pursuer and the reference orbit is $\mathbf{X}_{P0}$ and the position component is nonzero, then

$$\begin{aligned} \mathbf{X}_{Pr1} &= \mathbf{\Phi}_p\mathbf{X}_{P0}, \\ \mathbf{X}_{Pr2} &= \mathbf{\Phi}_p^2\mathbf{X}_{P0}, \end{aligned} \tag{8}$$

where $\mathbf{\Phi}_p^2 = \mathbf{\Phi}_p\mathbf{\Phi}$, $\mathbf{X}_{Pr1}$ and $\mathbf{X}_{Pr2}$ are position vectors.

According to (5) and Definition 1, we have

$$\begin{aligned} \mathbf{H}(\mathbf{y}_1)\mathbf{X}_{Pr1} &= \mathbf{H}(\mathbf{y}_1)\mathbf{X}_{Er1} = 0, \\ \mathbf{H}(\mathbf{y}_2)\mathbf{X}_{Pr2} &= \mathbf{H}(\mathbf{y}_2)\mathbf{X}_{Er2} = 0; \end{aligned} \tag{9}$$

namely,

$$\begin{aligned} \mathbf{H}(\mathbf{y}_1)(\mathbf{B}_r\mathbf{u}_E) &= \mathbf{H}(\mathbf{y}_1)\left(\mathbf{\Phi}_p\mathbf{X}_{P0}\right) = 0, \\ \mathbf{H}(\mathbf{y}_2)\left(\mathbf{\Phi}_p\mathbf{B}\mathbf{u}_E\right) &= \mathbf{H}(\mathbf{y}_2)\left(\mathbf{\Phi}_p^2\mathbf{X}_{P0}\right) = 0; \end{aligned} \tag{10}$$

According to the property of solution space, it is easy to find out the following relation:

$$\mathbf{B}_r\mathbf{u}_E = \alpha_1\left(\mathbf{\Phi}_p\mathbf{X}_{P0}\right), \tag{11}$$

$$\mathbf{\Phi}_p\mathbf{B}\mathbf{u}_E = \alpha_2\left(\mathbf{\Phi}_p^2\mathbf{X}_{P0}\right), \tag{12}$$

where α_i is nonzero real number.

Therefore, if (11) and (12) are proved to be held, the existence of completely unobservable maneuver can be proved. □

From (11), we have

$$\mathbf{u}_E = \alpha_1\left(\mathbf{B}_r^{-1}\mathbf{\Phi}_p\mathbf{X}_{P0}\right). \tag{13}$$

Substitute (13) into (12) and take the second step of (12) into consideration; then

$$\begin{aligned} \mathbf{\Phi}_p\left(\kappa_1\mathbf{\Phi} - \mathbf{B}\mathbf{B}_r^{-1}\mathbf{\Phi}_p\right)\mathbf{X}_{P0} &= 0, \\ \mathbf{\Phi}_p^2\left(\kappa_2\mathbf{\Phi} - \mathbf{B}\mathbf{B}_r^{-1}\mathbf{\Phi}_p\right)\mathbf{X}_{P0} &= 0, \end{aligned} \tag{14}$$

where $\kappa_i = \alpha_{i+1}/\alpha_i$.

Since $\mathbf{\Phi}_p$ is linear transformation and (14) do hold, the necessary and sufficient condition of the nonzero solution existence of $\mathbf{X}_{P0}$ is $\kappa_1 = \kappa_2 \equiv \kappa$ (the initial relative position is nonzero). Thus,

$$\mathbf{\Phi}_p\left(\kappa\mathbf{\Phi} - \mathbf{B}\mathbf{B}_r^{-1}\mathbf{\Phi}_p\right)\mathbf{X}_{P0} = 0. \tag{15}$$

In addition, if the nonzero solution of $\mathbf{X}_{P0}$ exists, according to the existence theorem of solutions of linear equation, we can acquire $\operatorname{rank}(\kappa\mathbf{\Phi} - \mathbf{B}\mathbf{B}_r^{-1}\mathbf{\Phi}_p) < 6$. Expand $(\kappa\mathbf{\Phi} - \mathbf{B}\mathbf{B}_r^{-1}\mathbf{\Phi}_p)$ and one can get

$$\begin{aligned} &\left(\kappa\mathbf{\Phi} - \mathbf{B}\mathbf{B}_r^{-1}\mathbf{\Phi}_p\right) \\ &= \begin{bmatrix} (\kappa - 1)\mathbf{\Phi}_{11} & (\kappa - 1)\mathbf{\Phi}_{12} \\ \kappa\mathbf{\Phi}_{21} - \mathbf{\Phi}_{22}\mathbf{\Phi}_{12}^{-1}\mathbf{\Phi}_{11} & (\kappa - 1)\mathbf{\Phi}_{22} \end{bmatrix}. \end{aligned} \tag{16}$$

Because $\operatorname{rank}(\mathbf{\Phi}) = 6$, it is easy to know that if and only if $\beta = 1$ the expression $\operatorname{rank}(\kappa\mathbf{\Phi} - \mathbf{B}\mathbf{B}_r^{-1}\mathbf{\Phi}_p) < 6$ can be held. Equation (15) is revised as

$$\mathbf{\Phi}_p\left(\mathbf{\Phi} - \mathbf{B}\mathbf{B}_r^{-1}\mathbf{\Phi}_p\right)\mathbf{X}_{P0} = 0. \tag{17}$$

Substitute $\kappa = 1$ into (16), then

$$\begin{aligned} &\begin{bmatrix} (\kappa - 1)\mathbf{\Phi}_{11} & (\kappa - 1)\mathbf{\Phi}_{12} \\ \kappa\mathbf{\Phi}_{21} - \mathbf{\Phi}_{22}\mathbf{\Phi}_{12}^{-1}\mathbf{\Phi}_{11} & (\kappa - 1)\mathbf{\Phi}_{22} \end{bmatrix} \\ &= \begin{bmatrix} 0 & 0 \\ \mathbf{\Phi}_{21} - \mathbf{\Phi}_{22}\mathbf{\Phi}_{12}^{-1}\mathbf{\Phi}_{11} & 0 \end{bmatrix}. \end{aligned} \tag{18}$$

Set $\mathbf{X}_{P0} = [\mathbf{X}_{Pr0} \quad \mathbf{X}_{Pv0}]^{\mathrm{T}}$. It can be seen from (17) and (18) that the value of $\mathbf{X}_{Pv0}$ makes no influence on the result of (17). When $\mathbf{X}_{Pr0}$ is arbitrary, the equalities do not always hold. Furthermore, (17) always holds only when $\mathbf{X}_{Pr0} = 0$. This condition goes with the antecedent hypothesis of *Theorem 2*; therefore *Theorem 2* is proved in this way. Unobservable maneuvers proofs have previously been developed for the bearings-only navigation problem in circular orbit [21]; here the related conclusion is extended to elliptical orbit successfully.

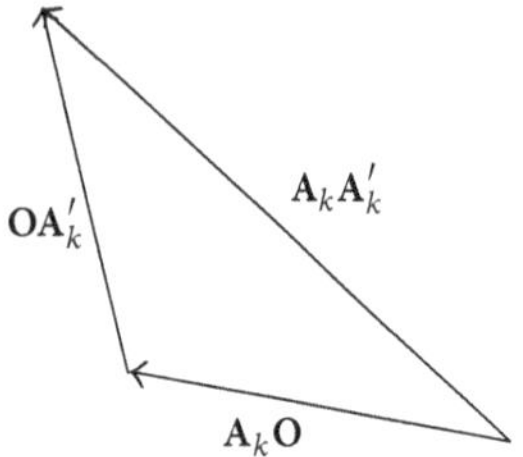

FIGURE 3: The components of measuring line of sight.

Though the completely unobservable maneuver is nonexistent, it is set under the situation of ideal measurement. In reality, if the angle measurements pursuer acquired after the evasive maneuver stays quite close to those of no maneuver and the difference approach to the measurement accuracy, pursuer could be unable to identify the evasive maneuver. Thus, these evasive maneuvers can be called approximate solution of the completely unobservable maneuver. It can be said that the difficulty of maneuver tracking for pursuer is increased when the difference of angle measurements the evasion caused is decreased. This is a new index to evaluate the superiority of an evasive maneuver.

4. Optimal Evasion Maneuvers Analysis

In this section, an analytical solution is derived based on the conclusion of Section 3 and meanwhile a numerical solution is given to prove the optimality of analytical solution.

4.1. Analytical Solution

4.1.1. Quantification of Space Geometry. To find the optimal evasive maneuver, it is necessary to quantify the relationship of angle measurements variation and space geometry of pursuer and evader. In Euclidean space, orthogonality between two vectors can be defined by using the notion dot product. Thus, two column unit vectors **a** and **b** are orthogonal when $\mathbf{a}^{\mathrm{T}} \cdot \mathbf{b} = 0$. On the contrary, if the scalar product of **a** and **b** is 1 or −1 ($\mathbf{a}^{\mathrm{T}} \cdot \mathbf{b} = 1$ or $\mathbf{a}^{\mathrm{T}} \cdot \mathbf{b} = -1$), then **a** and **b** are parallel.

From Figure 2 we know that $\mathbf{A}_k\mathbf{O}$ is the measuring line of sight of pursuer and the reference orbit, $\mathbf{A}_k\mathbf{A}'_k$ is the measuring line of sight of pursuer and evader after evasion. To each measuring time, the norm of $\mathbf{A}_k\mathbf{O}$ and $\mathbf{A}_k\mathbf{A}'_k$ is constant. Therefore, in order to decrease the variation of angle measurements caused by evasion, the included angle of $\mathbf{A}_k\mathbf{A}'_k$ and $\mathbf{A}_k\mathbf{O}$ must be as small as possible. As an extreme example, when the included angle of $\mathbf{A}_k\mathbf{A}'_k$ and $\mathbf{A}_k\mathbf{O}$ is 0 or $-\pi$, the evasive maneuver becomes the completely unobservable maneuver.

Since the norm of $\mathbf{A}_k\mathbf{A}'_k$ and $\mathbf{A}_k\mathbf{O}$ is definite value, if the evasive maneuver is optimal, the scalar product of $\mathbf{A}_k\mathbf{A}'_k$ and $\mathbf{A}_k\mathbf{O}$ should be maximum. It is seen from Figures 2 and 3 that the measuring line of sight $\mathbf{A}_k\mathbf{A}'_k$ can be expressed as

$$\mathbf{A}_k\mathbf{A}'_k = \mathbf{A}_k\mathbf{O} + \mathbf{O}\mathbf{A}'_k. \quad (19)$$

At k time, the scalar product of $\mathbf{A}_k\mathbf{A}'_k$ and $\mathbf{A}_k\mathbf{O}$ is

$$\begin{aligned}\mathbf{A}_k\mathbf{O}^{\mathrm{T}} \cdot \mathbf{A}_k\mathbf{A}' &= \mathbf{A}_k\mathbf{O}^{\mathrm{T}} \cdot \left(\mathbf{A}_k\mathbf{O} + \mathbf{O}\mathbf{A}'_k\right) \\ &= \mathbf{A}_k\mathbf{O}^{\mathrm{T}} \cdot \mathbf{A}_k\mathbf{O} + \mathbf{A}_k\mathbf{O}^{\mathrm{T}} \cdot \mathbf{O}\mathbf{A}'_k.\end{aligned} \quad (20)$$

It is easy to know that the value of $\mathbf{A}_k\mathbf{O}^{\mathrm{T}} \cdot \mathbf{A}_k\mathbf{O}$ is constant at each time; thus the scalar product of $\mathbf{A}_k\mathbf{A}'_k$ and $\mathbf{A}_k\mathbf{O}$ is decided by $\mathbf{A}_k\mathbf{O}^{\mathrm{T}} \cdot \mathbf{O}\mathbf{A}'_k$. Therefore, to minimize the variation of angle measurements at k time, the optimization at k time should minimize the following object:

$$J_k = -\mathbf{A}_k\mathbf{O}^{\mathrm{T}} \cdot \mathbf{O}\mathbf{A}'_k. \quad (21)$$

According to the relationship of evasive maneuver and space geometry, (21) can be revised as

$$J_k = -\left(\mathbf{\Phi}_P\mathbf{X}_{P0}\right)^{\mathrm{T}} \cdot \mathbf{B}_r\mathbf{u}. \quad (22)$$

During the approaching, the optimal evasion should minimize the object J_k at every moment; thus the object function of the whole approaching is as follows:

$$J = \sum_{k=0}^{N} J_k = -\sum_{k=0}^{N} \left(\mathbf{\Phi}_P\left(t_k, t_0\right)\mathbf{X}_{P0}\right)^{\mathrm{T}} \cdot \mathbf{B}_r\left(t_k, t_0\right)\mathbf{u}, \quad (23)$$

where N is the number of measurements.

Equation (23) shows that the evasive effectiveness is not only dependent on the maneuvers performed by **u** but also dependent on the position where they are executed. Since this equation is closed form, it allows for simple inclusion inside a global trajectory optimization scheme as an additional objective or independent objective, weighed by other objective such as relative distance or fuel consumption. This would permit a global trajectory optimizer to choose maneuvers that will increase the difficulty of maneuver tracking for pursuer. It is worth noting that (23) provides explicit solutions for any arbitrary set of initial conditions (nonzero initial relative position).

4.2. Algebraic Optimal Evasive Maneuver. To find the optimal evasive maneuver for a constant duration of measurement, the optimization variable of interest in (23) is the evasive maneuver **u**. The initial state $\mathbf{X}_{P0}$ is fixed when the evasive maneuver is executed. N is the number of measurements.

Since (23) is a linear function for evasive maneuver, it can only be minimized with respect to **u**. After the former quantification analysis above, the optimal evasive maneuver can be found based on (23). It is logical to limit the desired evasive maneuver **u** magnitude s in order to find the optimal evasive maneuver direction. Then, this constraint can be mathematically posed as follows:

$$\chi\left(\mathbf{u}\right) = \mathbf{u}^{\mathrm{T}}\mathbf{u} - s^2 = 0. \quad (24)$$

It should be noted that additional constraints could be considered here, if it is needed. Closed form solutions would be possible by the following steps with constraints mentioned above.

The constrained optimization problem transfers to minimize (23) with respect to the evasive maneuver $\mathbf{u}$, under the equality constraint proposed by (24). Since there is no inequality constraint in this problem, it can be converted to an equivalent unconstrained problem with the Lagrange multiplier technique used in [28]. The problem is converted to minimize the Lagrangian function

$$\Gamma(\mathbf{u}, \lambda) = J(\mathbf{u}) - \lambda \chi(\mathbf{u}) = 0, \tag{25}$$

where λ is the Lagrange multiplier corresponding to the equality constrain in (24).

The first-order optimality conditions are given by the derivatives of the Lagrangian function with respect to the optimization variables equal to zero, as well as the Lagrange multiplier. Thus, we have

$$\begin{aligned} \frac{\partial \Gamma}{\partial \mathbf{u}} &= -\sum_{k=0}^{N} \left(\mathbf{\Phi}_P(t_k, t_0)\mathbf{X}_{P0}\right)^{\mathrm{T}} \mathbf{B}_r(t_k, t_0) - 2\lambda \mathbf{u}^{\mathrm{T}} = 0, \\ \frac{\partial \Gamma}{\partial \lambda} &= \mathbf{u}^{\mathrm{T}}\mathbf{u} - s^2 = 0. \end{aligned} \tag{26}$$

According to (26), we have

$$\begin{aligned} \mathbf{u} &= -\frac{\sum_{k=0}^{N} \mathbf{B}_r^{\mathrm{T}}(t_k, t_0)\left(\mathbf{\Phi}_P(t_k, t_0)\mathbf{X}_{P0}\right)}{2\lambda}, \\ \lambda &= \pm\frac{1}{2s}\sqrt{\sum_{k=0}^{N} \left(\mathbf{\Phi}_P(t_k, t_0)\mathbf{X}_{P0}\right)^{\mathrm{T}} \mathbf{B}_r(t_k, t_0) \sum_{k=0}^{N} \mathbf{B}_r^{\mathrm{T}}(t_k, t_0)\left(\mathbf{\Phi}_P(t_k, t_0)\mathbf{X}_{P0}\right)}. \end{aligned} \tag{27}$$

Take the second-order optimality conditions into consideration in order to identify the stationary point corresponding to the minimum of the Lagrangian function.

$$\begin{aligned} &\frac{\partial^2 \Gamma}{\partial \mathbf{u}^2} = -2\lambda > 0 \longrightarrow \\ &\lambda < 0. \end{aligned} \tag{28}$$

According to (27) and (28), an algebraic expression for optimal evasive maneuver $\mathbf{u}_{\mathrm{opt}}$ can be obtained as follows:

$$\mathbf{u}_{\mathrm{opt}} = -s\frac{\sum_{k=0}^{N} \mathbf{B}_r^{\mathrm{T}}(t_k, t_0)\left(\mathbf{\Phi}_P(t_k, t_0)\mathbf{X}_{P0}\right)}{\sqrt{\sum_{k=0}^{N} \left(\mathbf{\Phi}_P(t_k, t_0)\mathbf{X}_{P0}\right)^{\mathrm{T}} \mathbf{B}_r(t_k, t_0) \sum_{k=0}^{N} \mathbf{B}_r^{\mathrm{T}}(t_k, t_0)\left(\mathbf{\Phi}_P(t_k, t_0)\mathbf{X}_{P0}\right)}}. \tag{29}$$

Equation (29) is the analytical expression of optimal evasive maneuver including the initial states $\mathbf{x}0$ and the expectable evasive maneuver magnitude s. Till now, the expression of the analytical solution is obtained, and it provides the optimal evasive maneuver with respect to relative space geometry and the angle measurements. The state transition $\mathbf{\Phi}_p$ and input transition $\mathbf{B}_r$ were proposed in Section 3.

4.3. Numerical Solution. In order to prove the optimality of analytical solution, an optimization model is established in this section. One of the most well-known evolutionary algorithms, GA, is employed to solve the optimization problem. The GA has been successfully applied in spacecraft trajectory optimization, for example, in designing low-thrust trajectories [29] and solving two different kinds of problems typical to astrodynamics [30].

Optimization Variables. Since the desired evasive maneuver ΔV magnitude is limited to s, in order to find the optimal direction of evasive maneuver, azimuth θ and elevation ε are selected as two optimization variables; namely,

$$\mathbf{D} = [\varepsilon, \theta]^{\mathrm{T}}. \tag{30}$$

In consideration of the space geometric relationship between evader and pursuer, the constraint conditions for an orbital evasion problem read

$$\begin{aligned} -2\pi &\le \varepsilon \le 2\pi, \\ -\pi &\le \theta \le \pi. \end{aligned} \tag{31}$$

In order to further simplify the problem, here we focus on the value range of the optimization variables. For two satellites in orbit, it is known from [19] that an increased difference of the system relative motion has a positive correlation with system observability. Therefore, if the initial orbit is coplanar, the evasive maneuver should better be coplanar, too.

Under such circumstance, new constraint conditions are as stated in the following:

$$\begin{gathered}-2\pi \leq \varepsilon \leq 2\pi, \\ |\theta| \leq \xi, \quad \xi = 0.01.\end{gathered} \tag{32}$$

Objective Function. Assume that $\overline{\mathbf{A}_k\mathbf{O}}$ and $\overline{\mathbf{OA}'_k}$ are the unit vectors of $\mathbf{A}_k\mathbf{O}$ and $\mathbf{OA}'_k$, separately. According to Section 4.1, it is easy to know that the closer to 1 the value of $\overline{\mathbf{A}_k\mathbf{O}} \cdot \overline{\mathbf{OA}'_k}$ is, the smaller the variation of angle measurements is. During the measurement, when the sum of $\overline{\mathbf{A}_k\mathbf{O}} \cdot \overline{\mathbf{OA}'_k}$ is max, the evasive maneuver is optimal. Thus, objective function is as follows:

$$\max_{\mathbf{u}} f(\mathbf{u}) = \sum_{k=0}^{N} \overline{\mathbf{A}_k\mathbf{O}} \cdot \overline{\mathbf{OA}'_k}, \tag{33}$$

where $\overline{\mathbf{A}_k\mathbf{O}}$ and $\overline{\mathbf{OA}'_k}$ are equal to the unit vector of $\mathbf{\Phi}_p(t_k, t_0)\mathbf{X}_{P0}$ and $\mathbf{B}_r(t_k, t_0)\mathbf{u}$.

In order to use GA, the objective function is revised as

$$\min_{\mathbf{u}} f(\mathbf{u}) = -\sum_{k=0}^{N} \overline{\mathbf{A}_k\mathbf{O}} \cdot \overline{\mathbf{OA}'_k}. \tag{34}$$

Thus, the GA optimization model is established. Through the results comparison of analytical and numerical solutions, the optimality of analytical solution can be proved.

5. Simulation

Simulation results are presented in this section. Consider an illustrative example: evader is in a HEO and semimajor axis is 45485189 m, eccentricity is 0.713, inclination of orbit is 1.10187 rad, right ascension of ascending node (RAAN) is 0.84489 rad, argument of perigee is 4.7022 rad, and true anomaly is 1.4856 rad. Assume that the initial time is $t_0 = 0$, and initial relative states $\mathbf{X}_{P0}$ at t_0 is

$$\begin{aligned}\mathbf{X}_{P0} = [&-85204.071\,\mathrm{m}, \\ &-47965.164\,\mathrm{m}, 0, 16.845\,\mathrm{m/s}, 4.306\,\mathrm{m/s}, 0.0]^{\mathrm{T}}.\end{aligned} \tag{35}$$

The pursuer uses optical camera to get relative measurements. The measurement frequency of the optical camera is assumed as 0.1 Hz. The magnitude of evasive impulse is fixed to 3 m/s (namely, $s = 3$ m/s). The optimization parameters of GA are as follows: 100 for population size; 30 for maximum generations number; 0.90 for crossover probability; and 0.08 for mutation probability. Constraint conditions are chosen as those in Section 4.2.

If the evasion is aimed at 101 angle measurements, then the measuring time is 1000 s. It is easy to know that the minimum value of objective function should be −101. Thus, the optimization results are obtained in Figure 4.

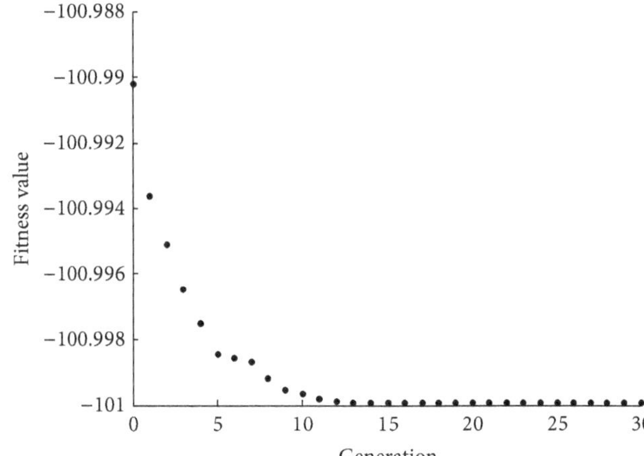

Figure 4: The optimization results of GA.

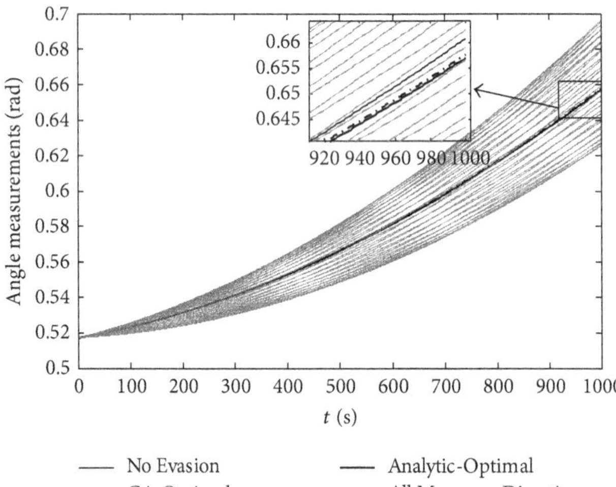

Figure 5: Angle measurements in 1000 s.

It can be seen from Figure 4 that the numerical solution is −100.9998. After 15 generations, the results reach convergence. The direction of optimal evasive maneuver is $\mathbf{D}_{V\text{opt}} = [-2.4031, 0]^{\mathrm{T}}$. Take correlated parameters into (29), the analytical solution is acquired and the direction of analytical evasive maneuver is $\mathbf{D}_{V\text{analytical}} = [-2.4219, 0]^{\mathrm{T}}$.

The measurement effectiveness of the optimal evasive maneuvers is validated by comparing the GA-optimal and analytical-optimal evasive maneuvers to the propagation of 80 different maneuver directions, covering ε from -2π to 2π. The angle measurements during the previous 1000 s are shown in Figure 5.

Set the angle measurements of reference orbit as reference measurements. The angle measurements with an evasive maneuver at each time minus reference measurements become the variation of angle measurements. The variation of angle measurements during the previous 1000 s is given in Figure 6.

As seen from Figure 6, the variation of angle measurements caused by the GA-optimal and analytical-optimal

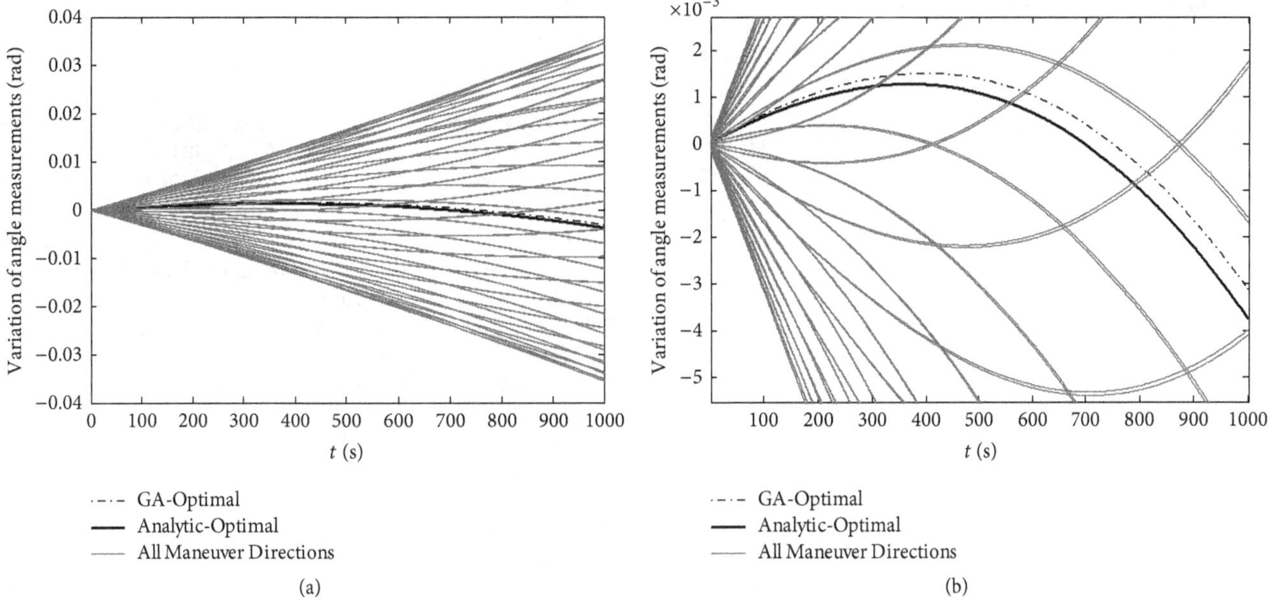

FIGURE 6: The variation of angle measurements due to optimal maneuvers (a) and zoom (b).

solutions are much smaller than other maneuver directions. Although the analytical-optimal variation of angle measurements at the end time is bigger than that of GA-optimal, the sum of the variation is optimal for the whole measuring time. It can be seen that the variation of angle measurements caused by the analytical-optimal solution is no more than 0.001 rad for the previous 800 s, which means that the pursuer will even not find out the evasion if the measurement accuracy is not better than 0.001 rad.

In order to show that the proposed analytical and numerical maneuvers provide the most guidance error in the filter estimate, a Kalman filter is used as the navigation filter. The simulation conditions are the same with the simulation in Figures 4–6, and the initial guidance error is assumed to be nonzero. The simulation step is set to 5 s. The guidance error of the optimal evasive maneuvers is validated by comparing the GA-optimal and analytical-optimal evasive maneuvers to the propagation of 60 different maneuver directions, covering ε from -2π to 2π. The guidance errors in the two directions of orbit plane are shown in Figures 7 and 8.

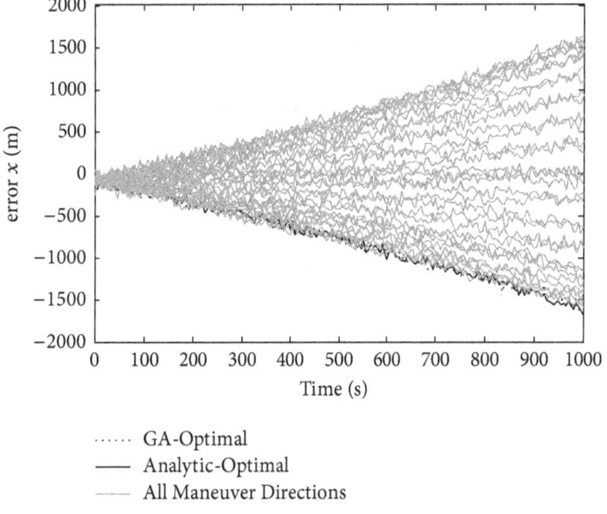

FIGURE 7: Navigation errors caused by maneuvers (Dircetion x).

As Figures 7 and 8 show, the analytical and numerical maneuvers provide the most improvement of guidance error. Though the guidance error caused by the analytical maneuvers is better than that of numerical maneuvers, the difference is minor. The simulation results show that the analytical analysis in Section 4 is effective. Moreover, since the solution is analytical, it will have a potential application in engineering utilization.

6. Conclusions

An analytical optimal evasion strategy is proposed for an evading satellite against a noncooperative rendezvous spacecraft. This work extends the research objective to not only orbit debris but also spacecraft with maneuver ability. Through analysis of relative space geometry, the completely unobservable maneuver is defined and proved to be nonexistent. An analytical closed-form solution is proposed to compute optimal evasive maneuvers for angles-only navigation. Based on this analytical method, an optimal evasive maneuver can be quickly obtained. Since the previous evasion strategies required numerical optimization which a lot of time is needed, the method in this work should be an effective improvement compared with the previous evasion state of art. Moreover, in order to decline the navigation accuracy of pursuer, the relative space geometry is analyzed and quantified. Though the derivation is under the situation of HEO, it is also adaptable for orbital evasion problem in circular orbit.

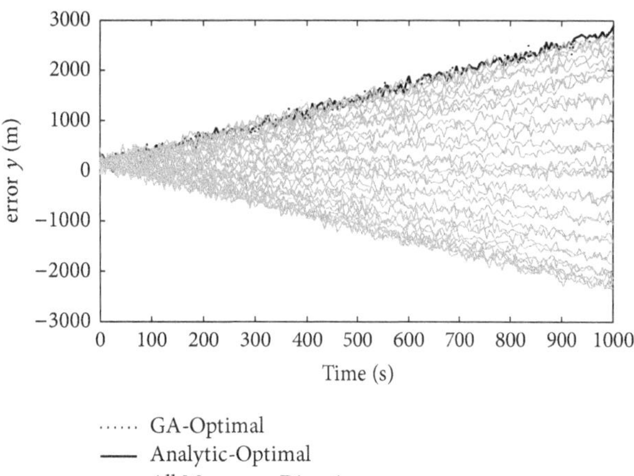

FIGURE 8: Navigation errors caused by maneuvers (Dircetion y).

This work proves that the angle measurements and navigation accuracy can be used in orbital evasion problem, which is often neglected in previous research. This research proposes a new research method that means a potential step closer to engineering utilization.

Conflicts of Interest

The authors declare that they have no conflict of interests.

Acknowledgments

Project is supported by the National Natural Science Foundation of China (Grant no. 11572345).

References

[1] J. Shinar, Y. Rotsztein, and E. Bezner, "Analysis of three-dimensional optimal evasion with linearized kinematics," *Journal of Guidance, Control, and Dynamics*, vol. 2, no. 5, pp. 353–360, 1979.

[2] I. Forte, A. Steinberg, and J. Shinar, "The effects of non–linear kinematics in optimal evasion," *Optimal Control Applications and Methods*, vol. 4, no. 2, pp. 139–152, 1983.

[3] N. Sánchez-Ortiz, M. Belló-Mora, and H. Klinkrad, "Collision avoidance manoeuvres during spacecraft mission lifetime: risk reduction and required ΔV," *Advances in Space Research*, vol. 38, no. 9, pp. 2107–2116, 2006.

[4] T. Shima, "Optimal cooperative pursuit and evasion strategies against a homing missile," *Journal of Guidance, Control, and Dynamics*, vol. 34, no. 2, pp. 414–425, 2011.

[5] K. Chan, "Spacecraft maneuvers to mitigate potential collision threats," in *Proceedings of the Astrodynamics Specialist Conference and Exhibit (AIAA/AAS '02)*, Monterey, Calif, USA, August 2002.

[6] H.-D. Kim and H.-J. Kim, "Optimal collision avoidance maneuver to maintain a LEO station keeping," in *Proceedings of the 61st International Astronautical Congress (IAC '10)*, pp. 3056–3060, International Astronautical Federation (IAF), Prague, Czech Republic, October 2010.

[7] J. Shinar and D. Steinberg, "Analysis of optimal evasive maneuvers based on a linearized two-dimensional kinematic model," *Journal of Aircraft*, vol. 14, no. 8, pp. 795–802, 1977.

[8] I. Forte, A. Steinberg, and J. Shinar, "The effects of non-linear kinematics in optimal evasion," *Optimal Control Applications and Methods*, vol. 4, no. 2, pp. 139–152, 1983.

[9] B. D. Kelly and S. D. Picciotto, "Probability based optimal collision avoidance maneuvers," AIAA 2005-6775, 2005.

[10] R. P. Patera, "General method for calculating satellite collision probability," *Journal of Guidance, Control, and Dynamics*, vol. 24, no. 4, pp. 716–722, 2001.

[11] R. P. Patera and G. E. Peterson, "Space vehicle maneuver method to lower collision risk to an acceptable level," *Journal of Guidance, Control, and Dynamics*, vol. 26, no. 2, pp. 233–237, 2003.

[12] R. P. Patera, "Satellite collision probability for nonlinear relative motion," *Journal of Guidance, Control, and Dynamics*, vol. 26, no. 5, pp. 728–733, 2003.

[13] C. Bombardelli, "Analytical formulation of impulsive collision avoidance dynamics," *Celestial Mechanics and Dynamical Astronomy*, vol. 118, no. 2, pp. 99–114, 2014.

[14] S.-C. Lee, H.-D. Kim, and J. Suk, "Collision avoidance maneuver planning using GA for LEO and GEO satellite maintained in keeping area," *International Journal of Aeronautical and Space Sciences*, vol. 13, no. 4, pp. 474–483, 2012.

[15] E. H. Kim, H. D. Kim, and H. J. Kim, "Optimal solution of collision avoidance maneuver with multiple space debris," *Journal of Space Operations—Letters*, vol. 9, no. 3, pp. 20–31, 2012.

[16] A. D. de Jesus and R. R. de Sousa, "Processing optimized for symmetry in the problem of evasive maneuvers," *Computational & Applied Mathematics*, vol. 34, no. 2, pp. 521–534, 2015.

[17] S. E. Hammel and V. J. Aidala, "Observability Requirements for Three-Dimensional Tracking via Angle Measurements," *IEEE Transactions on Aerospace and Electronic Systems*, vol. 21, no. 2, pp. 200–207, 1985.

[18] S. C. Nardone and V. J. Aidala, "Observability criteria for bearings-only target motion analysis," *Institute of Electrical and Electronics Engineers. Transactions on Aerospace and Electronic Systems*, vol. 17, no. 2, pp. 162–166, 1981.

[19] D. A. Vallado, "Evaluating gooding angles-only obit determination of space based space surveillance measurements," in *Proceedings of the AAS Born Symposium*, Boulder, Colo, USA, May 2010.

[20] D. C. Woffinden and D. K. Geller, "Observability criteria for angles-only navigation," *IEEE Transactions on Aerospace and Electronic Systems*, vol. 45, no. 3, pp. 1194–1208, 2009.

[21] J. Grzymisch and W. Fichter, "Observability criteria and unobservable maneuvers for in-orbit bearings-only navigation," *Journal of Guidance, Control, and Dynamics*, vol. 37, no. 4, pp. 1250–1259, 2014.

[22] J. Grzymisch and W. Fichter, "Analytic optimal observability maneuvers for in-orbit bearings-only rendezvous," *Journal of Guidance, Control, and Dynamics*, vol. 37, no. 5, pp. 1658–1664, 2014.

[23] Y. Dateng, Y. Luo, J. Zicheng, and G. Tang, "Multi-objective evolutionary optimization of evasive maneuvers including observability performance," in *Proceedings of the IEEE Congress on Evolutionary Computation (CEC '15)*, pp. 603–610, Sendai, Japan, May 2015.

[24] D. C. Woffinden and D. K. Geller, "Optimal orbital rendezvous maneuvering for angles-only navigation," *Journal of Guidance, Control, and Dynamics*, vol. 32, no. 4, pp. 1382–1387, 2009.

[25] W. Fehse, *Automated Rendezvous and Docking of Spacecraft*, Cambridge University Press, Cambridge, UK, 2003.

[26] K. Alfriend, S. R. Vadali, and P. Gurfil, *Spacecraft Formation Flying: Dynamics, Control and Navigation*, vol. 2, Butterworth-Heinemann, Oxford, UK, 2009.

[27] K. Yamanaka and F. Ankersen, "New state transition matrix for relative motion on an arbitrary elliptical orbit," *Journal of Guidance, Control, and Dynamics*, vol. 25, no. 1, pp. 60–66, 2002.

[28] S. Boyd and L. Vandenberghe, *Convex optimization*, Cambridge University Press, Cambridge, UK, 2004.

[29] G. A. Rauwolf and V. L. Coverstone-Carroll, "Near-optimal low-thrust orbit transfers generated by a genetic algorithm," *Journal of Spacecraft and Rockets*, vol. 33, no. 6, pp. 859–862, 1996.

[30] G. Taini, L. Amorosi, A. Notarantonio, and G. B. Palmerini, "Applications of genetic algorithms in mission design," in *Proceedings of the IEEE Aerospace Conference*, 902, 890 pages, IEEE, Big Sky, Mont, USA, March 2005.

4

Experimental Validation of Fly-Wheel Passive Launch and On-Orbit Vibration Isolation System by Using a Superelastic SMA Mesh Washer Isolator

Seong-Cheol Kwon,[1] Mun-Shin Jo,[2] and Hyun-Ung Oh[1]

[1] *Space Technology Synthesis Laboratory, Department of Aerospace Engineering, Chosun University, 375 Seosuk-dong, Dong-gu, Gwangju 501-759, Republic of Korea*
[2] *Mechanical Design Team, Hanwha Systems, 304 Cheoin-gu, Yongin 449-886, Republic of Korea*

Correspondence should be addressed to Hyun-Ung Oh; ohu129@chosun.ac.kr

Academic Editor: Paolo Tortora

On-board appendages with mechanical moving parts for satellites produce undesirable micro-jitters during their on-orbit operation. These micro-jitters may seriously affect the image quality from high-resolution observation satellites. A new application form of a passive vibration isolation system was proposed and investigated using a pseudoelastic SMA mesh washer. This system guarantees vibration isolation performance in a launch environment while effectively isolating the micro-disturbances from the on-orbit operation of jitter source. The main feature of the isolator proposed in this study is the use of a ring-type mesh washer as the main axis to support the micro-jitter source. This feature contrasts with conventional applications of the mesh washers where vibration damping is effective only in the thickness direction of the mesh washer. In this study, the basic characteristics of the SMA mesh washer isolator in each axis were measured in static tests. The effectiveness of the design for the new application form of the SMA mesh washer proposed in this study was demonstrated through both launch environment vibration test at qualification level and micro-jitter measurement test which corresponds to on-orbit condition.

1. Introduction

The quality of high-resolution images obtained from earth observation satellites can be degraded by undesirable micro-disturbances induced by on-board appendages that have mechanical moving parts, such as reaction wheel assembly (RWA) [1], control moment gyro [2], cryogenic cooler [3], and gimbal-type antenna [4]. Among the various vibration disturbance sources, micro-jitter caused by the RWA is known to have the greatest effect on the performance of the high-resolution optical payloads. The major causes of the disturbances from the RWA are rotor imbalance, bearing imperfections, motor disturbances, and motor driven errors [5]. Recently, micro-jitter level is becoming quietly lower by high precise balancing of the spinning rotor of the RWAs. However, these micro-jitters may still seriously affect the image quality of high-resolution observation satellites as the acceptable amplitudes of the micro-jitters are also becoming much lower to meet strict mission requirements.

To ensure the performance of observation satellites, it is important task to isolate the micro-jitters from the RWAs. This isolation can be accomplished by mounting the RWAs on a vibration isolator with low stiffness by attenuating the transmitted vibration to the main payload structures, which are sensitive to the micro-jitters. Kamesh et al. [6] proposed a low-frequency flexible platform that consisted of folded continuous beams for passive RWA vibration isolation. The effectiveness of the design was confirmed through tests conducted by mounting the RWA on several configurations of the flexible platform. The results indicated that this passive isolation platform is capable of mitigating the disturbance vibration by serving as an isolation mount, not only for RWAs but also for other disturbance sources in spacecraft.

Davis et al. [7] proposed a passive viscous-damped strut (D-strut) with a very low fundamental frequency of 1.5 Hz to isolate disturbance forces and torques from a Honeywell HM-1800 RWA with a frequency band from 2 to 300 Hz. In addition, it has been widely used in various space programs including the Hubble mission. Zhang et al. [8] proposed a new vibration isolation system that contains a multistrut vibration isolation platform with multiple tuned mass dampers to achieve RWA micro-jitter isolation. The numerical simulation results demonstrated that the RWA disturbances were effectively attenuated by the newly proposed vibration isolation system while simultaneously validating its safety performance when the speed of the RWA crossed the resonance frequency.

Oh et al. [9] proposed a RWA variable-damping isolator using a biometal fiber valve with the characteristics of low power consumption, which can select the suitable damping according to the rotating speed of the RWA in order to minimize the transmitted force to the main payload structure. Additionally, the effectiveness of the design was demonstrated by the micro-jitter measurement test of a dummy RWA supported by the isolator on an air-floating type micro-disturbance measurement device [10].

Several types of passive RWA vibration isolation systems [11–14], including the aforementioned applications, provide excellent micro-jitter isolation performance. However, the structural safety of the RWA cannot be guaranteed under the severe vibration conditions experienced in a launch environment when supported by a low-stiffness isolator. If the isolator supporting the RWA is rigidly fixed by a holding and release mechanism during launch and then released in an orbit, the structural safety problems can be readily solved. However, this approach increases the system complexity, reduces its reliability, and increases the mass of the total system. In addition, if there is a problem in activating the release mechanism, then the micro-vibrations cannot be expected anymore. Therefore, Oh et al. [15] developed a passive launch and on-orbit vibration isolation system (PLOVIS) for the purpose of vibration attenuation of a cryogenic cooler under both launch and on-orbit vibration environments. The effectiveness of the isolator was demonstrated by launch environment tests and a micro-vibration measurement test of the cooler. Oh et al. [16, 17] also proposed a strategy to use a compressed shape memory alloy (SMA) mesh washer on the existing cooler isolation system [15] to attenuate the micro-vibration induced by cryocooler operation in on-orbit environment. Additionally, the SMA washer effectively enhanced the vibration attenuation performance during severe launch environments when compared to the conventional isolation system [15]..

In this study, we investigate the possibility of achieving a novel launch and on-orbit vibration isolation system which guarantees the vibration isolation performance in a launch environment without the requirement of a holding and release mechanism, while effectively isolating the micro-jitter disturbances from the RWAs on-orbit. For this, we focused on a compressed mesh washer that exploits the pseudoelasticity of SMA wire, as proposed by Youn et al. [18]. The largely deformed SMA mesh washer can recover its original shape without plastic deformation upon unloading because of its pseudoelastic or superelastic SMA behavior, which is caused by stress-induced phase transformation from austenite to martensite. The large damping of the mesh washer is also one of the great advantages of the SMA mesh washer. The strategy to achieve the design goal of this study is to use the radial direction of the ring-type SMA mesh washer as a main axis to support the vibration sources. This strategy is in contrast with the applications of conventional mesh washers [16–20], in which vibration damping is effective only in the thickness direction of the washer.

In the previous study, the effectiveness of the new application form of the SMA mesh washer using a radial direction of the ring-type mesh washer was investigated [21, 22]. Kwon et al. proposed a novel pseudoelastic gear that utilized the ring-type SMA mesh washer as a spring-blade to bridge the output shaft of the motor with an existing gear wheel [21]. This application was researched and developed for attenuating the micro-jitter induced by stepper motor activation of the two-axis gimbal antenna. This application makes it possible to achieve a low rotational stiffness and high damping characteristics, which are the main contributors to the micro-jitter isolation capability. In addition, the pseudoelasticity of the SMA mesh washer guaranteed the structural safety of the gear itself even under unexpected circumstances such as an overdriving torque condition. Its effectiveness was verified through the micro-jitter measurement tests using a gimbal-type antenna.

Oh et al. proposed a whole antenna isolation system employing a ring-type SMA mesh washer that supports the gimbal-type antenna [22]. This application makes it possible to achieve a relatively much lower rotational stiffness of the mesh washer than in the axial direction. The rotational axis of the azimuth gear wheel of the antenna corresponds to the main jitter sources induced by stepper motor activation and an imperfect intermeshed harmonic-drive-gear configuration. Therefore, the weak rotational stiffness of the mesh washer isolator enhances the performance of the micro-jitter isolation. However, the effectiveness of the vibration isolation performance under a launch environment, where the vibration level is quietly much higher than the on-orbit micro-vibration condition, was not verified in the previous study.

In this study, we verified the feasibility of the new application form of a superelastic SMA mesh washer by applying it to a RWA vibration isolation system. The objective of the design is to guarantee the structural safety of both the jitter sources and the isolator itself in severe launch environments without requiring an additional holding and release mechanism and while effectively isolating the RWA-induced micro-jitter. To measure the basic characteristics of the ring-type SMA mesh washer isolator, which is different application form with the conventional washers, we performed a static test at various compressive and elongation ranges of the isolator, in both the axial and lateral directions. Based on the static test results, we measured the basic characteristics of the isolator in each axis. To confirm the micro-jitter isolation capability by using the new form of the compressed SMA mesh washer isolator, a micro-jitter measurement test of the RWA was performed on a Kistler table with and without the isolation system. In this paper, launch vibration environment tests, such as sine,

FIGURE 1: Pseudoelastic compressed SMA mesh washer isolator [21].

random vibrations, and shock tests, were also performed at the qualification level. These test results indicated that the launch and on-orbit vibration can be significantly attenuated by using the newly proposed application of the compressed SMA mesh washer. The isolation system guarantees both the structural safety of the RWA during the launch environment and the micro-jitter isolation of RWA during on-orbit conditions.

2. Passive Launch and On-Orbit Vibration Isolator Using a SMA Mesh Washer

2.1. Basic Characteristics of Superelastic SMA Mesh Washer. To achieve a passive launch and on-orbit vibration isolator that does not require a hold and release mechanism, we focused on the superelastic behavior of the compressed SMA mesh washer as shown in Figure 1 [21]. The SMA mesh washer was manufactured by pressing a knitted wire mesh by using a mold of the required shape. The SMA mesh washer used in this study was fabricated from SE508 superelastic SMA wire with diameter of 0.2 mm. The outer diameter, inner diameter, and thickness of the SMA mesh washer were 16 mm, 8 mm, and 5 mm, respectively. The density and mass of the mesh washer were 1.7 g/cm^3 and 1.88 g. The density was 50% lower than that of the mesh washer used by Youn et al. [18] and Oh et al. [16, 17, 22].

Figure 2 shows an example of the representative static test results of the SMA mesh washer shown in Figure 1 during compressive loading and unloading in the thickness direction of the washer [21]. The test results indicate that a largely deformed SMA mesh washer in the thickness direction with a displacement of 2.5 mm can recover its original shape without plastic deformation upon unloading due to its pseudoelastic or superelastic SMA characteristics. This behavior is a material effect in which the original shape restores itself upon unloading after any large deformation over 10%, yet without being plastically deformed. This is one of the great features of the superelastic SAM mesh washer, which are rarely observed in ordinary metal materials. In addition, the SMA mesh washer also provided better damping capacity than any conventional metal washer isolator [18]. Youn et al. [18] proposed to use the SMA mesh washer as a pyroshock isolator instead of using a general metal washer. However,

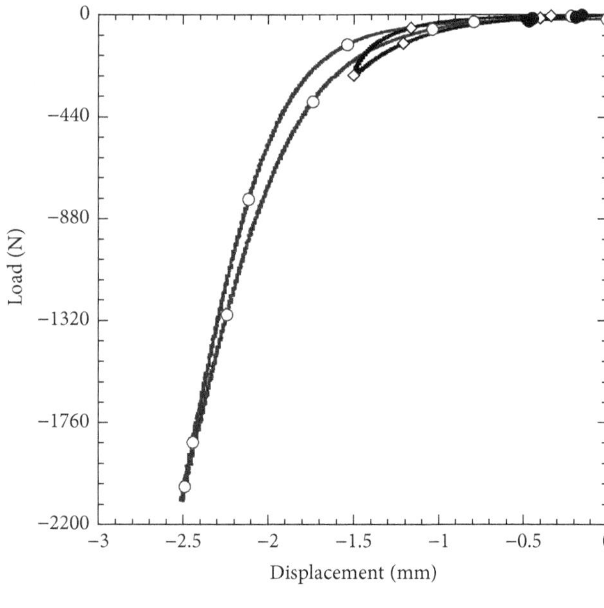

FIGURE 2: Representative load-displacement relationships for a SMA mesh washer along thickness direction [21].

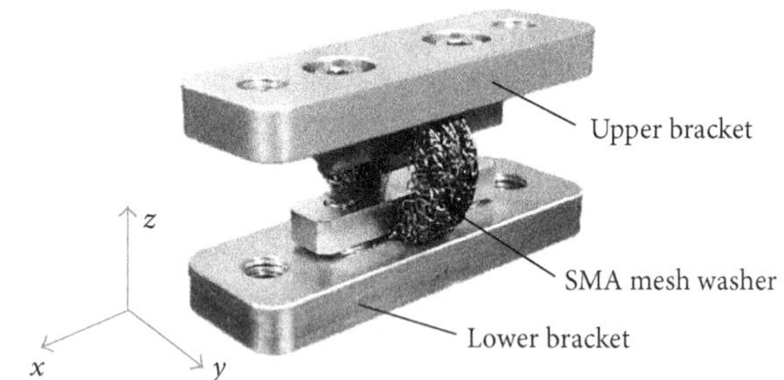

FIGURE 3: New application form of the ring-type SMA mesh washer isolator.

this application is effective only in the thickness direction of the washer. In addition, the characteristics of the washer may be changed by implementing a fastening torque in a real assembly process to satisfy the torque requirement. It is difficult to apply the conventional mesh washer for RWA micro-jitter isolation to achieve objectives of this study.

To overcome the drawbacks discussed above, we proposed a new application form of the SMA mesh washer, as shown in Figure 3. Unlike the applications of a conventional mesh washer, in which vibration damping is effective only in the thickness direction of the washer, this application uses a radial direction of the ring-type mesh washer as a main axis to support the RWA. This application achieved a relatively much lower rotational stiffness for the mesh washer than using the axial direction. This lower rotational stiffness may contribute to micro-jitter isolation as the rotational axis corresponds to the main micro-jitter sources of the RWA.

FIGURE 4: Representative static loading test configuration of the ring-type SMA mesh washer along with x-axis.

Static loading tests were performed to confirm the basic characteristics of the ring-type SMA mesh washer isolator on each axis. Then, the output data obtained from the tests was used to determine the accommodation of the ring-type SMA mesh washer isolator on the RWA. Figure 4 shows the representative test set-up for the static loading test of the ring-type SMA mesh washer isolator along the x-axis. During the test, three repeated cycles of elongation and contraction loading were applied to the washer with various displacement ranges based on increments of ±1 mm along the x-, y-, and z-axes, respectively. The test conditions for the temperature and strain rate were 20°C and 20 mm/min, respectively.

Figures 5(a) and 5(b) represent the load-displacement relationships obtained from the static loading test of the SMA mesh washer isolator. In this figure, the results are plotted on the x-, y-, and z-axes for displacement ranges of ±1 mm and ±3 mm. The test results indicated the mesh washer completely recovered its original shape without any plastic deformation in the maximum displacement range of ±3.0 mm. The hysteresis induced by ring-type SMA mesh washer during the loading-unloading cycle indicates that the SMA mesh washer experienced better energy absorption because of its pseudoelastic shape memory effects and the energy dissipation characteristics of the mesh isolator itself. The area enclosed by hysteretic curves is different for each axis, which means that the isolator has different damping capacities.

The static test results along the z-axis of the isolator showed substantial variation in the slopes in accordance with the deflection range of the mesh washer. The slope of the isolator decreases as the deflection of the washer increases. This is especially apparent along the z-axis. This phenomenon might be related to the direction of the woven pseudoelastic SMA fibers that comprise the mesh. The SMA mesh washer used in this study was fabricated through a pressing process of the knitted pseudoelastic SMA wire mesh, which is then stacked in layers to produce the mesh washer.

Therefore, the geometry of the mesh washer causes it to be more susceptible to delamination between the stacked layers of the compressed washer during loading in the z-axis than during loading in the other axes. This is the main reason for the substantial variation in stiffness with regard to the deflection range of the mesh washer. However, plastic deformation, or any permanent damage due to local delamination, was not observed during the static test along the z-axis. Additionally, the washer experiencing repeated large deflections of ±3 mm during static loading tests completely recovered its original shape without any plastic deformation of the washer due to the pseudoelastic SMA behavior.

From the test results, we calculated the equivalent damping and stiffness values. The equivalent damping can be achieved by calculating the area enclosed by the hysteresis curves and the equivalent stiffness of the isolator. In the calculation for obtaining the equivalent damping, an equivalent linearization method where the nonlinear phenomena in slopes and hysteresis are regarded as linear ones in accordance with the energy-balancing principle is used [23]. To assess the damping capacity of the isolator along each axis, the equivalent linear damping coefficient of ζ_{eq} is calculated using the following equation:

$$\zeta_{eq}(a_0) = \frac{\Delta E(a_0)}{2\pi a_0^2 K_{eq}}, \quad (1)$$

where ΔE is the area enclosed by the hysteresis curve and a_0 is the displacement amplitude. The equivalent stiffness values of K_{eq} for various displacement ranges are calculated by linear fitting method by using the slopes of the load-displacement relationships shown in Figure 5.

Figure 6 shows the estimated equivalent stiffness and damping values of K_{eq} and ζ_{eq} under various displacement ranges along each axis. The K_{eq} along the z-axis indicates the highest values compared to the other axes. For instance, the stiffness value with a ±1 mm deflection of the isolator in the z-axis is 9.4 times larger than the value obtained from the x-axis, which shows the lowest stiffness value. However, it can be observed that the stiffness in the z-axis relatively largely declines with an increase in deflection range, compared to the other axes, owing to the aforementioned reason for the direction of the woven SMA fibers that are meshed together. The calculated equivalent damping values of ζ_{eq} with respect to the deflection ranges of the isolator in each axis are also plotted in Figure 6. The results indicate that the maximum value of ζ_{eq} is 0.4 in x-axis when the displacement range of the isolator reaches ±2 mm, which is greater than those for the other two axes. In addition, the values of ζ_{eq} in x-axis increase with an increase of the displacement range from ±1 mm to ±2 mm and then decrease when the displacement range of the isolator is ±3 mm. This axis shows slightly different phenomena than the other two axes; however, the trend is nearly equivalent to the results obtained from Figure 13 of [23].

The estimated equivalent stiffness and damping values of the ring-type SMA mesh washer shown in Figure 6 are much smaller than the results obtained from Figures 5 and 6 of [22], even though their trends are almost all the same. This decrease in the stiffness and damping values is because the density of the SMA mesh washer employed in this study is 50% less than that used in the previous study. However, once we consider the jitter isolation methodology for the

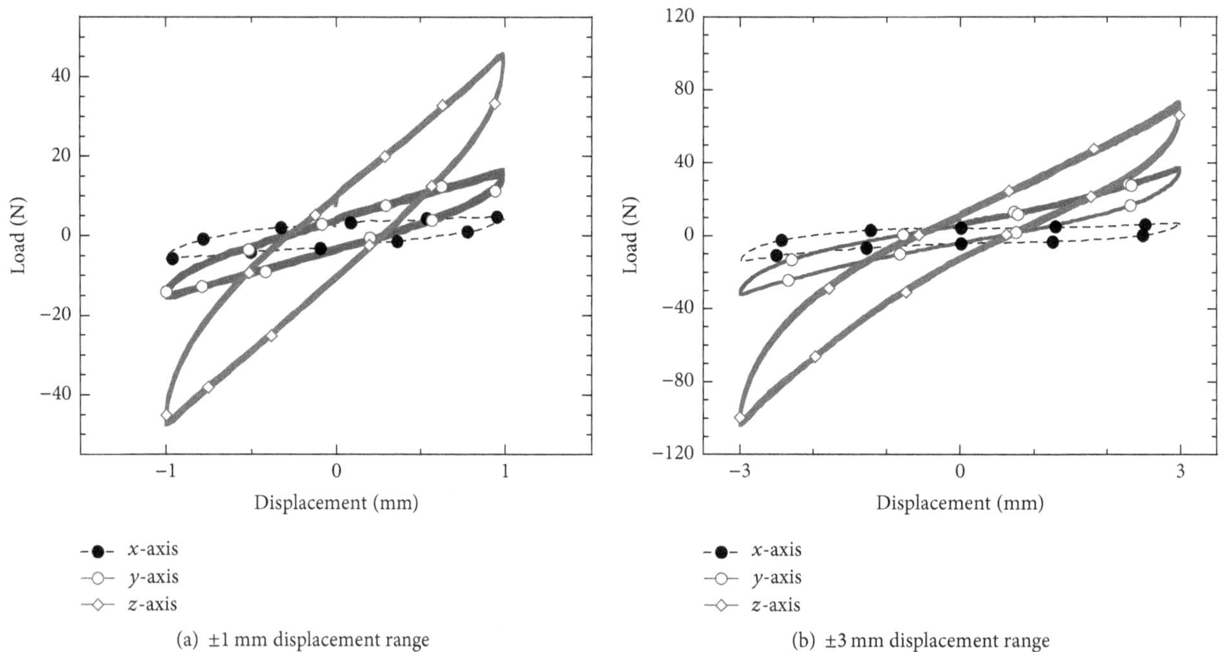

FIGURE 5: Load-displacement relationships for displacement ranges of 1 mm and 3 mm on each axis ((a): x-axis, (b): y-axis, (c): z-axis).

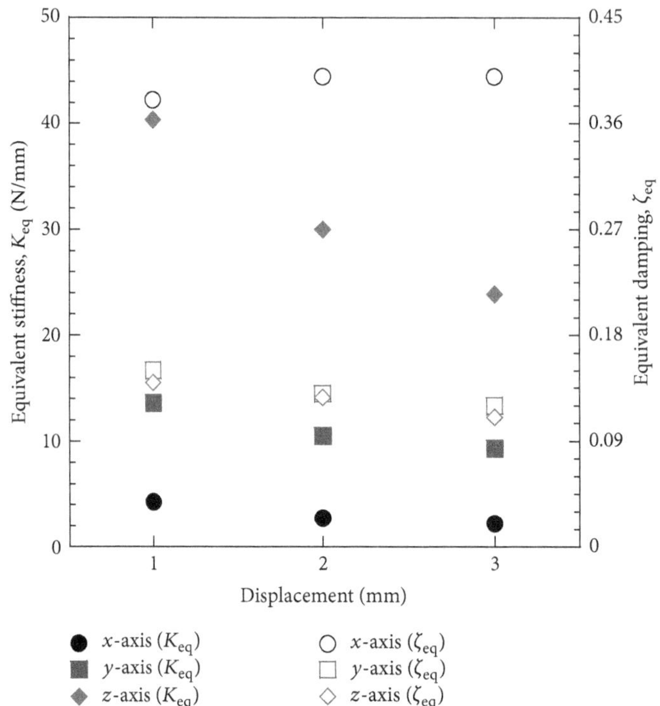

FIGURE 6: Estimated values of the equivalent stiffness (K_{eq}) and damping (ζ_{eq}) of the isolator on each axis with respect to displacement ranges.

RWA-induced micro-jitter, a lower stiffness value is always desirable.

2.2. RWA Passive SMA Mesh Washer Isolator. In this study, to verify the effectiveness of the isolator design, which guarantees vibration attenuation in a launch environment while effectively isolating the micro-disturbances in an on-orbit environment, RWA for small satellite application was considered as a vibration isolation target for the experimental investigation. The functionality of the RWA considered in this study was verified through the STSAT-2C (Science and Technology Satellite 2C, South Korean Satellite launched in

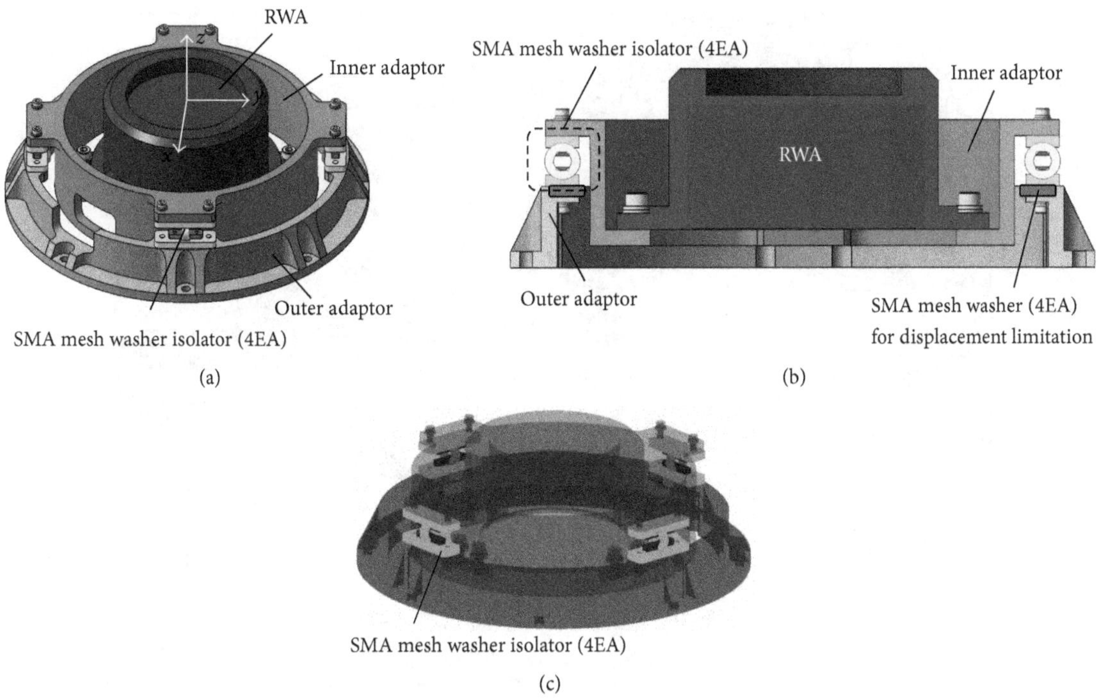

FIGURE 7: Configuration of the RWA combined with the ring-type SMA mesh washer passive isolation system ((a): isometric view, (b): sectional view, and (c): real configuration of the RWA isolation system).

2013) program as the experimental demonstration payload. The mass of the RWA is 1.38 kg; the dimensions are 168 mm × 168 mm × 70 mm; the wheel inertia is 0.001143 kgm^2; and the maximum reaction torque is 11.1 mNm. In this study, a verification model having the same specifications as the RWA flight model was used to verify the isolator proposed in this study.

Figure 7 shows the configuration of a RWA combined with the ring-type SMA mesh washer passive isolation system. To implement the mesh washer isolator onto the RWA, inner and outer adapters were additionally manufactured and integrated into the RWA as shown in Figure 7(a) for the purpose of validation of isolator design as modification of the existing RWA hardware was not allowed anymore.

In case of compatibility of the proposed ring-type SMA mesh washer isolator against different types of RWA, several things should be trade-off studied, that is, the number of ring-type SMA mesh washer isolators and its arrangement with regard to the design requirement of the RWA. Such trade-off study could be carried out based on the basic characteristics of the ring-type SMA mesh washer isolator obtained from static loading tests as shown in Figure 6.

The ring-type isolators are integrated with the RWA through the inner adapter such that they are on the center of the gravity plane of the RWA as shown in Figure 7(b). This allows us to establish a system that does not require an additional holding and release mechanism for the RWA because this greatly minimizes the additional movement of the RWA that is a result of the rotational moments of the RWA. Figure 7(c) shows the accommodation of the SMA mesh washer isolators. The accommodation of the isolators was decided based on the static test results where the RWA isolation system has the 1st eigenfrequency of approximately 18 Hz in x-y plane for the RWA-induced micro-jitter isolation. Four SMA mesh washers are installed on the outer adapter and are used as a mechanical protector to limit the movements of the RWA within 3.0 mm in x-y plane under launch vibration environment because the stiffness of the isolator in x- and y-axis is much lower than the z-axis. This part makes additional contribution of attenuation of the mechanical loads transmitted to the RWA when the inner adapter bumps into outer adapter at around the resonance frequency bands of the RWA isolation system under severe launch environments.

2.3. Numerical Investigation of the RWA Isolation System. Based on the basic characteristic test results of the ring-type SMA mesh washer as shown in Figure 6, it can be noticed that the mechanical behavior of the RWA isolation system varied with regard to the given vibration environments, that is, launch vibration and on-orbit micro-vibration. Under the harsh launch vibration environments, the RWA isolation system experiences large displacements for all axes within 3 mm due to existence of the mechanical protector. In this condition, once we consider the mechanical property of the ring-type SMA mesh washer and its arrangement in the RWA, the estimated equivalent stiffness of the RWA isolation system along with x-y plane is approximately 24,400 N/m. Based on this value and the mass of the RWA, the estimated 1st eigenfrequency of the RWA isolation system along x-y plane is 21.1 Hz under the launch vibration environment. In the same manner, the estimated 1st eigenfrequency of the RWA isolation system along z-axis is 40.3 Hz.

FIGURE 8: Configuration of the launch vibration test.

On the other hand, under the on-orbit micro-vibration environment, the rotational direction of the RWA is corresponded to the x-axis of the ring-type SMA mesh washer. In this condition, the calculated equivalent stiffness of the RWA isolation system, obtained from the case when the ring-type SMA mesh washer is deflected by 1 mm along x-axis, is 17,200 Nm. Thus the expected 1st eigenfrequency of the RWA isolation system is 17.76 Hz under the on-orbit micro-vibration environment. Based on this value, it can be noticed that the RWA isolation system can achieve effective micro-vibration isolation capability through the frequency decoupling methodology.

3. Experimental Validation of RWA Passive Launch and On-Orbit Isolation System

3.1. Launch Vibration Test Results. To verify the effectiveness of the isolator design for the launch environment, sine, random vibration tests, and shock tests were performed. In the test, a dummy RWA was used because the test objective was to qualify the isolator design under qualification launch loads. Figure 8 shows the launch vibration test configurations for the x-, y- and z-axes. The dummy RWA mounted on the isolator had a connection with the vibration shaker through the use of an adaptor. The vibration input from the vibration shaker was obtained from the reference accelerometer. The vibration responses from the RWA isolation system were then obtained from the 3-axis accelerometer placed on the center of gravity of the RWA.

Figures 9(a), 9(b), and 9(c) show the sine vibration test specifications for each axis and the output acceleration measured by the 3-axis accelerometer on the RWA isolation system. In the case of the x-axis excitation shown in Figure 9(a), the highest acceleration at the RWA was 7.2 g at 18 Hz, which was originated from the translational mode of the RWA isolation system along the x-axis. This value is almost similar to the estimated 1st eigenfrequency of 21.1 Hz in Section 2.3. In addition, 2nd highest response was followed by around 23 Hz, which was induced by the rotational mode of the RWA isolation system along the z-axis. In the resonance frequency ranges from 18 to 23 Hz, the estimated displacement of the RWA isolation system was 4.3 mm along the in-plane direction. However, the isolation system cannot exceed 3.0 mm along the x- or y-axis due to the mechanical protector on the outer adapter that limits the movements of RWA within 3.0 mm as shown in Figure 7(b). Therefore, the inner adaptor bumped into outer adaptor. However, the mechanical vibration loads transmitted to the RWA during the bump between the adaptors were outstandingly dissipated due to the significant damping characteristic of the SMA mesh washer installed on the outer adaptor. This phenomenon contributed towards the reduction of the maximum peak acceleration at the RWA, allowing the design requirement of the RWA to be further mitigated. Alternately, the acceleration responses were greatly reduced after the resonance frequency higher than 23 Hz because the displacement of the RWA isolation system was reduced below 0.52 mm. Hence, the inner and outer adaptors were no longer bumped into each other. In addition, the frequency decoupling effect due to supporting the RWA with the low-stiffness isolator contributed towards the reduction of the acceleration response for the higher frequency ranges from 24 to 100 Hz. In case of the y-axis excitation as shown in Figure 9(b), the highest acceleration at the RWA was 9.2 g

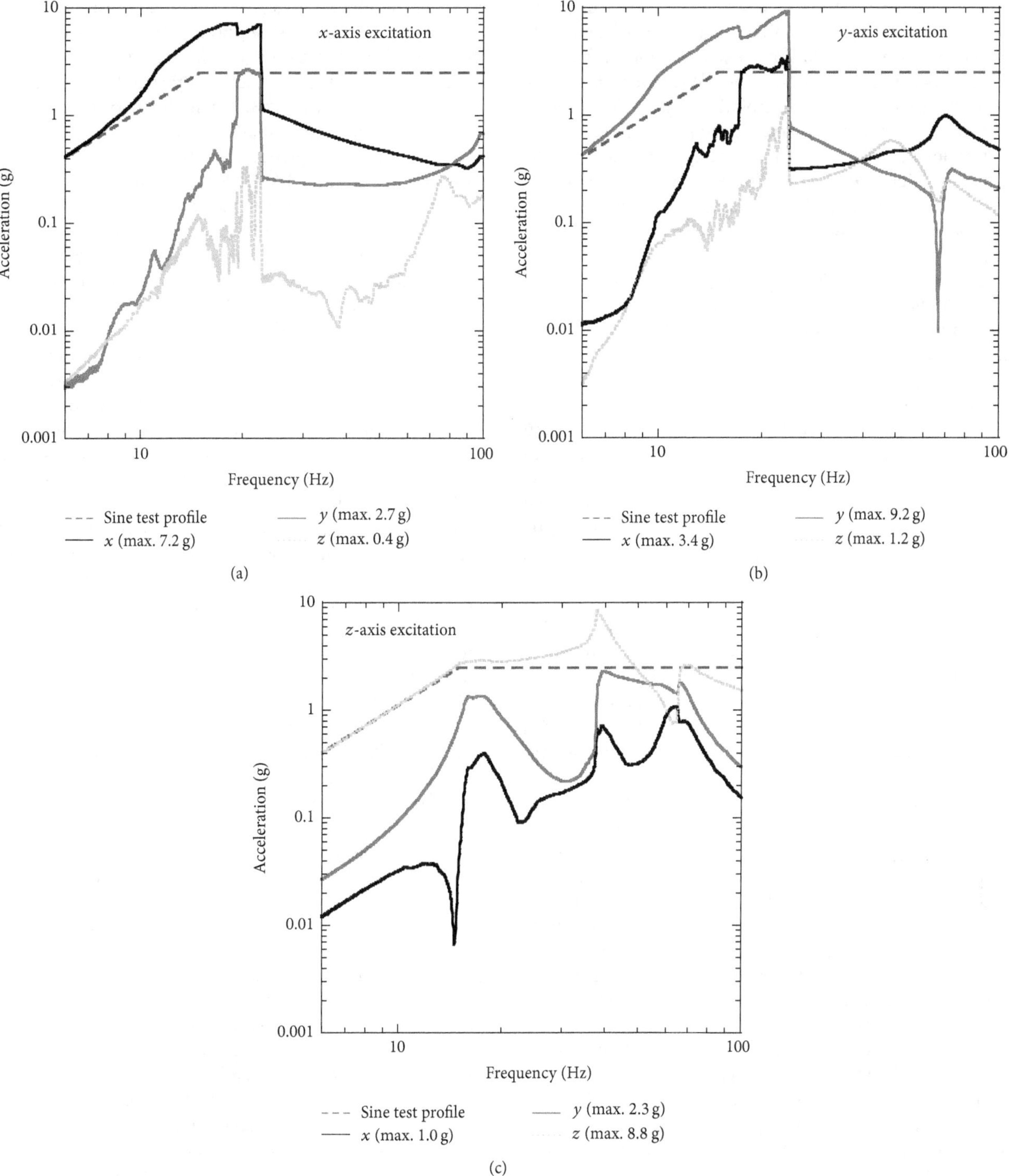

FIGURE 9: Results for sine vibration tests of the RWA isolation system with respect to each axis excitation ((a): x-axis, (b): y-axis, and (c): z-axis).

at 24 Hz. In addition, the overall tendency showed similar results when compared to the results obtained from the x-axis excitation. This was due to the symmetrical design of the SMA mesh washer isolator in the x-y plane as shown in Figure 7(c). For the case of the z-axis excitation as shown in Figure 9(c), the highest acceleration of 8.8 g was observed at 38 Hz which is also almost identical to the estimated 1st eigenfrequency of the RWA isolation system along z-axis. In this condition, the estimated deformation of the RWA isolation system along the z-axis was 1.5 mm. The resonance frequency observed at 38 Hz was relatively higher than the 1st eigenfrequency of the RWA isolation system in the x- and y-axes, around 18–23 Hz.

This was because the stiffness of the SMA isolator in the z-axis was relatively higher than the x- and y-axes, as shown in Figure 5. However, even if the z-axis excitation from the vibration shaker is exerted onto the RWA isolation system, a rotational mode of the RWA isolation system along the x- and y-axes was mainly observed during the vibration test. This rotational mode was identified at approximately 70 Hz, as shown in Figure 9(c). A similar phenomenon can also be observed in the random vibration test results described later. This behavior of the RWA isolation system can be introduced by the fact that the movements of the RWA occurred not only in the z-axis but also in the y-z or x-z planes due to accommodations of the SMA mesh washer isolator.

From the above sine vibration test results, it can be seen that the highest acceleration response obtained from each of the x-, y-, and z-axes, when corresponded axis excitation was induced to the RWA isolation system, did not exceed the RWA's design load of 42.1 g, guaranteeing the structural safety of the RWA and isolator itself under sine vibration loads without requiring an additional launch-locking device. However, this design approach cannot be achieved using stainless-steel wire mesh as this material exhibited plastic deformation after a compressive displacement of 0.5 mm [21]. In addition, because of the relatively higher stiffness value of the steel washer within the elastic range, the jitter isolation capability might be degraded because the isolation capability heavily depends on the stiffness value. For these reasons, the pseudoelastic SMA mesh washer is used in this study.

Figures 10(a), 10(b), and 10(c) show the random vibration test specifications for each axis and the output responses with the same vibration test configuration, which is shown in Figure 8. For comparison of the vibration reduction capabilities of the SMA mesh washer isolator, a random vibration test where the RWA was rigidly mounted onto the adaptor was performed as well. The representative results when the y-axis excitation was induced to the RWA without the isolator are shown in Figure 11. In here, the 1st eigenfrequency of the RWA was observed at around 200 Hz, indicating that the vibration loads were amplified to that resonance frequency without isolation and directly transmitted to the RWA. Furthermore, the overshoot at approximately 1000 Hz was also sharply amplified. This overshoot occurred primarily due to the structural resonance of the mechanical support equipment. The vibration response in the y-axis was 175.5 grms. This response level may result in structural problems to the RWA and thereby potentially bring up the structural safety of the RWA during harsh launch vibration environments, especially if the design load of the RWA becomes stricter.

However, the output response in the y-axis, when the same axis excitation was induced, was significantly decreased to 0.51 grms by establishing the SMA mesh washer isolator on the RWA as shown in Figure 10(b). Here, it can be noticed that the severe random vibration levels were isolated by a factor of 344 when compared to the rigid condition. This superior vibration isolation capability was due to lowering the 1st eigenfrequency of the RWA from 200 Hz to 20 Hz. The same vibration isolation methodology was observed in the sine vibration test results, as shown in Figure 9. However, the vibration energy dissipation methodology by a bump between the inner adaptor and the mechanical protector, as experienced during the sine vibration tests, was not seen during the random vibration test as the deformation of the RWA isolation system was too small. Practically, the estimated deformation in the y-axis, when the corresponding axis random excitation was induced, was approximately 0.17 mm at the resonance frequency of 20 Hz. In this condition, the rotational mode of the RWA isolation system along the x-axis was also observed at approximately 70 Hz. On the other hand, a similar behavior of the RWA isolation system in the x-axis was seen in the overall frequency bands when the x-axis excitation was induced. This behavior was due to the symmetric configurations of the RWA isolation system in the x-y plane.

The response along the z-axis, when the z-axis excitation was induced to the RWA isolation system, indicated slightly different results compared to the x- or y-axis responses. Because the stiffness of the SMA mesh washer isolator in the z-axis was relatively higher than those of the x- and y-axis, therefore, the 1st eigenfrequency of the RWA isolation system along the z-axis was observed at around 30 Hz. In contrast, as indicated in basic characteristic results, the equivalent damping in the z-axis was relatively lower than for the other axes. Therefore, the peak occurring at 1st eigenfrequency was relatively higher than the other axes. However, this was no longer of concern in regard to the structural safety of the RWA because the estimated acceleration response at the resonance frequency did not exceed the design load of the RWA. Additionally, the random vibration input level of 14.1 grms was significantly decreased to 2.89 grms.

Figure 12 shows the representative frequency response spectrum of the RWA isolation system obtained from the low level random sweep (LLRS) tests along the y-axis before and after the launch vibration tests referred to above. This test was performed to investigate the structural safety of the RWA isolation system by comparing the dynamic response of the RWA isolation system before and after the launch vibration tests. In general, to satisfy the test requirement of the vibration test, no variations of 1st eigenfrequency less than 5% and amplitudes less than 3 dB should be observed from the LLRS test. The results of LLRS indicate that the 1st eigenfrequency of the RWA isolation system along the y-axis was observed at around 18–20 Hz, which is the same as the results obtained from the launch vibration tests. This result also indicated that no variations of frequencies and amplitudes before and after the launch vibration tests occurred. Hence, the structural safety of the isolator itself can be guaranteed in spite of being subjected to harsh launch vibration loads. Therefore, it is expected that no performance degradation for the micro-jitter isolation in the on-orbit environment occurs.

From this test result, it can be expected that a significantly deformed SMA mesh washer, occurring in a launch environment, can recover its original shape without plastic deformation upon unloading in a 0-g environment owing to its pseudoelastic behavior, as shown in the static test results. These facts are great advantages to use the pseudoelastic SMA mesh washer in this study as a smart adaptive system that can be used in both a launch environment and on-orbit micro-jitter isolation without requiring an additional launch lock device.

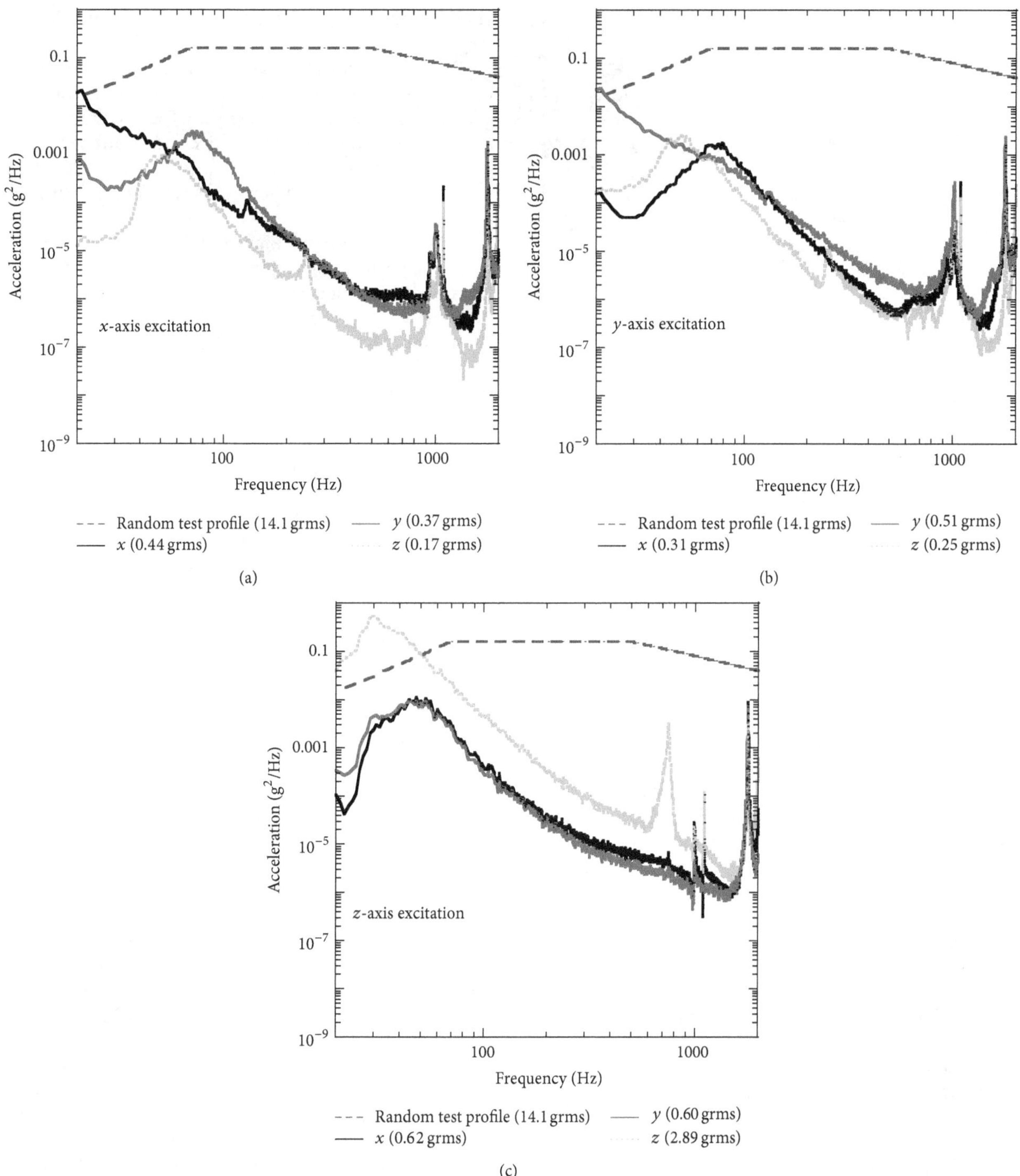

FIGURE 10: Results for random vibration tests of the RWA isolation system with respect to each axis excitation ((a): x-axis, (b): y-axis, and (c): z-axis).

On the other hand, the ring-type SMA mesh washer is subjected to substantial cyclic deformations during launch vibration environments; thus a functional fatigue such as degradations of superelasticity or shape recovery effect can be a concern. In general, the formation and growth of microscopic defects such as dislocations of the superelastic SMA material are caused by cyclic loadings. These dislocations are formed and aggregated mainly near the interfaces between austenite and stress-induced martensite phases and the interfaces between the martensite variants with different crystallographic orientations [24]. Such dislocations in the SMA material result in occurrence of residual strain in the SMA itself and the residual strain gradually increases as the loading cycle increases. Besides, the hysteresis area becomes lower as

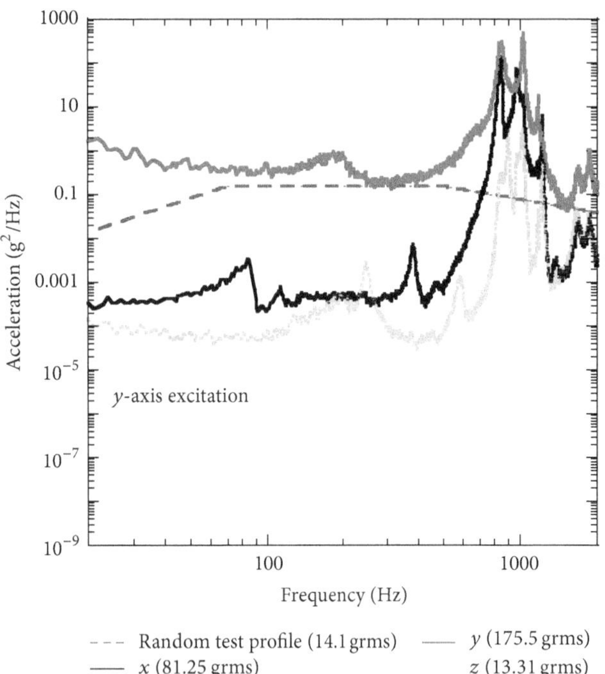

FIGURE 11: Representative results for random vibration tests of the RWA without isolation system when the y-axis excitation was induced.

the loading cycle increases [25]. This means that the damping capability of the superelastic SMA is decreased if the cyclic loadings are repeatedly accumulated in the SMA material. However, once we investigate the static loading test result for ±3 mm as shown in Figure 5(b), no phase transformation phenomenon from austenite to martensite can be observed under given displacement limits of ±3 mm which is the maximum allowable deflection of the RWA isolation system due to a mechanical protector. Therefore, considering the main fatigue failure mechanism of the superelastic SMA such as the interfaces between austenite and stress-induced martensite phases, it can be revealed that no severe failure mechanism acts on the ring-type SMA mesh washer during launch vibration environments. Moreover, it can be also noticed from the LLRS test result, shown in Figure 12, that no variations of frequencies and amplitudes before and after the launch vibration tests occurred. Hence, the fatigue effect can be negligible to the current design of the RWA isolation system proposed in this study.

Figure 13 shows the shock test results for the x-, y-, and z-axes when each shock excitation along the corresponding axis was induced. The maximum shock response spectrum (SRS) on the RWA based on a maximum SRS input of 793 g was 3.3 g in the y-axis, thereby indicating that the shock level transmitted to the RWA was greatly attenuated by the factor of 240. This superior attenuation capability was due to the frequency decoupling principal that occurred when mounting the RWA with a low-stiffness isolator. The 1st eigenfrequency of the RWA isolation system in the x- and y-axes was observed at around 20 Hz. This was similar when compared with the results from the sine and random vibration tests for the identical axis, including the z-axis responses.

These test results indicate that the isolation system using the pseudoelasticity of the SMA mesh washer is effective for guaranteeing the structural safety of a RWA without requiring an additional launch-locking device and greatly reducing the harsh vibration loads transmitted to the RWA that enables mitigating the strict design requirements of the RWA.

3.2. Micro-Jitter Measurement Test Results. Generally, it has been known that micro-jitter disturbances induced by the RWA activation can be divided into three categories based on their origin: rotor imbalance, ball bearing imperfection, and motor driven disturbances [12]. In here, it has also been known that the rotor imbalance of the RWA has the most significant effect on micro-jitter disturbances and generates disturbances at the same frequency of rotation, which is referred to as fundamental harmonic (H1). At some wheel speed, this fundamental harmonic can be significantly amplified through frequency coupling with the structural modes of the RWA. This effect will seriously impact the image quality of a high-resolution observation satellite. Alternately, irregularities in ball bearings, motor, lubrication, and so forth generate disturbances that usually occur at integral and fractional multiples of the fundamental harmonic frequency as sub- and superharmonics (H0.5, H2, H2.7, H3, etc.). However, these disturbances from the irregularities are usually significantly smaller than those of the fundamental harmonics.

To investigate the micro-jitter isolation performance when employing a new application of the SMA mesh washer as proposed in this study, a micro-jitter measurement test, with and without an isolation system, was performed on the Kistler table. Figure 14 shows the micro-jitter measurement test set-up with and without the RWA isolation system. The RWA was installed on the Kistler table through inner and outer adaptor, and four isolators bridged the inner and outer adaptor with a low stiffness, respectively. The test set-up without the isolation system was implemented by rigidly connecting those inner and outer adaptor, such that the micro-jitter from the RWA was directly transmitted to the Kistler table without isolation. In the micro-vibration measurement test of the RWA, the 0 g device compensating for the 1 g effect during the ground test was not applied to simulate zero gravity on-orbit environment. Therefore, the mesh washer was subjected to prestressed conditions owing to the RWA mass, including the inner adaptor for a total of 2 kg. In this condition, the estimated deflection of the RWA isolation system along the direction of gravity due to the RWA with an adaptor was 0.12 mm. However, the estimated deflection might not have a considerable effect on the eigenfrequency shift of the isolation system between 1 g and 0 g conditions because the stiffness variation of the mesh washer due to RWA mass was negligibly small. During the test, the free-rundown method, which decreased the wheel speed naturally by turning off the operation command, was adopted to investigate the impact of the micro-jitter of the RWA, such that the wheel speed of the RWA was varied from 5000 to 0 rpm in the test.

Figures 15(a) and 15(b) show examples of waterfall plots of the RWA disturbance forces and torques without the isolation

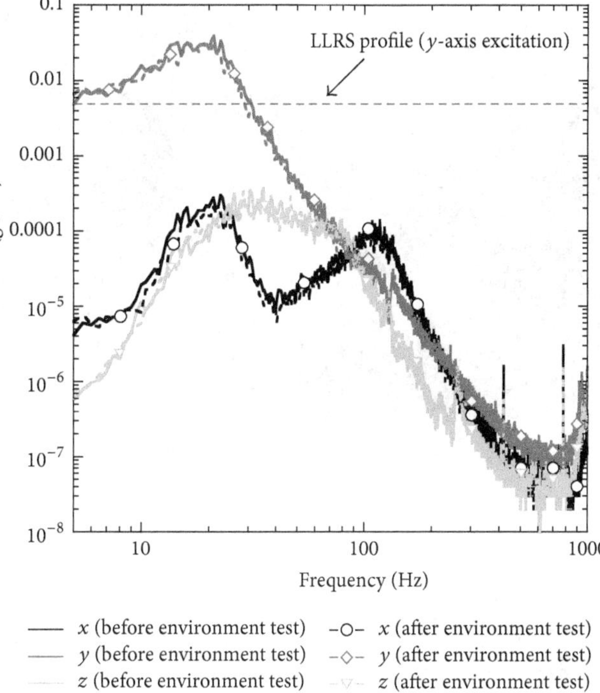

FIGURE 12: Frequency response spectrum of the RWA isolation system obtained from the low level random sweep (LLRS) test on the y-axis before and after the launch vibration tests.

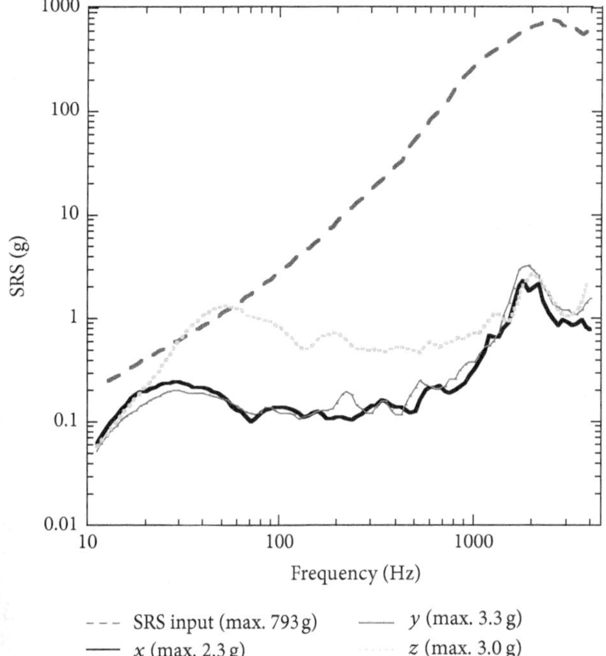

FIGURE 13: Results for shock test of RWA isolation system on each axis excitation.

system. From these results, it can be seen that the fundamental harmonic at the 5000 rpm has the largest amplitude and gradually decreases as the wheel speed decreases. In here, the amplitude of the fundamental harmonic is proportional to the square of the wheel speed [26]. Aside from the fundamental harmonic, a number of superharmonics were also observed in all axes with relatively lower amplitudes compared to the fundamental harmonic.

The disturbance, which appears at frequency of 130 Hz with same amplitude throughout the entire test speed ranges, mainly comes from the electrical power supply equipment. In practice, this electrical noise cannot be avoidable during the test even if some efforts were implemented towards reducing this noise [26]. However, the electrical noise is much smaller when compared to the harmonic disturbances; therefore, it can be ignored. Meanwhile, the peak disturbances occurring at around 280 Hz and 420 Hz primarily result from the frequency coupling effect between the RWA structural modes and the harmonic disturbances. These peak disturbances were observed in all axes.

Figures 16(a) and 16(b) show waterfall plots of the RWA disturbance forces and torques with the isolation system. The test was conducted under the same conditions as the without isolation test to impartially assess the micro-jitter isolation performance of the RWA isolation system proposed in this study. From this result, it can be seen that the fundamental harmonic has a different shape and its maximum amplitudes in all waterfall plots are considerably reduced when compared to the results of the rigid condition. For example, in the case of the x-axis, the maximum peak value occurring at the fundamental harmonic without the isolation is approximately 3.9 N at 80 Hz, which corresponds to a maximum wheel speed of 5000 RPM. However, this value is effectively attenuated to 1.8 N by the factor of 2.16 at the same wheel speed condition due to the SMA isolation system. In fact, the main feature of

FIGURE 14: Test set-up for micro-jitter measurement of RWA with and without isolation system.

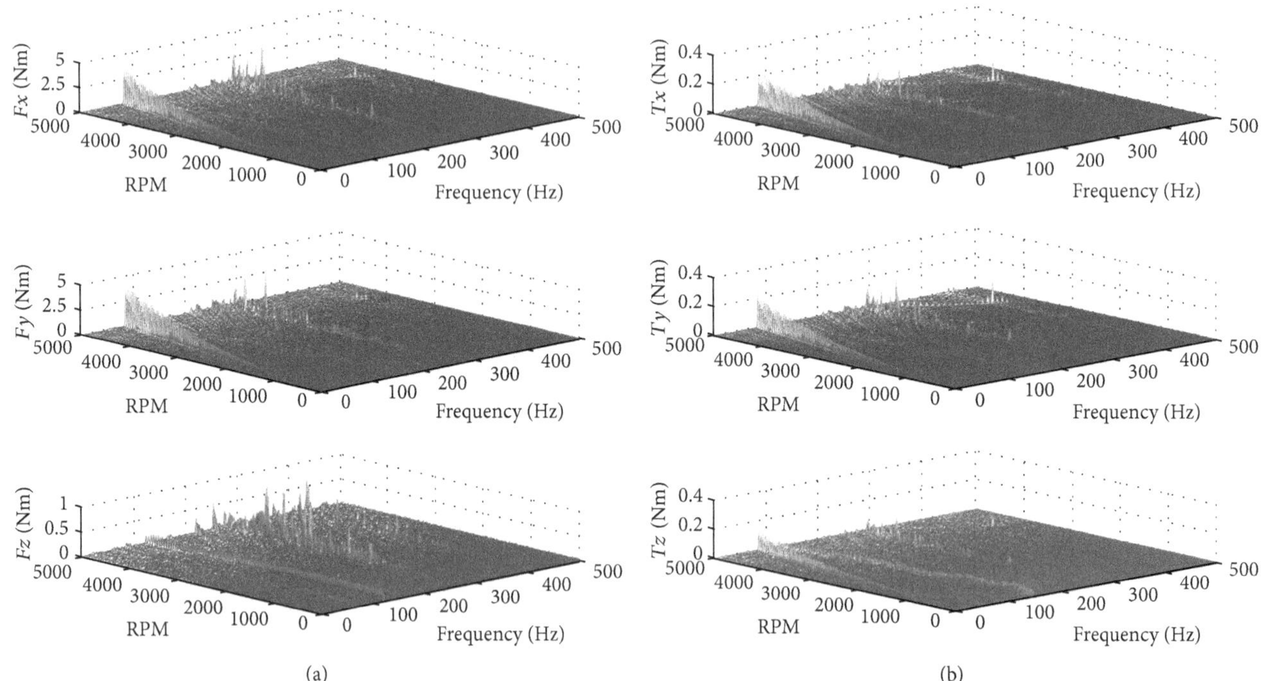

FIGURE 15: Waterfall plots of disturbance forces and torques for RWA activation without isolation ((a): disturbance forces and (b): disturbance torques).

the RWA-induced micro-jitter isolation, when utilizing the proposed application form of SMA mesh washer, is to put the eigenfrequency of the isolation system towards a lower frequency region than that of the without isolation condition to effectively reduce higher frequency disturbances. The 1st eigenfrequency of the RWA with the isolation system is observed in a low-frequency range at around 20 Hz in the x- and y-axes, which is the same as the results obtained from the launch vibration tests. These frequencies are much lower than the eigenfrequency of a RWA without isolation system, which is roughly 280 Hz. Therefore, the harmonic disturbances induced by the RWA activation as well as the structural modes of the RWA itself are effectively isolated on the basis of the 1st eigenfrequency of the isolation system.

However, some amplitudes are still observed in the waterfall plots because the fundamental harmonic of the RWA and the 1st eigenfrequency of the isolation system are coupled with each other at a specific wheel speed region, which corresponds to approximately 1700 RPM. These peaks are not avoidable in the current system because the isolation methodology proposed in this study is based on the passive approach of frequency decoupling. However, the maximum peak force and torque level at the resonance frequency of isolation system, 20 Hz, are also effectively attenuated due to the significant damping characteristics of the SMA mesh washer as shown in Figure 6.

On the other hand, the 1st eigenfrequency of the RWA isolation system along the z-axis is not clearly observed in

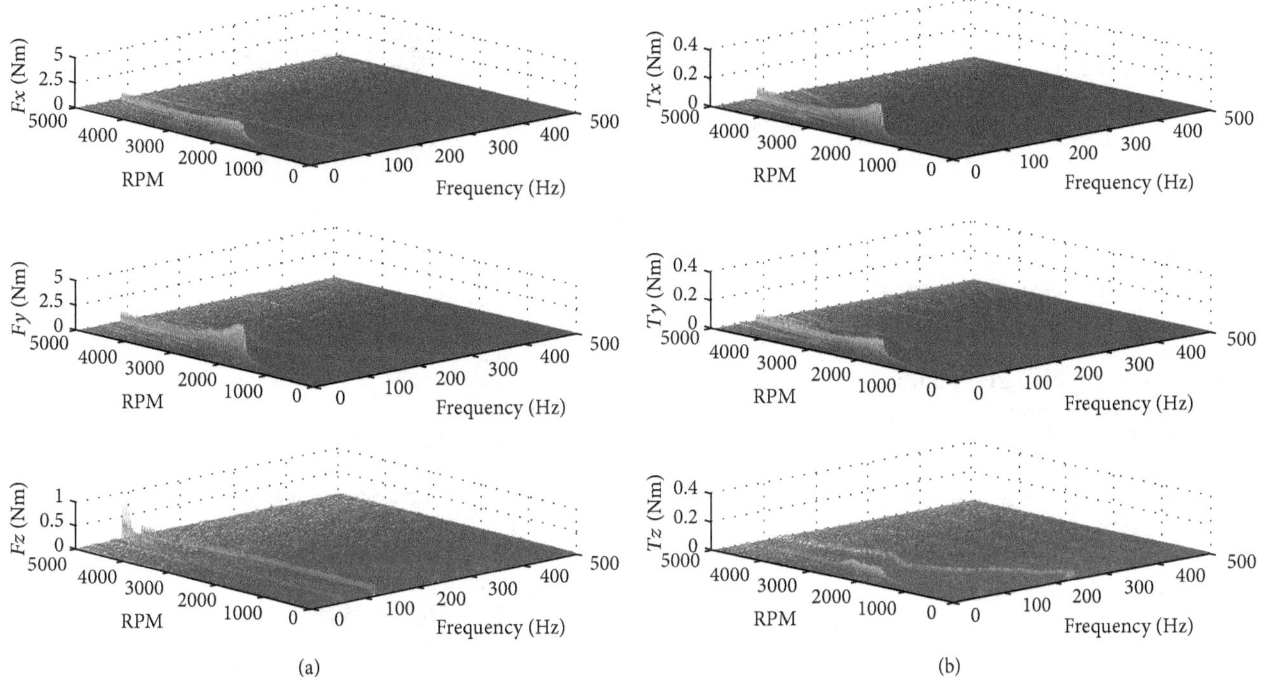

FIGURE 16: Waterfall plots of disturbance forces and torques for RWA activation with isolation ((a): disturbance forces and (b): disturbance torques).

the waterfall plot because the main jitter contributor of the RWA is the rotational movement in the x-y plane. Therefore, the micro-jitter disturbances along the z-axis are less critical when compared to the x- or y-axis. In practice, the maximum peak value with the isolation system in the z-axis is approximately 0.3 N at the maximum wheel speed and this value is almost negligible. However, from the basic characteristic and launch vibration tests results, it can be assumed that it existed at around 40 Hz and this value is included in the fundamental harmonic frequency ranges of the RWA. Therefore, the fundamental harmonic disturbances are slightly amplified when coupled with the 1st eigenfrequency of the isolation system along the z-axis. However, higher harmonics including structural modes of the RWA are significantly isolated due to the frequency decoupling. For instance, the maximum peak value of 1.7 N at 280 Hz without isolation system decreases to nearly 0 N with the isolation system.

In terms of position sensitivity of the ring-type SMA mesh washer on the micro-jitter isolation performance, the equivalent stiffness and damping values along x- and y-axes, which are corresponded to the main excitation axis of the RWA, are not significantly varied as the displacement changes, as shown in Figure 6. This means that the dynamic characteristic of the RWA isolation system does not significantly change with regard to the alignment shift of the ring-type SMA mesh washer. Thus, stable micro-jitter isolation performance can be expected regardless of position sensitivity.

For a more detailed investigation with regard to the effectiveness of the isolation system, the total disturbances reduction ratio obtained with and without the isolation system is calculated and summarized in Table 1. This table shows the ratio of maximum and mean peak reduction and the standard deviation reduction for each axis. From the results, it can be noticed that the maximum peak reduction ratios of 59.32% and 42.92% for forces and torques are observed on the Fx and Ty, respectively. The Fx result also shows the highest peak mean reduction ratio of 58.94% while the highest value of 53.41% for torque is obtained from Ty. With regard to the standard deviation reduction ratio for forces and torques, all axes showed considerable reduction ratios compared to the rigid condition. This reduction occurred because the high frequency disturbances were all significantly isolated due to the implementation of the isolation system. In here, the main contribution of micro-jitter isolation mainly results not from the pseudoelastic behavior of SMA effects but from the relatively lower stiffness of the isolator. The micro-jitter measurement test results indicate that the application of the SMA mesh washer isolator is effective in reducing RWA-induced micro-jitter when mounting the RWA with a low rotational stiffness isolator, which is identical to the main jitter sources induced by the rotational movement of the RWA. The SMA isolator proposed in this study is also effective in the z-axis even though its stiffness is relatively higher than those for the x- and y-axis.

4. Conclusion

In this study, a novel application form of the ring-type SMA mesh washer isolator that can be used in both launch vibration and on-orbit micro-jitter environments without requiring a launch lock device was evaluated. The RWA for a small satellite application was considered as a vibration isolation target for the purpose of experimentally investigating the SMA mesh washer isolator as proposed in this study. The

TABLE 1: Total disturbance reduction ratio, with and without isolators.

Reduction rate	F_x	F_y	F_z	T_x	T_y	T_z
Peak reduction (%)						
Max.	59.32	23.65	44.37	−0.04	42.93	30.12
Mean	58.944	37.01	55.64	42.72	53.41	19.49
STD reduction (%)	52.05	22.49	58.71	15.34	42.41	32.92

basic characteristics of the SMA mesh washer isolator were demonstrated through the static loading tests.

The effectiveness of the isolator design in launch environments was assessed through qualification level sine, random vibration, and shock tests. The results of the LLRS before and after the launch vibration tests indicated that the structural safety of the RWA isolation system subjected to harsh vibration loads was guaranteed without any variations of frequencies and amplitudes. The micro-jitter measurement test results indicated that the RWA-induced micro-jitter harmonics including RWA structural modes were outstandingly isolated due to shifting the 1st eigenfrequency of the RWA isolation system towards a lower frequency.

Conflicts of Interest

The authors declare that there are no conflicts of interest regarding the publication of this paper.

Acknowledgments

This research was supported by National Research Foundation of Korea (NRF) funded by the Ministry of Science, ICT and Future Planning (MSIP) (NRF-2015R1A2A2A01003672).

References

[1] K. KOMATSU and H. UCHIDA, "Microvibration in spacecraft," *Mechanical Engineering Reviews*, vol. 1, no. 2, pp. SE0010–SE0010, 2014.

[2] X. Li and W. Cheng, "Test and analysis of micro-vibrations generated by large control moment gyroscope," *International Journal of Control and Automation*, vol. 7, no. 2, pp. 297–308, 2014.

[3] S. Riabzev, A. Veprik, H. Vilenchik, and N. Pundak, "Control of dynamic disturbances produced by a pulse tube refrigerator in a vibration-sensitive instrumentation," *Cryogenics*, vol. 49, no. 1, pp. 7–11, 2009.

[4] D. K. Kim and H. T. Choi, "Velocity optimization method of X-band antenna for jitter attenuation," in *Proceedings of the The 21st International Congress on Sound and Vibration*, Beijing, China, 2014.

[5] Z. Zhang, G. S. Aglietti, and W. Zhou, "Microvibrations induced by a cantilevered wheel assembly with a soft-suspension system," *AIAA Journal*, vol. 49, no. 5, pp. 1067–1079, 2011.

[6] D. Kamesh, R. Pandiyan, and A. Ghosal, "Passive vibration isolation of reaction wheel disturbances using a low frequency flexible space platform," *Journal of Sound & Vibration*, vol. 331, no. 6, pp. 1310–1330, 2012.

[7] P. Davis, D. Cunningham, and J. Harrell, "Advanced 1.5 Hz passive viscous isolation system," in *Proceedings of the 35th AIAA/ASME/ASCE/AHS/ASC Structures, Structural Dynamics, and Materials Conference*, Hilton Head, SC, USA, 1994.

[8] Y. Zhang, Z. Guo, H. He, J. Zhang, M. Liu, and Z. Zhou, "A novel vibration isolation system for reaction wheel on space telescopes," *Acta Astronautica*, vol. 102, pp. 1–13, 2014.

[9] H.-U. Oh, K. Izawa, and S. Taniwaki, "Development of variable-damping isolator using bio-metal fiber for reaction wheel vibration isolation," *Smart Materials and Structures*, vol. 14, no. 5, pp. 928–933, 2005.

[10] H.-U. Oh, S. Taniwaki, N. Kinjyo, and K. Izawa, "Flywheel vibration isolation test using a variable-damping isolator," *Smart Materials and Structures*, vol. 15, no. 2, pp. 365–370, 2006.

[11] K. J. Pendergast and C. J. Schauwecker, "Use of a passive reaction wheel jitter isolation system to meet the Advanced X-ray Astrophysics Facility imaging performance requirements," in *Proceedings of the 1998 Conference on Space Telescopes and Instruments V. Part 1 (of 2)*, pp. 1078–1094, March 1998.

[12] Z. Zhang, G. S. Aglietti, W. Ren, and D. Addari, "Microvibration analysis of a cantilever configured reaction wheel assembly," *Advances in Aircraft and Spacecraft Science*, vol. 1, no. 4, pp. 379–398, 2014.

[13] W.-Y. Zhou, G. S. Aglietti, and Z. Zhang, "Modelling and testing of a soft suspension design for a reaction/momentum wheel assembly," *Journal of Sound and Vibration*, vol. 330, no. 18-19, pp. 4596–4610, 2011.

[14] L. P. Davis, J. F. Wilson, R. E. Jewell, and J. J. Roden, *Hubble Space Telescope Reaction Wheel Assembly Vibration Isolation System*, NASA Marshall Space Flight Centre, 1986.

[15] H.-U. Oh, K.-J. Lee, and M.-S. Jo, "A passive launch and on-orbit vibration isolation system for the spaceborne cryocooler," *Aerospace Science and Technology*, vol. 28, no. 1, pp. 324–331, 2013.

[16] H.-U. Oh, S.-C. Kwon, and S.-H. Youn, "Characteristics of spaceborne cooler passive vibration isolator by using a compressed shape memory alloy mesh washer," *Smart Materials and Structures*, vol. 24, no. 1, Article ID 015009, 2015.

[17] S.-C. Kwon, S.-H. Jeon, and H.-U. Oh, "Performance evaluation of spaceborne cryocooler micro-vibration isolation system employing pseudoelastic SMA mesh washer," *Cryogenics*, vol. 67, pp. 19–27, 2015.

[18] S.-H. Youn, Y.-S. Jang, and J.-H. Han, "Compressed mesh washer isolators using the pseudoelasticity of SMA for pyroshock attenuation," *Journal of Intelligent Material Systems and Structures*, vol. 21, no. 4, pp. 407–421, 2010.

[19] H.-K. Jeong, J.-H. Han, S.-H. Youn, and J. Lee, "Frequency tunable vibration and shock isolator using shape memory alloy wire actuator," *Journal of Intelligent Material Systems and Structures*, vol. 25, no. 7, pp. 908–919, 2014.

[20] S.-H. Youn, Y.-S. Jang, and J.-H. Han, "Development of a three-axis hybrid mesh isolator using the pseudoelasticity of a shape memory alloy," *Smart Materials and Structures*, vol. 20, no. 7, Article ID 075017, 2011.

[21] S.-C. Kwon, S.-H. Jeon, and H.-U. Oh, "Performance investigation of a novel pseudoelastic SMA mesh washer gear wheel with micro-jitter attenuation capability," *Smart Materials and Structures*, vol. 25, no. 5, Article ID 055004, 2016.

[22] H.-U. Oh, S.-H. Jeon, T.-H. Kim, and Y.-H. Kim, "Experimental feasibility study for micro-jitter attenuation of stepper-actuated X-band antenna-pointing mechanism by using pseudoelastic SMA mesh washer," *Smart Materials and Structures*, vol. 24, no. 4, Article ID 045010, 2015.

[23] X. Yan and J. Nie, "Study of a new application form of shape memory alloy superelasticity," *Smart Materials and Structures*, vol. 12, no. 6, pp. N14–N23, 2003.

[24] G. Kang and D. Song, "Review on structural fatigue of NiTi shape memory alloys: pure mechanical and thermo-mechanical ones," *Theoretical and Applied Mechanics Letters*, vol. 5, no. 6, pp. 245–254, 2015.

[25] A. Isalgue, H. Soul, A. Yawny, and C. Auguet, "Functional fatigue recovery of superelastic cycled NiTi wires based on near 100°C aging treatments," in *Proceedings of the 10th European Symposium on Martensitic Transformations (ESOMAT '15)*, vol. 33, September 2015.

[26] G. Park, D.-O. Lee, and J.-H. Han, "Development of multi-degree-of-freedom microvibration emulator for efficient jitter test of spacecraft," *Journal of Intelligent Material Systems and Structures*, vol. 25, no. 9, pp. 1069–1081, 2014.

5

Effective Computational Approach for Prediction and Estimation of Space Object Breakup Dispersion during Uncontrolled Reentry

Deok-Jin Lee[1], **Eun Jung Choi**[2], **Sungki Cho**[2], **Jung-Hyun Jo**[2], **and Tae Soo No**[3]

[1]*School of Mechanical & Automotive Engineering, Kunsan National University, Gunsan 54150, Republic of Korea*
[2]*Center for Space Situational Awareness, Korea Astronomy and Space Science Institute, Daejeon 305-348, Republic of Korea*
[3]*Department of Aerospace Engineering, Chonbuk National University, Jeonju 54896, Republic of Korea*

Correspondence should be addressed to Deok-Jin Lee; deokjlee@kunsan.ac.kr

Academic Editor: Hikmat Asadov

This paper provides an effective approach for the prediction and estimation of space debris due to a vehicle breakup during uncontrolled reentry. For an advanced analysis of the time evolution of space debris dispersion, new efficient computational approaches are proposed. A time evolution of the dispersion of space pieces from a breakup event to the ground impact time is represented in terms of covariance ellipsoids, and in this paper, two covariance propagation methods are introduced. First, a derivative-free statistical linear regression method using the unscented transformation is utilized for performing a covariance propagation. Second, a novel Gaussian moment-matching method is proposed to compute the estimation of the covariance of a debris dispersion by using a Gauss-Hermite cubature-based numerical integration approach. Compared to a linearized covariance propagation method such as the Lyapunov covariance equation, the newly proposed Gauss-Hermite cubature-based covariance computation approach could provide high flexibilities in terms of effectively representing an initial debris dispersion and also precisely computing the time evolution of the covariance matrices by utilizing a larger set of sigma points representing debris components. In addition, we also carry out a parametric study in order to analyze the effects on the accuracy of the covariance propagation due to modeling uncertainties. The effectiveness of the newly proposed statistical linear regression method and the Gauss-Hermite computational approach is demonstrated by carrying out various simulations.

1. Introduction

In the past 50 years, over 16,000 metric tons of man-made space objects and unexpected asteroids have entered the Earth's atmosphere. Although most of them burn while entering the atmosphere, some of the debris have survived to impact the Earth's surface and do expose a risk to people and property [1, 2]. To prevent these casualties, researchers have given considerable attention to the research area of reentry estimation of space objects. Usually, the starting altitude of a reentry object will be at around 120 km and then, breakup event happens at 78 km. However, the problem of estimation of space debris dispersion during the atmospheric reentry is still quite difficult due to unknown effects and uncertainties. Especially, the effects of atmospheric uncertainties are very difficult to determine since the atmospheric density and drag coefficient undergoes large fluctuations while decaying at that altitude [3]. In addition the characteristics of uncontrolled space objects including the ballistic coefficients are largely insufficient to model the reentry trajectory. This is because the exact physical parameters of the objects like mass, area of cross section, shape, and dimensions are not available. Thus, the prediction of a reentry trajectory is generally estimated based on the statistical and stochastic method [4]. As shown in Figure 1, an estimated position and velocity vector of a reentering object is propagated by using a numerical integration model up to the breakup point at an altitude of 78 km. After the breakup occurred, the debris dispersion could be modeled by using various stochastic estimation approaches [5]

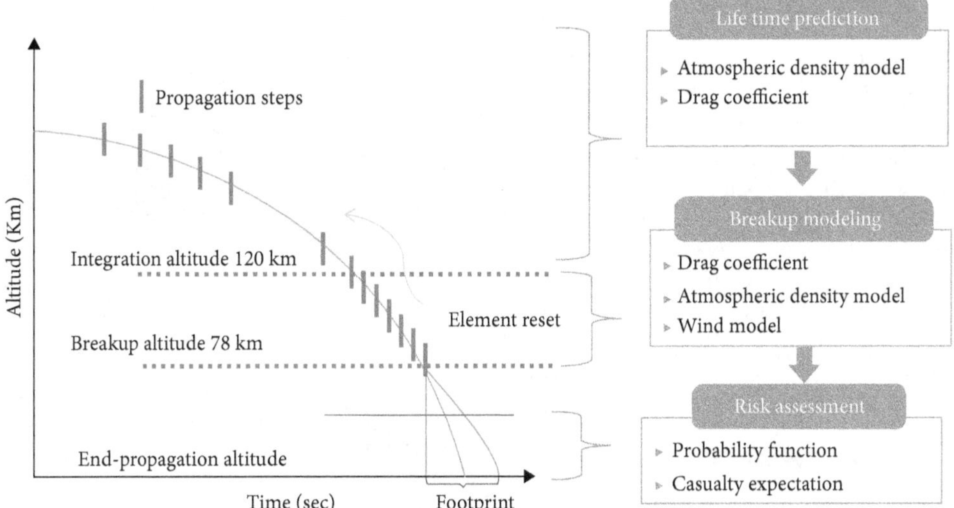

FIGURE 1: Methodology for designing a reentry trajectory of an uncontrolled space object.

Several approaches for the estimation of the reentry time and impact locations have been proposed over the last decades [4–8]. For instance, Tardy and Kluever have estimated the states of orbital debris by using an optimal estimation method, and then Monte-Carlo simulation was carried out to predict the impact location with a final covariance matrix [5]. Reyhanoglu and Alvarado have proposed a Lyapunov equation-based covariance propagation method to predict the time evolution of debris trajectories [7]. In that method, a concept of positional probability ellipsoids is employed for the visualization of the time evolution of the debris dispersion. However, the Lyapunov equation-based covariance propagation method is based on the linearization of the translational equations of motion of the reentering space debris and the equations of the atmospheric reentry is subject to unknown uncertainties; thus, a high degree of neglected and truncated nonlinearities could lead to a degraded estimation of the debris dispersion [8, 9].

In order to compensate for the drawbacks of the Lyapunov-based estimation of the debris dispersion, alternative solutions could be employed by increasing the order of the Taylor-series expansion of the nonlinear system or by using an advanced numerical technique [9, 10]. These efficient alternatives include the unscented transformation [11], the divided-difference numerical integration [12–14], and the cubature quadrature integration [15]. As discussed in [11], the unscented transformation is able to capture the higher-order moments caused by the nonlinear transform better than the Taylor-series-based approximations used in the Lyapunov-based covariance propagation in [5]. The unscented transformation-based estimation technique is called the sigma point estimator [16, 17] in the sense that the estimator works on the principle that a set of deterministically sampled sigma points is used to parameterize the mean and covariance of the Gaussian random variables without the linearization step. The classification of the sigma point method is also interpreted as a special case of a weighted statistical linear regression (SLR) [18]. Even though the statistical linear regression approach and the unscented transformation approach could represent the estimation of the debris reentry trajectories, the estimation accuracy of the debris dispersion could be limited, because they capture the debris dispersion with the second-moment covariance using a fixed small number of sigma points extracted from a predefined covariance ellipsoid. Therefore, it is necessary to design an advanced approach which could include more precise distribution information of the debris dispersion and take into account the nonlinearities due to uncertain nonlinear motion of the reentering debris.

The main contributions of this paper are twofold. First, a precise and effective Gaussian cubature transformation-based [19, 20] estimation approach is proposed to compute the time evolution of the debris dispersion by using a moment-matching technique. The moment-matching formulation enables usage of many precise numerical integration methods such as Gauss-Hermite quadratures and cubature rules for multidimensional states [21]. In the Gauss-Hermite integration approach, a much larger set of sigma points and weights could be chosen compared to those of the Lyapunov covariance and the statistical regression approach such that with polynomial integrand the numerical approximation becomes exact and leads to more precise representation of the statistical moments (mean and covariance) of the distribution of a debris dispersion. Second, various parametric studies are carried out to analyze the effects of modeling uncertainties due to unpredictable atmospheric density and drag coefficients as well as wind effects near to the ground. For the analysis of the effects of the density model affecting the accuracy of the reentry prediction, four difference density models were used in the prediction of reentry trajectory. Specifically, NRLMSISE-00 [22], CIRA-72 [23], U.S. standard 1976, and exponential models [24] are compared with several different values of drag coefficient describing the unknown shape of breakup debris. During the atmospheric reentry, a time-varying empirical wind model is utilized to take into account a realistic environment, and it is based on HWM-07 (Horizon Wind Model 2007) [24, 25].

The remainder of this paper is organized as follows: Section 2 provides a description of the formulation of the reentry translational equations of motion. Section 3 provides two new approaches for the estimation and prediction of the debris dispersion due to the breakup during the reentry which is introduced by propagating the covariances of positional probabilistic ellipsoids. Section 4 shows performance comparisons of the proposed estimation approaches, that is, the statistical linear regression with the unscented transformation and the Gauss-Hermite cubature-based numerical approximation are demonstrated. Lastly, Section 5 provides a discussion of the simulation results.

2. Reentry Equations of Motion

2.1. Coordinate System. Let $\mathbf{x} = [x_1, x_2, x_3]^T$ denote the topocentric horizon coordinate system at the breakup instant. The origin of this frame is at (θ_0, ϕ_0) on the Earth's surface, where (θ_0, ϕ_0) denotes the initial longitude and latitude. The x_1x_2 plane is the local horizon, which is the plane tangent to the sphere at the origin. x_3 is normal to this plane directed towards the zenith (i.e., the coordinate x_3 is the geometric altitude). The x_1 axis is directed eastward and the x_2 axis points north. The $x_1x_2x_3$ frame is also referred to as the ENU (east-north-up) coordinate [24] as shown in Figure 2.

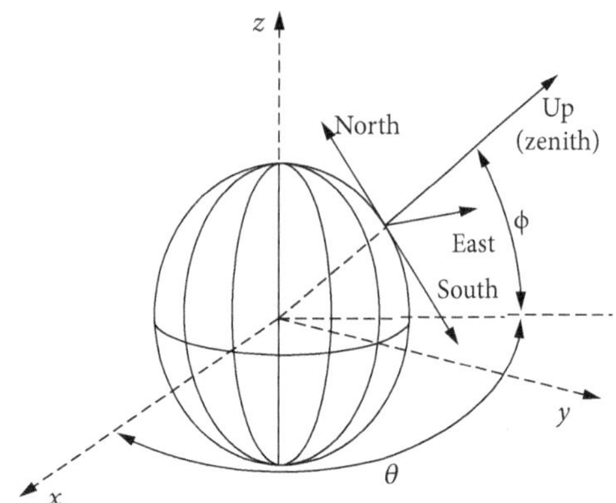

FIGURE 2: ENU (east-north-up) coordinate system for reentry equations of motion.

2.2. Nonlinear Translational Equations of Motion. The governing equation of motion for an uncontrolled space object entering into the atmosphere perturbed by the atmospheric drag uncertainty but ignoring the lifting force can be described by the following equations with position $\mathbf{r}$ and velocity $\mathbf{v}$ with their corresponding initial conditions $\mathbf{r}(t_0)$ and $\mathbf{v}(t_0)$ [5].

$$\begin{aligned}\dot{\mathbf{r}}(t) &= \mathbf{v}(t),\\ \dot{\mathbf{v}}(t) &= \mathbf{a}_D(t) + \mathbf{a}_G(t) - 2\boldsymbol{\omega}\times\mathbf{v}(t) - \boldsymbol{\omega}\times[\boldsymbol{\omega}\times(\mathbf{r}(t) + R_e\hat{\mathbf{e}}_3)] + \boldsymbol{\xi}(t),\end{aligned} \tag{1}$$

where $\mathbf{r} = [x_1, x_2, x_3]^T$ and $\mathbf{v} = [v_1, v_2, v_3]^T$ denote the position and velocity vectors, respectively, with respect to the ENU coordinate frame. $R_e = 6378 \times 10^3$ m is the Earth's radius, $\mathbf{e}_3 = [0, 0, 1]^T$ is the third standard unit vector in $\mathscr{R}^3$ and $\boldsymbol{\xi}$ is the random acceleration vector due to modeling uncertainties and disturbances, and it is assumed that the noise vector is a zero-mean Gaussian process, $E[\boldsymbol{\xi}(t)\boldsymbol{\xi}(t)^T] = \delta(t-s)\mathbf{Q}(t)$, where $\mathbf{Q}(t)$ is the positive definite process noise covariance matrix. $\boldsymbol{\omega} = [\omega_1, \omega_2, \omega_3]^T$ denotes the angular velocity vector of the ENU coordinates with respect to the ECEI (Earth-centered-Earth-fixed inertial) frame and each term of the angular velocity is computed by (θ_0, ϕ_0) if the origin of the ENU topocentric horizon coordinate system is given $[\omega_1, \omega_2, \omega_3]^T = [0, \omega_e\cos\phi_0, \omega_e\sin\phi_0]^T$, where $\omega_e = 72.9217 \times 10^{-6}$ rad/s is the Earth's rotation rate. $\mathbf{a}_G$ denotes the gravitational acceleration, $\mathbf{a}_G = g\hat{\mathbf{e}}_3$, and based on the inverse square gravity model, the gravitational acceleration is calculated by

$$g = g_e\left(\frac{R_2}{R_e + x_3}\right)^2, \tag{2}$$

where g_e denotes the gravitational acceleration at zero altitude. $\mathbf{a}_D$ is the instantaneous acceleration due to the atmospheric density and it is assumed to be opposed to the direction of motion and proportional to the atmospheric density ρ and the velocity squared [24].

$$\begin{aligned}\mathbf{a}_D(t) &= -\frac{1}{2\beta}\rho|\mathbf{v}_{\text{rel}}(t)|\mathbf{v}_{\text{rel}}(t),\\ \beta &= \frac{m_s}{C_dA},\end{aligned} \tag{3}$$

where β is the ballistic drag coefficient, C_d is the aerodynamic drag coefficient, A is the cross-sectional area of the satellite in a plane normal to the relative velocity vector, and m_s is the mass of the reentering space object. $\mathbf{v}_{\text{rel}} = \mathbf{v} - \mathbf{w}$ denotes the relative air velocity vector, where $\mathbf{w}$ is the wind velocity vector, and $v_{\text{rel}} = \|\mathbf{v}_{\text{rel}}\| = \sqrt{\mathbf{v}_{\text{rel}}^T\mathbf{v}_{\text{rel}}}$ is the magnitude of the relative air velocity vector. For the atmospheric density, an analytical exponential atmosphere model is utilized as

$$\rho(x_3) = \rho_e \exp\left(\frac{-x_3}{H}\right), \tag{4}$$

where $\rho_e = 1.752\,\text{kg/m}^3$ and $H = 6.7 \times 10^3$ m.

Now, after defining an augmented state vector $\mathbf{x}(t) \equiv [\mathbf{r}^T(t)\,\mathbf{v}^T(t)]^T$, then the nonlinear differential equation could be written by

$$\dot{\mathbf{x}}(t) = \mathbf{f}(\mathbf{x}(t)) + \mathbf{G}(t)\boldsymbol{\eta}(t), \tag{5}$$

where

$$\mathbf{f}(\mathbf{x}(t)) \equiv \begin{bmatrix}\mathbf{v}(t)\\ \mathbf{a}_D(t) + \mathbf{a}_G - 2\boldsymbol{\omega}\times\mathbf{v}(t) - \boldsymbol{\omega}\times[\boldsymbol{\omega}\times(\mathbf{r}(t) + R_e\hat{\mathbf{e}}_3)]\end{bmatrix}, \tag{6}$$

and $\boldsymbol{\eta}(t) \equiv [\mathbf{0}_{3\times1}^T \, \boldsymbol{\xi}^T(t)]^T$ is the augmented process noise vector, and $\mathbf{G}(t) = [\mathbf{0}_{3\times3}\mathbf{I}_{3\times3}]^T$.

2.3. State-Space Nonlinear Equations of Motion. It is assumed that the initial nominal state of a reentering space vehicle right before the breakup is given by propagating an orbital decay reentry trajectory until 78 km in ECEF (Earth-centered-Earth-fixed). After the coordinate transformation from the ECEF to ENU frame, the initial position of a breakup event is given by geometric altitude z_0, longitude θ_0, and latitude ϕ_0. The initial position (θ_0, ϕ_0) which locates the origin of the ENU frame on the Earth's surface is given by

$$(x_1(t_0), x_2(t_0), x_3(t_0)) = (0, 0, z_0). \quad (7)$$

Then, the nonlinear translational equations of motion in (1) can be rewritten in the form of a state-space representation [5] for numerical integration as follows:

$$\begin{aligned}
\dot{x}_1(t) &= v_1(t), \\
\dot{x}_2(t) &= v_2(t), \\
\dot{x}_3(t) &= v_3(t), \\
\dot{v}_1(t) &= -\frac{\rho}{2\beta} v_{\text{rel}}(t)(v_1(t) - w_1) - 2(\omega_2 v_3(t) - \omega_3 v_2(t)) \\
&\quad + \omega_e^2 x_1(t) + \xi_1(t), \\
\dot{v}_2(t) &= -\frac{\rho}{2\beta} v_{\text{rel}}(t)(v_2(t) - w_2) - 2\omega_3 v_1(t) - \omega_2\omega_3(R_e + x_3(t)) \\
&\quad + \omega_3^2 x_2(t) + \xi_2(t), \\
\dot{v}_3(t) &= -\frac{\rho}{2\beta} v_{\text{rel}}(t)(v_3(t) - w_3) - g(x_3) + 2\omega_2 v_1(t) \\
&\quad - \omega_2\omega_3 x_2(t) + \omega_2^2(x_3(t) + R_e) + \zeta_3(t),
\end{aligned} \quad (8)$$

where $\mathbf{w} = [w_1, w_2, w_3]^T$ is the wind velocity vector.

3. Algorithms for Estimation of Debris Dispersion

In this section, two new approaches for the estimation of debris dispersion due to a space vehicle breakup are proposed. The estimation of debris dispersion is represented in terms of a covariance propagation which describes probabilistic ellipsoids from a nominal reentry trajectory. First, a statistical linear regression using the unscented transformation is introduced to compute the time evolution of the covariance information. Second, a Gaussian moment-matching method which calculates the first and second moments of the probabilistic distribution of the debris dispersion by using a Gauss-Hermite cubature numerical approximation is proposed. Figure 3 depicts the difference between the conventional Lyapunov-based covariance approach and the unscented transformation-based covariance computation. In this work, detailed derivation for the covariance computation using the Lyapunov equation is not included but it can be referred to [5].

3.1. Statistical Linear Regression Approach for Estimation of Debris Dispersion. In general, statistical linear error propagation is more accurate than the error propagation by first-order Taylor-series approximation. For a statistically linearized error propagation, consider a nonlinear mapping with a set of weighted sigma points $\{\mathcal{X}_i, \mathcal{W}_i\}, \quad i = 1, \dots, r$ that are selected to line on the principle component axes of the state covariance with a weight [18].

$$\mathcal{Z}_i = \mathbf{h}(\mathcal{X})_i. \quad (9)$$

Then, the first- and second-order statistics of the transformed points, mean, and covariance, are computed from the sigma point $\mathcal{X}_i$ and transformed sigma point $\mathcal{Z}_i$:

$$\begin{aligned}
\hat{\mathbf{x}} &\triangleq \sum_{i=1}^{r} \mathcal{W}_i \mathcal{X}_i, \\
\mathbf{P}^{\mathcal{X}\mathcal{X}} &\triangleq \sum_{i=1}^{r} \mathcal{W}_i [\mathcal{X}_i - \hat{\mathbf{x}}][\mathcal{X}_i - \hat{\mathbf{x}}]^T, \\
\mathbf{P}^{\mathcal{X}\mathcal{Z}} &\triangleq \sum_{i=1}^{r} \mathcal{W}_i [\mathcal{X}_i - \hat{\mathbf{x}}][\mathcal{Z}_i - \hat{\mathbf{z}}]^T.
\end{aligned} \quad (10)$$

Before applying a statistical linear regression method, it is necessary to transform the continuous nonlinear differential equation in (6) into a discrete-time nonlinear difference equation using an approximate method, such as the Euler method, or numerical Runge-Kutta approach. In this section, it is assumed that a nonlinear discrete-time equation is given in the form [11]

$$\mathbf{x}_k = \mathbf{f}_k(\mathbf{x}_k, \boldsymbol{\zeta}_k, t_k), \quad (11)$$

where $\mathbf{x}_k \in \mathbb{R}^n$ is the $n \times 1$ state vector. It is assumed that the noise vector $\boldsymbol{\zeta}_k \in \mathbb{R}^q$ is the $q \times 1$ state noise vector and is a zero-mean Gaussian process satisfying

$$E\left\{\boldsymbol{\zeta}_k \boldsymbol{\zeta}_j^T\right\} = \delta_{kj}\mathbf{Q}_k, \quad \forall k, j, \quad (12)$$

where $\mathbf{Q}_k$ is the process noise covariance matrix. In this paper, we utilize the unscented transformation mapping technique to derive the first- and second-order moment statistics. First, in order to generate a set of weighted sigma points, the state vector $\mathbf{x}_k \in \mathbb{R}^n$ is redefined as an augmented state vector $\mathbf{x}_k^a \in \mathbb{R}^{n+q}$ along with noise variables and the corresponding augmented covariance matrix on the diagonal is reconstructed by

$$\begin{aligned}
\mathbf{x}_k^a &= \begin{bmatrix} \mathbf{x}_k \\ \boldsymbol{\zeta}_k \end{bmatrix}, \\
\mathbf{P}_k^a &= \begin{bmatrix} \mathbf{P}_k & \mathbf{0} \\ \mathbf{0} & \mathbf{Q}_k \end{bmatrix},
\end{aligned} \quad (13)$$

where $\mathbf{x}_k^a$ is the augmented state vector with the dimension $n_a = n + q$. Then, the set of scaled symmetric sigma points

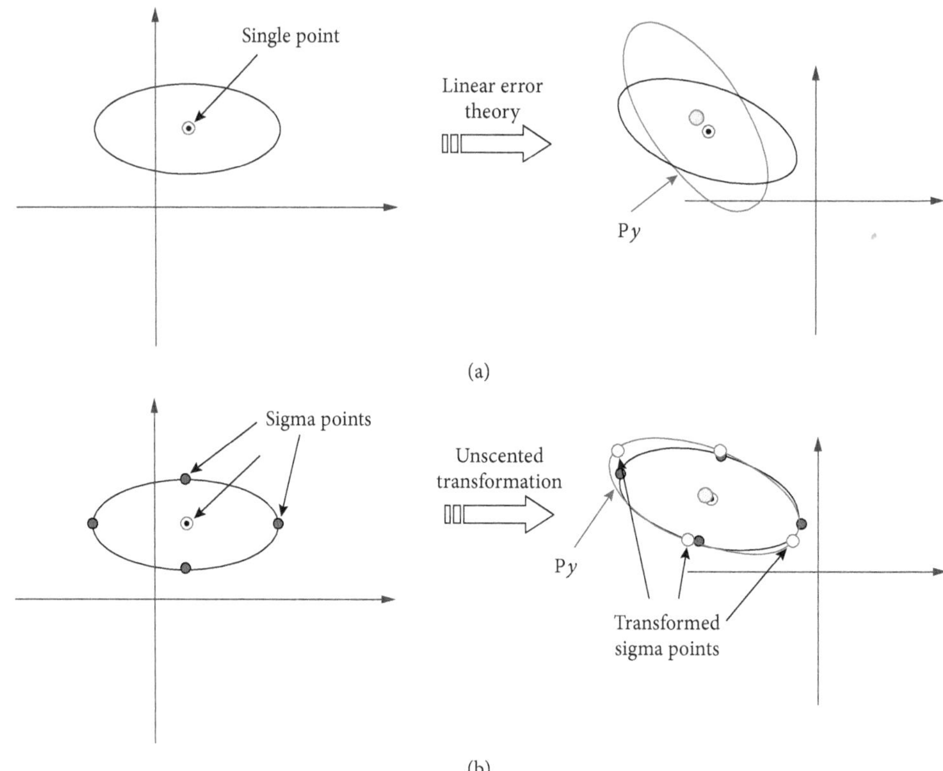

FIGURE 3: (a) Lyapunov equation-based covariance propagation. (b) Statistical linear regression-based covariance propagation method—unscented transformation approach.

$\{\mathcal{X}^a_{i,k}\}_{i=0}^{2n_a} = \{[(\mathcal{X}^x_{i,k})^T(\mathcal{X}^{\varsigma}_{i,k})^T]^T\}_{i=0}^{2n_a}$ with the augmented state vector and covariance matrix is constructed by [11].

$$\begin{aligned}
\mathcal{X}^a_{0,k} &= \hat{\mathbf{x}}^a_k,\\
\mathcal{X}^a_{i,k} &= \hat{\mathbf{x}}^a_k + \left(\sqrt{(n_a+\lambda)\mathbf{P}^a_k}\right)_i, \quad i = 1,\ldots,n_a,\\
\mathcal{X}^a_{i,k} &= \hat{\mathbf{x}}^a_k - \left(\sqrt{(n_a+\lambda)\mathbf{P}^a_k}\right)_i, \quad i = n_a+1,\ldots,2n_a,
\end{aligned} \tag{14}$$

where $\lambda = \alpha^2(n_a+\kappa) - n_a$ is a scaling parameter with the constant parameter $0 \le \alpha \le 1$ and κ providing an extra degree of freedom for fine-tuning the higher-order moment ($\kappa = 3 - n_a$ for a Gaussian distribution). The term $(\sqrt{(n_a+\lambda)\mathbf{P}^a_k})_i$, is the ith column or row vector of the weighted square root of the scaled covariance matrix $(n_a+\lambda)\mathbf{P}^a_k$. The corresponding weights for the mean and covariance are defined by

$$\begin{aligned}
W_i^{(m)} &= \frac{\lambda}{(n_a+\lambda)}, \quad i = 0,\\
W_i^{(m)} &= \frac{1}{\{2(n_a+\lambda)\}}, \quad i = 1,\ldots,2n_a,\\
W_i^{(c)} &= \frac{\lambda}{(n_a+\lambda)+(1-\alpha^2+\beta)}, \quad i = 0,\\
W_i^{(c)} &= \frac{1}{\{2(n_a+\lambda)\}}, \quad i = 1,\ldots,2n_a,
\end{aligned} \tag{15}$$

where β is a third parameter for incorporating extra higher-order effects, and $\beta = 2$ is the optimal value for Gaussian distribution.

As for the prediction step for the next time $k+1$, the predicted state estimate vector $\hat{\mathbf{x}}_{k+1|k}$ and its predicted covariance matrix $\mathbf{P}_{k+1|k}$ are computed from the propagated sigma point vectors as

$$\boldsymbol{\mathcal{X}}^x_{k+1} = \mathbf{f}(\boldsymbol{\mathcal{X}}^x_k, \boldsymbol{\mathcal{X}}^w_k, k), \tag{16}$$

$$\hat{\mathbf{x}}_{k+1|k} = \sum_{i=0}^{2n_a} W_i^{(m)} \mathbf{f}(\mathcal{X}^x_k, \mathcal{X}^w_k, k), \tag{17}$$

$$\mathbf{P}_{k+1|k} = \sum_{i=0}^{2n_a} W_i^{(c)} \left[\mathcal{X}^x_{i,k+1} - \hat{\mathbf{x}}_{k+1|k}\right]\left[\mathcal{X}^x_{i,k+1} - \hat{\mathbf{x}}_{k+1|k}\right]^T, \tag{18}$$

where $\mathcal{X}^x_{i,k}$ is a weighted sigma point vector of the first n elements of the ith augmented sigma point vector $\mathcal{X}^a_{i,k}$ and $\mathcal{X}^w_{i,k}$ is a weighted sigma point vector of the next q elements of $\mathcal{X}^a_{i,k}$, respectively.

3.2. Gaussian Moment-Matching Approach for Estimation of Debris Dispersion. Even though the statistical linear regression (SLR) using the unscented transformation approach could represent the propagation of the debris dispersion due to a breakup event, the SLR-based approach estimates the debris dispersion with the second-moment covariance using a finite small number of sigma points. The statistical regression approach is not a truly global approximation and it does not work well with nearly singular covariances, that is, nearly deterministic systems with small covariances. For compensating the drawbacks, it is necessary to consider a more realistic approach which can precisely capture the initial debris dispersion and also take into account the neglected nonlinearities from the statistical linear regression. In this section, the Gaussian cubature transformation approach [10] is proposed to estimate the debris dispersion with a moment-matching technique by using a set of transformed sigma lattice points which represent breakup components in the debris dispersion (shown in Figure 4).

Consider the transformation of the state $\mathbf{x}$ and the transformed random variable $\mathbf{y} = \mathbf{g}(\mathbf{x}) + \mathbf{q}$ where $\mathbf{x} \sim \mathcal{N}(\mathbf{m}, \mathbf{P})$ and $\mathbf{q} \sim \mathcal{N}(\mathbf{0}, \mathbf{Q})$, then the moment-matching-based Gaussian approximation to the joint distribution of $\mathbf{x}$ and $\mathbf{y}$ is represented as [19]

$$\begin{pmatrix} \mathbf{x} \\ \mathbf{y} \end{pmatrix} \sim \mathcal{N}\left(\begin{pmatrix} \mathbf{m} \\ \widehat{\boldsymbol{\mu}}_M \end{pmatrix}, \begin{pmatrix} \mathbf{P} & \mathbf{C}_M \\ \mathbf{C}_M^T & \mathbf{S}_M \end{pmatrix} \right), \tag{19}$$

where

$$\begin{aligned} \widehat{\boldsymbol{\mu}}_M &= \int \mathbf{g}(\mathbf{x}) \mathcal{N}(\mathbf{x} \mid \mathbf{m}, \mathbf{P}) d\mathbf{x}, \\ \mathbf{S}_M &= \int (\mathbf{g}(\mathbf{x}) - \widehat{\boldsymbol{\mu}}_M)(\mathbf{g}(\mathbf{x}) - \widehat{\boldsymbol{\mu}}_M)^T \mathcal{N}(\mathbf{x} \mid \mathbf{m}, \mathbf{P}) d\mathbf{x} + \mathbf{Q}, \\ \mathbf{C}_M &= \int (\mathbf{x} - \mathbf{m})(\mathbf{g}(\mathbf{x}) - \widehat{\boldsymbol{\mu}}_M)^T \mathcal{N}(\mathbf{x} \mid \mathbf{m}, \mathbf{P}) d\mathbf{x}, \end{aligned} \tag{20}$$

and $\mathbf{x}$ is assumed to be Gaussian with the distribution, $\mathbf{x} \sim \mathcal{N}(\mathbf{m}, \mathbf{P})$, $\mathbf{m}$ is the mean, and $\mathbf{P}$ is the covariance.

The debris dispersion can be modeled with the first $\widehat{\boldsymbol{\mu}}_M$ and second moments $\mathbf{S}_M$ of the original distribution of $\mathbf{x}$ and the transformed distribution $\mathbf{y} = \mathbf{g}(\mathbf{x}) + \mathbf{q}$, and this leads to the problem of forming Gaussian approximation to $(\mathbf{x}, \mathbf{y})$ by directly approximating the integrals. The advantage of the moment-matching formation lies in that it enables usage of well-known numerical integration methods such as Gauss-Hermite quadrature [20], cubature rules [19], and central difference methods [13, 14]. In this section, the integrals in (20) can be evaluated practically by the Gauss-Hermite cubature-based numerical integration method as shown in Figures 4 and 5. By generalizing the change of variable idea, we can form approximation to multidimensional integrals of the form in (19). First, it is needed to decompose the covariance, $\mathbf{P} = \sqrt{\mathbf{P}}\sqrt{\mathbf{P}}^T$, where $\sqrt{\mathbf{P}}$ is the Cholesky factor of the covariance. Now, a new integration variable is defined by [19]

$$\mathbf{x} = \mathbf{m} + \sqrt{\mathbf{P}}\boldsymbol{\sigma}, \tag{21}$$

where $\boldsymbol{\sigma} = \boldsymbol{\sigma}^{(i_1,\ldots,i_n)}$ is a n-dimensional vector with one-dimensional unit sigma point $\sigma^{(i_k)}$ at element k. Then, the Gaussian integral in (20) can be rewritten as

$$\int \mathbf{g}(\mathbf{x}) \mathcal{N}(\mathbf{x} \mid \mathbf{m}, \mathbf{P}) d\mathbf{x} = \int \mathbf{g}\left(\mathbf{m} + \sqrt{\mathbf{P}}\boldsymbol{\sigma}\right) \mathcal{N}(\boldsymbol{\sigma} \mid 0, \mathbf{I}) d\boldsymbol{\sigma}. \tag{22}$$

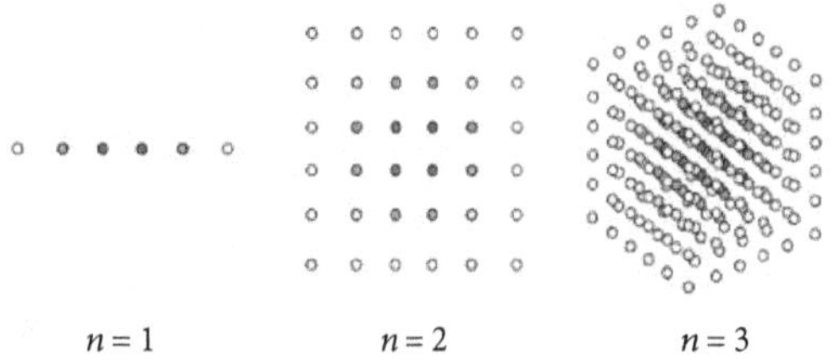

FIGURE 4: In Gauss-Hermite cubature integration rule, the lattice points in dimension $n = 1,2,3$ are required to perform p product rule integral approximation, that is, p^n cubature points. The color marks the weight of the cubature points [19].

The integral in (22) is formulated in terms of the multidimensional unit Gaussian $\mathcal{N}(\mathbf{g} \mid 0, \mathbf{I})$ and can be written as an iterated integral over one-dimensional Gaussian distributions. Each of the one-dimensional integrals can be approximated with the Gauss-Hermite quadrature as

$$\begin{aligned} &\int \mathbf{g}\left(\mathbf{m} + \sqrt{\mathbf{P}}\boldsymbol{\sigma}\right) \mathcal{N}(\boldsymbol{\sigma} \mid 0, \mathbf{I}) d\boldsymbol{\sigma} \\ &= \int \cdots \int \mathbf{g}\left(\mathbf{m} + \sqrt{\mathbf{P}}\boldsymbol{\sigma}\right) \mathcal{N}(\sigma_1 \mid 0, 1) d\sigma_1 \times \cdots \times \mathcal{N}(\sigma_n \mid 0, 1) d\sigma_n \\ &\approx \sum_{i_1,\ldots,i_n} W^{(i_1)} \times \cdots \times W^{(i_n)} \mathbf{g}\left(\mathbf{m} + \sqrt{\mathbf{P}}\sigma^{(i_1,\ldots,i_n)}\right), \end{aligned} \tag{23}$$

where $W^{(i_k)}$, $k = 1, \ldots, n$ are simply the corresponding one-dimensional Gauss-Hermite weights with pth-order approximation, which can be calculated by

$$W^{(i)} = \frac{p!}{p^2 \left[H_{p-1}\left(\sigma^{(i)}\right)\right]^2}, \tag{24}$$

and $H_p(x)$ is a Hermite polynomial of order p given by

$$H_p(x) = (-1)^p \exp\left(\frac{x^2}{2}\right) \frac{d^p}{dx^p} \exp\left(\frac{-x^2}{2}\right). \tag{25}$$

The unit sigma points $\sigma^{(i)}$, $i = 1, \ldots, p$ is calculated by finding the roots of the Hermite polynomial $H_p(x)$. The extension of the Gauss-Hermite quadrature rule to an n-dimensional cubature rule by using the product rule lattice approach yields a rather good numerical integration method. Based on the Gauss-Hermite cubature rule, the multidimensional weights are calculated as the products of one-dimensional weights

$$\begin{aligned} W^{(i_1,\ldots,i_n)} &= W^{(i_1)} \times \cdots \times W^{(i_n)} \\ &= \frac{p!}{p^2 \left[H_{p-1}\left(\sigma^{(i_1)}\right)\right]^2} \times \cdots \times \frac{p!}{p^2 \left[H_{p-1}\left(\sigma^{(i_n)}\right)\right]^2}. \end{aligned} \tag{26}$$

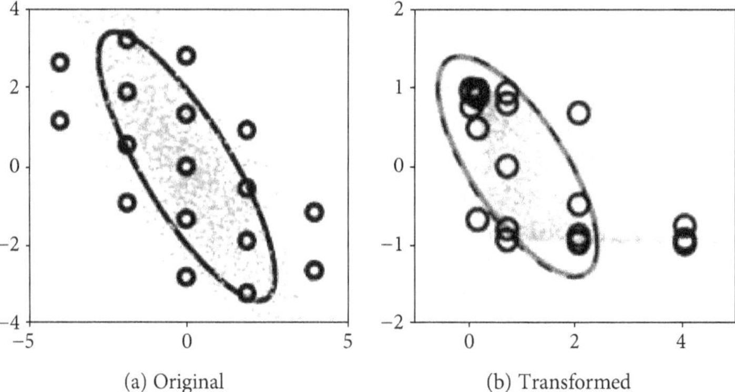

FIGURE 5: Illustration of a fifth-order Gauss-Hermite cubature-based approximation to a nonlinear transformation. The covariance of the true distribution is presented by the dashed line and the solid line is the approximation [10].

Also, the multidimensional unit sigma points can be given as Cartesian product of the one-dimensional unit sigma points

$$\boldsymbol{\sigma}^{(i_1,\ldots,i_n)} = \begin{pmatrix} \sigma^{(i_1)} \\ \vdots \\ \sigma^{(i_n)} \end{pmatrix}. \tag{27}$$

It is noted that the number of sigma points required for n-dimensional integral with pth-order rule is p^n, which could become unfeasible when the number of dimensions grows which is indicated in Figure 4.

Based on the previous Gauss-Hermite cubature integration approach, an additive form of the multidimensional Gauss-Hermite cubature integral technique is used to represent the covariance prediction of a debris dispersion. It is assumed that the posterior density function p^n is known. First, the unit sigma point $\sigma^{(i)}$, $i = 1, \ldots, p$ is calculated by finding the roots of the Hermite polynomial $H_p(x)$. Based on the Gauss-Hermite cubature rule, the multidimensional weights are calculated as the products of one-dimensional weights:

$$\begin{aligned} W^{(i_1,\ldots,i_n)} &= W^{(i_1)} \times \cdots \times W^{(i_n)} \\ &= \frac{p!}{p^2\left[H_{p-1}\left(\sigma^{(i_1)}\right)\right]^2} \times \cdots \times \frac{p!}{p^2\left[H_{p-1}\left(\sigma^{(i_n)}\right)\right]^2}, \end{aligned} \tag{28}$$

where one-dimensional weight $W^{(i_k)}$ is given by using (26). Now, the sigma points can be generated by

$$\mathcal{X}^i_{k|k} = \mathbf{m}_{k|k} + \sqrt{\mathbf{P}_{k|k}}\boldsymbol{\sigma}^i, \quad i = 1, \ldots, p^n, \tag{29}$$

where $\mathbf{m}_{k|k}$ is the mean vector of the current debris state and $\mathbf{P}_{k|k}$ is the covariance matrix representing the debris dispersion at the event of the breakup time, k, and the unit sigma points $\boldsymbol{\sigma}^{(i_1,\ldots,i_n)}$ are defined in (27). Next, the sigma points can be propagated by using the nonlinear discrete equation in (11) as

$$\mathcal{X}^i_{k+1|k} = \mathbf{f}\left(\mathcal{X}^i_{k|k}, k\right), \quad i = 1, \ldots, p^n. \tag{30}$$

As for the prediction step for the next time $k+1$, the predicted state vector $\mathbf{m}_{k+1|k}$ of the debris trajectory and its predicted covariance matrix $\mathbf{P}_{k+1|k}$ representing the prediction of the debris dispersion are computed by using the propagated sigma point vectors in (30) as

$$\mathbf{m}_{k+1|k} = \sum_{i=1}^{p^n} W^{(i)} \mathcal{X}^i_{k+1|k}, \tag{31}$$

$$\mathbf{P}_{k+1|k} = \sum_{i=1}^{p^n} W^{(i)} \left[\mathcal{X}^i_{k+1|k} - \mathbf{m}_{k+1|k}\right] \left[\mathcal{X}^i_{k+1|k} - \mathbf{m}_{k+1|k}\right]^T + \mathbf{Q}_{k+1}, \tag{32}$$

where the weight $W^{(i)}$ is defined in (28) and $\mathbf{Q}_{k+1}$ is the process noise covariance representing the uncertainties of truncated modeling errors and unknown errors.

The multidimensional Gaussian-Hermite cubature-based covariance prediction approach which represents the debris dispersion and the distribution in time can be interpreted as a special form of a Monte-Carlo integration approach.

4. Simulation Results

In this section, first, various parametric studies are carried out to analyze the effects of uncertainties in the atmospheric density and drag coefficient along with wind effects. Then, the performance of the proposed estimation techniques are investigated by comparing two covariance propagation methods; the unscented transformation approach and the Gauss-Hermite cubature integration method.

4.1. Parametric Study for Analysis of Effects of Modeling Uncertainties. Atmospheric drag most strongly influences the motion of reentry objects near the Earth. When it comes to determining accurately atmospheric drag, the values of atmospheric density and drag coefficient are the main factors to the trajectory in the sense that the information on the specific shape and quantity of objects are generally unknown. In light of this reason, the different values of the drag coefficient are selected to see the how the drag coefficient affects the reentry point and the probabilistic ellipsoid of debris dispersion caused by the breakup event in this study. In this

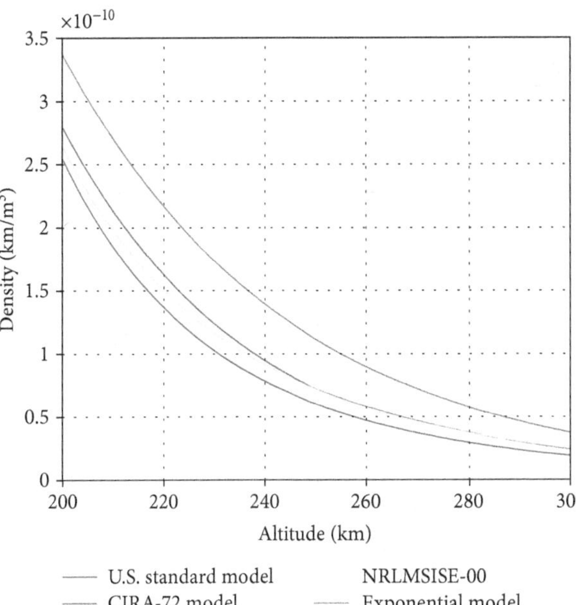

FIGURE 6: Density profiles for density models.

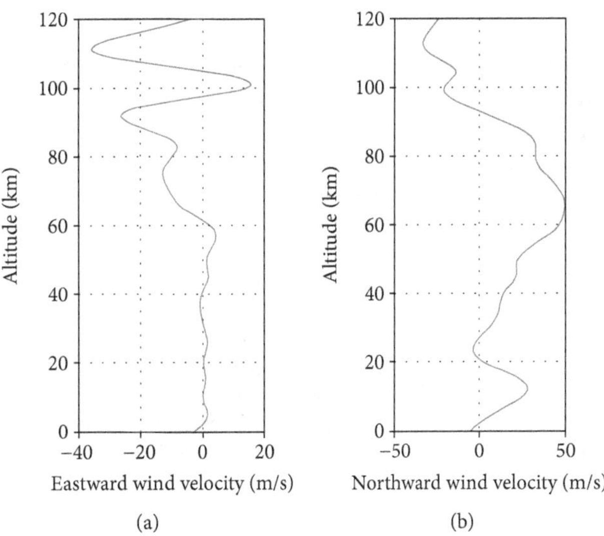

FIGURE 7: Wind profiles for different altitudes.

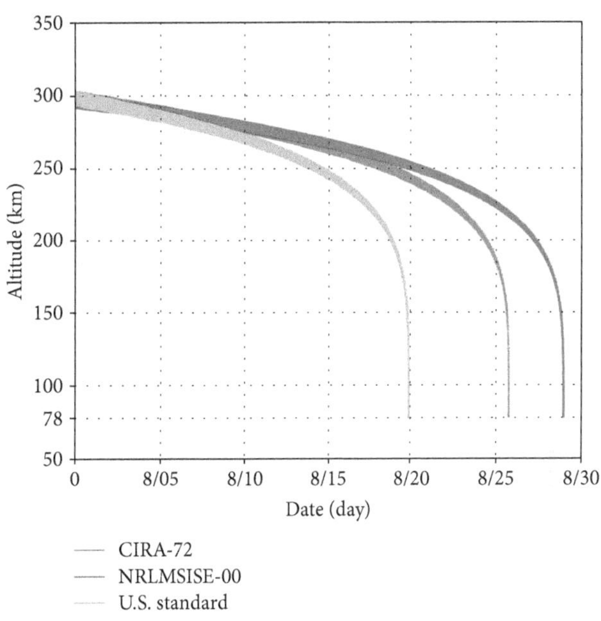

FIGURE 8: Decay of altitude with different atmospheric models.

study, four different atmospheric density models including CIRA-72, Exponential model, U.S. standard model, and NRLMSISE-00 are used to analyze the effects of the atmospheric density model onto the reentry trajectory. Figure 6 shows density profiles corresponding to each density model as a function of altitude. Since wind also influences the reentry trajectory near to the ground, it is necessary to design a mathematical wind profile model. In this study, HWM-07 (Horizontal Wind Model 2007) [24] is employed to consider a more accurate reentry model. Figure 7 shows the velocity of eastward and westward winds with respect to the history of altitude.

In order to analyze the effects of the density model and the drag coefficient, two simulation examples are investigated. The first example considers the lifetime prediction of uncontrolled space objects with different atmospheric density models and drag coefficients. The satellite under consideration has the following orbit parameters: $a = 6678\,\text{km}$, $e = 1.0 \times 10^{-5}$, $i = 28.5^\circ$, $\omega = 0^\circ$, and $\Omega = 0^\circ$, and epoch time is defined as August 1, 2015. Figure 8 shows that the orbital decay of the space object until 78 km is different with respect to atmospheric models. Using the U.S. standard model, the decay of altitude is faster than any other density models. On the other hand, the NRLSMSISE-00 model predicts the decay of altitude of the reentry object with the slowest drop among the three models.

Table 1 presents the reentry latitude as well as longitude of different atmospheric models at the 78 km altitude and reentry time. Compared to the STK LTP (lifetime prediction) tool [26], it is shown that the results of reentry time are similar. Figure 9 indicates the orbital decay of space objects with a different drag coefficient. As the drag coefficient increases, it is shown that the space object falls faster.

On the other hand, a second example is illustrated to show the effects of different density models and drag coefficient onto the debris nominal trajectory as well as the time evolution of the debris dispersion from 78 km at the breakup event to impact ground instance. For the simulation study, the number of 200 samples is drawn to model the initial debris breakup dispersion with the assumption of a Gaussian distribution. It is assumed that the initial breakup state is given by $\mathbf{x}(t_0) = [0 \quad 0 \quad 78,000\,\text{m} \quad 709.9\,\text{m/s} \quad 0\,\text{m/s} \quad -124.0\,\text{m/s}]$ with the breakup altitude at 78 km. The total simulation time was about 32 min. An initial breakup dispersion is generated with an incremental speed $\Delta\mathbf{v}$ uniformly added to the nominal velocity $\mathbf{v}(t_0)$ in every direction by creating a spherical grid of $32 \times 32 = 1024$ points, that is, $\mathbf{v}_{i,j} = \mathbf{v}(t_0) + \Delta\mathbf{v}_{i,j}$ [7]. So, with each 32 points for longitude and latitude,

Table 1: Terminal data of lifetime prediction with different density models.

Density model	Reentry latitude	Reentry longitude	Reentry time	STK LTP tool
CIRA-72	−38.936°	−345.39°	Aug. 26 (01:15)	Aug. 25 (18:21)
U.S. standard	−27.306°	−224.34°	Aug. 28 (15:44)	Aug. 27 (09:02)
NRLMSISE-00	−8.302°	−319.91°	Aug. 20 (05:31)	Aug. 19 (03:22)

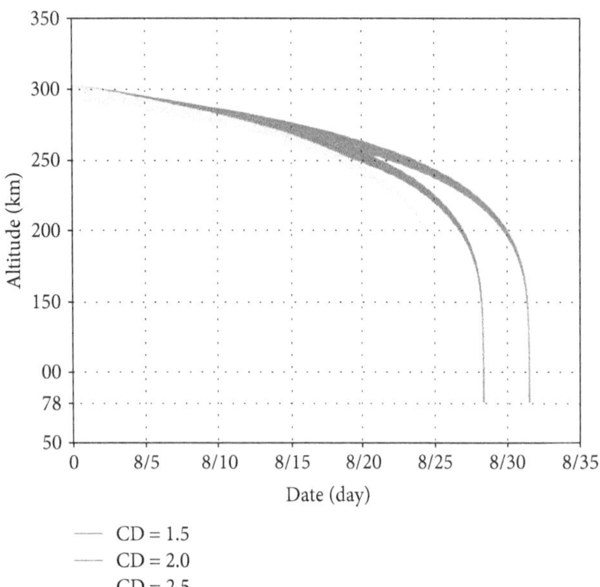

Figure 9: Decay of altitude with different drag coefficients.

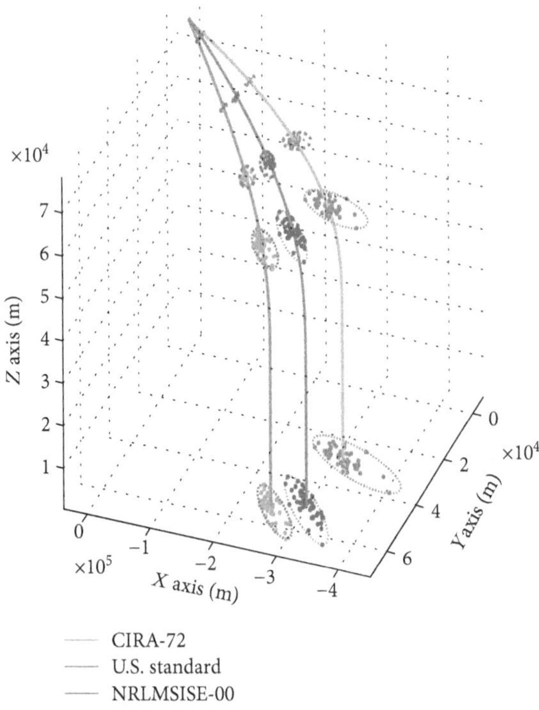

Figure 10: Time evolution of probability ellipsoid with different density models.

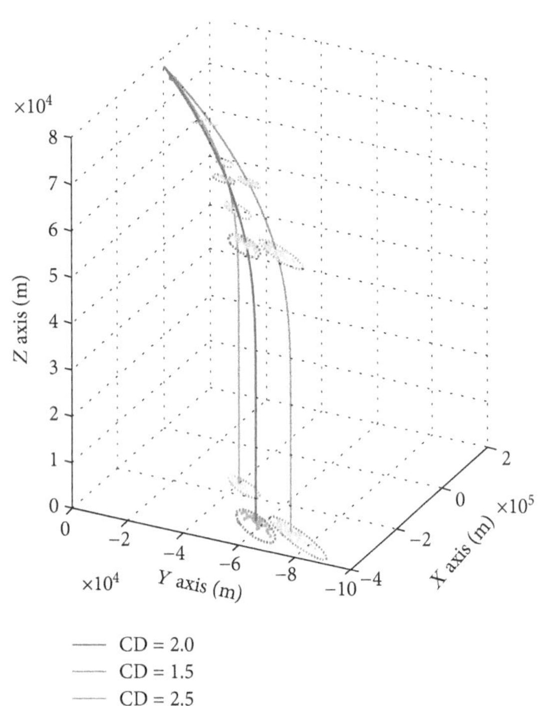

Figure 11: Time evolution of probability ellipsoid with different drag coefficients.

respectively, initial velocities in the initial debris dispersion are created by

$$\mathbf{v}_{i,j} = \mathbf{v}(t_0) + \Delta v \begin{bmatrix} \cos\phi_i \cos\theta_j \\ \cos\phi_i \sin\theta_j \\ \sin\phi_i \end{bmatrix}, \quad i, j = 1, 2, \ldots, 32. \tag{33}$$

The values of each latitude and longitude are taken with $\phi_i \in [-\pi/2, \pi/2)$ and $\theta_j \in [0, 2\pi)$, respectively, where $i, j = 1, 2, \ldots, 32$ and $\Delta v = 100$ m/s is an incremental speed.

Figures 10 and 11 show the positional ellipsoid trajectory of breakup dispersion with different atmospheric density models and drag coefficients. As can be seen, even though each case of breakup event occurred at the same position, the final location of the ground impact point is very different depending on the type of the density model and the drag coefficient. Figures 12 and 13 show the final location and size of the positional ellipsoid on the ground in detail. For the case of different atmospheric models, the CIRA-72 model generated the largest impact area of dispersion debris, while the NRLSMSISE-00 model showed a smaller one, which

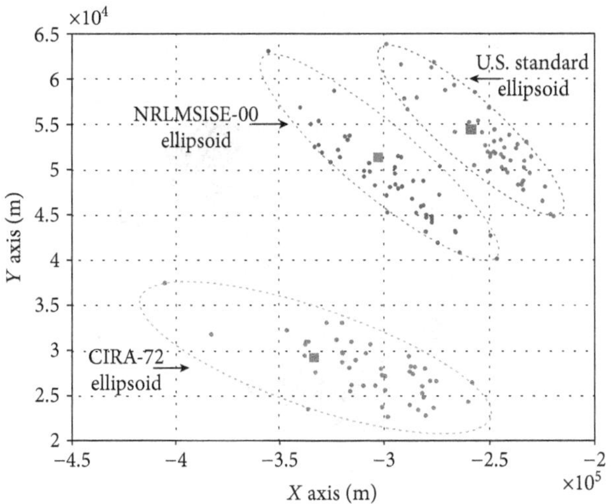

FIGURE 12: Impact footprint (density models).

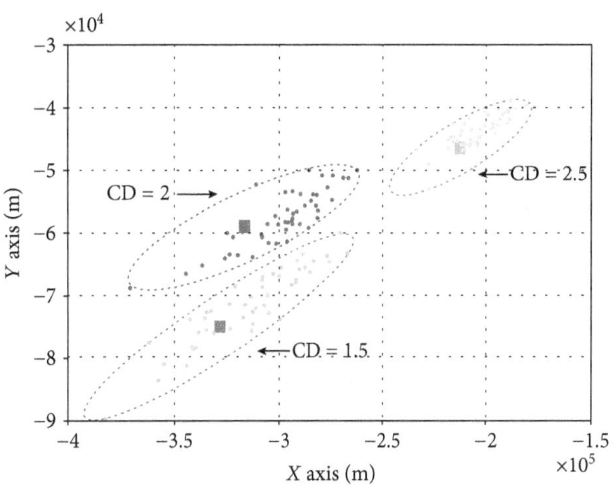

FIGURE 13: Impact footprint (drag coefficients).

indicates that the CIRA-72 model is subject to a lot of uncertainty errors compared to that of the NRLMSISE-00 models. Table 2 describes the footprint statistics including impact areas. For the case of a different drag coefficient, the size of the final positional ellipsoid of debris dispersion becomes smaller as the values of drag coefficients increase. It is indicated that a higher drag coefficient value leads to a faster fall to the ground with a short evolution. As can be shown in Table 3, the positional ellipsoidal area of CD = 1.5 is almost twice larger than that of CD = 2.5. It is seen that a properly estimated drag coefficient is one of the main factors in the propagation of the probabilistic positional ellipsoid.

4.2. Statistical Computation of Debris Dispersion with Covariance Estimation. In this section, a detailed analysis of the estimation of debris dispersion due to the breakup event during the reentry is made by using one of the proposed covariance propagation methods, the unscented transformation technique. Note that the positional variation at the breakup instant is small enough to be neglected but the variation of the debris velocity becomes relatively big. Therefore, it is assumed that an initial breakup dispersion could be generated by adding an incremental speed $\delta\mathbf{v}_{i,j}$ to the initial nominal velocity $\mathbf{v}(t_0)$ in every direction, $\mathbf{v}_{i,j} = \mathbf{v}(t_0) + \delta\mathbf{v}_{i,j}$ like in (33). Now, based on the initial breakup model with an incremental velocity dispersion as above, an initial debris dispersion at the event of the breakup time t_0 is remodeled with a Gaussian distribution before covariance propagation for the time evolution of the debris dispersion in (34).

$$\mathbf{P}(t_0) = 10^6 \operatorname{diag}\left(\begin{bmatrix} 10^{-10} & 10^{-10} & 10^{-10} & 0.0025\,(\mathrm{m/s})^2 & 0.0025\,(\mathrm{m/s})^2 & 0.0053\,(\mathrm{m/s})^2 \end{bmatrix}\right). \quad (34)$$

Note that the initial breakup dispersion generated with an incremental speed $\Delta\mathbf{v}$ uniformly added to the nominal velocity $\mathbf{v}(t_0)$ in every direction could enter the inside of the initial covariance and thus the initial debris dispersion at the breakup event is simulated as shown in Figure 14.

Since the probabilistic distribution of the debris dispersion is Gaussian, the center of the ellipsoid becomes the breakup point and the dispersion is represented by an ellipsoid. Uncertainty due to unknown and neglected acceleration terms is characterized by using the process noise covariance, that is, $E\{\boldsymbol{\eta}(t)\boldsymbol{\eta}(\varsigma)^T\} \equiv \mathbf{Q}(t)\delta(\tau - \varsigma)$ in (12).

$$\mathbf{Q}(t) = 10^6 \operatorname{diag}\left(\begin{bmatrix} 0 & 0 & 0 & 10^{-16} & 10^{-16} & 10^{-16} \end{bmatrix}\right). \quad (35)$$

It is assumed that the breakup altitude is 78 km and the initial breakup position is located at $\mathbf{x}(t_0) = [0 \;\; 0 \;\; 78{,}000\,\mathrm{m} \;\; 709.9\,\mathrm{m/s} \;\; 0\,\mathrm{m/s} \;\; -124.0\,\mathrm{m/s}]$ as seen in Figure 15. The total simulation time was about 32 minutes. Based on the initial covariance in (34), the process noise covariance matrix in (35), and other initial conditions, the initial covariance estimate of the debris dispersion could be propagated in time by using the one of the covariance propagation techniques such as the Gauss-Hermite cubature or the unscented transformation approach. Before propagating the covariance of the debris dispersion, the unscented transformation technique needs to generate a set of sigma points which also represents a set of initial debris fragments at the breakup event. Now, a set of scaled symmetric sigma points $\{\mathcal{X}^a_{i,k}\}_{i=0}^{2n_a} = \{[(\mathcal{X}^x_{i,k})^T (\mathcal{X}^\varsigma_{i,k})^T]^T\}_{i=0}^{2n_a}$ with the augmented state vector and covariance matrix is constructed by

TABLE 2: Footprint statistics for various atmospheric density models.

Density model	Impact latitude	Impact longitude	Impact area
CIRA-72	−3.007°	−240.2°	1522 km^2
U.S. standard	−27.306°	−224.34°	619.2 km^2
NRLMSISE-00	−8.302°	−319.91°	913.8 km^2

TABLE 3: Footprint statistics for various drag coefficients.

Drag coefficient	Impact latitude	Impact longitude	Impact area
CD = 1.5	12.23°	−137.3°	1093 km^2
CD = 2.0	−7.452°	−347.3°	916.5 km^2
CD = 2.5	−11.96°	−118.3°	547.3 km^2

$$\begin{aligned} \mathcal{X}^a_{0,0} &= \hat{\mathbf{x}}^a_0, \\ \mathcal{X}^a_{i,0} &= \hat{\mathbf{x}}^a_0 + \left(\sqrt{(n_a+\lambda)\mathbf{P}^a_0}\right)_i, \quad i = 1, \dots, n_a, \\ \mathcal{X}^a_{i,0} &= \hat{\mathbf{x}}^a_0 - \left(\sqrt{(n_a+\lambda)\mathbf{P}^a_0}\right)_i, \quad i = n_a+1, \dots, 2n_a, \end{aligned} \tag{36}$$

where the term, $(\sqrt{(n_a+\lambda)\mathbf{P}^a_0})_i$, is the ith column or row vector of the weighted square root of the scaled covariance matrix $(n_a+\lambda)\mathbf{P}^a_0$, and the augmented initial covariance is given by

$$\begin{aligned} \mathbf{x}^a_0 &= \begin{bmatrix} \mathbf{x}_0 \\ \zeta_0 \end{bmatrix}, \\ \mathbf{P}^a_0 &= \begin{bmatrix} \mathbf{P}(t_0) & \mathbf{0} \\ \mathbf{0} & \mathbf{Q}(t)\Delta t \end{bmatrix}. \end{aligned} \tag{37}$$

After the generation of the initial sigma points, the covariance is propagated by using (16), (17), and (18).

Figure 16 depicts the 3-dimensional positional probability ellipsoids corresponding to a confidence interval of 99.99% with a sequence of time instants. The breakup event during the reentry happened at the altitude of 78 km, and the predicted trajectory and the positional covariance ellipsoid of the debris fragments are well depicted with the usage of the proposed breakup dispersion estimation method. As can be shown, the positional probability ellipsoid representing the debris dispersion increases quickly within 2.4 minutes, and after 5 minutes the ellipsoid increases gradually until it impacts the ground. This phenomenon is understood by the fact that after 3 minutes from the breakup event the motion becomes a sharp fall and becomes nearly constant in motion.

This phenomenon is verified in Figure 17 which shows the plots of the standard deviation of the positional covariance matrices in each direction over time. As can be expected from the time evolution of the debris dispersion in Figure 16, the variance sharply increases within 3 minutes but gradually changes after that time, that is, the falling motion 3 minutes after the breakup event.

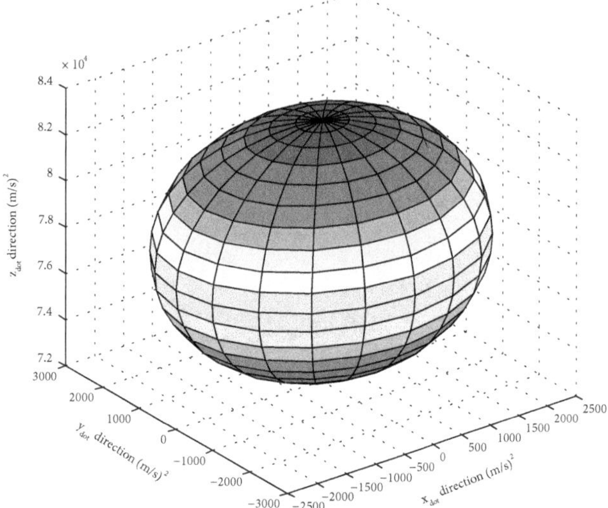

FIGURE 14: Initial estimate of debris dispersion.

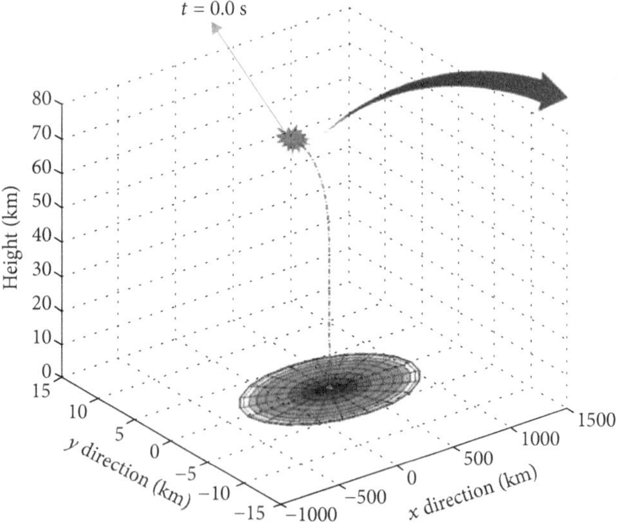

FIGURE 15: Breakup point and impact footprint.

For a better analysis of the first few minutes, a detailed view is illustrated in Figure 18 which depicts the change of the initial debris dispersion in terms of the ellipsoid from the breakup event to 100 seconds. The covariance ellipsoid contains the debris fragments which are broken into all directions with the magnitude of the initial covariance matrix. For the first 100 seconds, the dispersion grows fast and becomes 10 times larger than the initial breakup ellipsoid. This indicates that right after the breakup event there are lots of uncertainties acting on the motion of the instantaneous debris dispersion and also the motion is highly nonlinear. A commonly used metric for covariance matrices is the square root of the trace of the covariance ellipsoid as seen in Figure 17.

The simulation is made with different ballistic coefficients using the empirical density and wind models. Figure 19 shows the plot of flight time versus ballistic coefficient. In

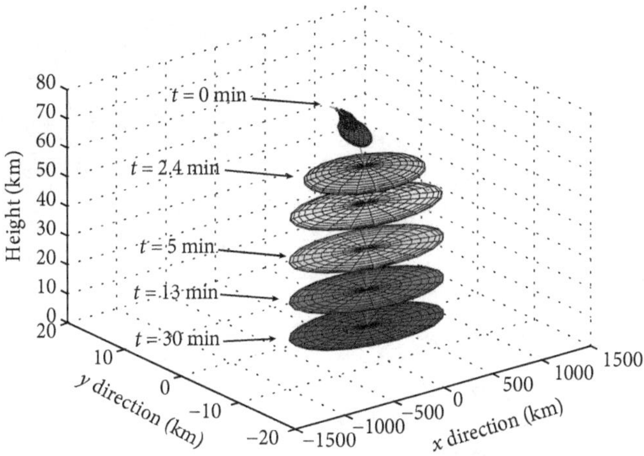

FIGURE 16: Dispersion probability ellipsoids for a sequence of time and impact footprint on the ground.

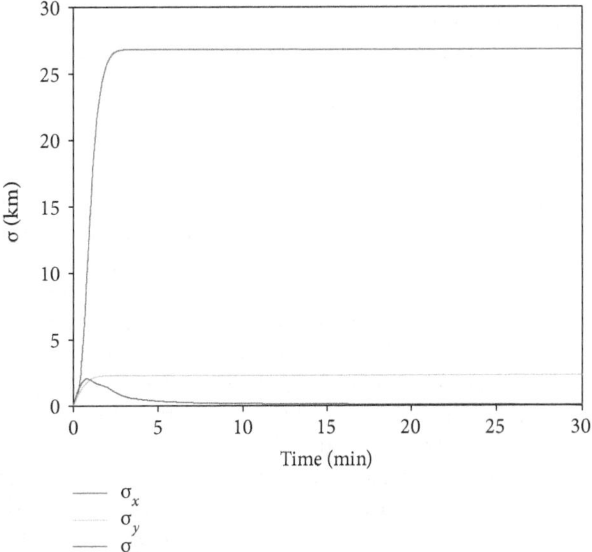

FIGURE 17: Magnitude of positional covariances in each direction (x, y, and z).

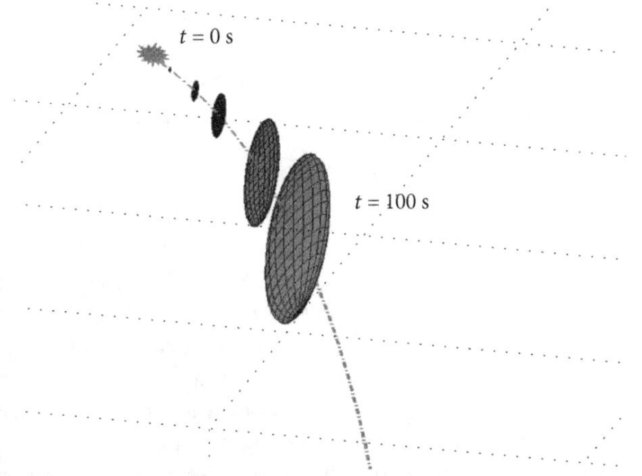

FIGURE 18: Initial dispersion of ellipsoids after the breakup event within 100 seconds.

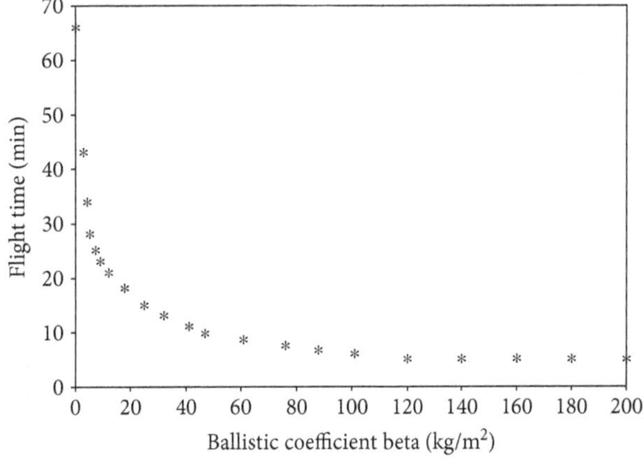

FIGURE 19: Plot of the empirical ballistic coefficient as a function of flight time [7].

Figure 19, for the variance of the ballistic coefficient in terms of altitude, an empirical density profile obtained from the MSISE-00 density model [22] is used. In addition, in order to include the effects of wind near the ground, a wind model is developed by using a piecewise cubic spline method through a curve fitting of real wind data. Figure 20 represents the wind profiles used in the simulation in the eastern and northern direction as a function of different altitudes. The wind model used in this work is based on HWM-07 (Horizon Wind Model 2007) [7, 24]. Figure 21 illustrates the effects of the wind strength onto the time variation of the dispersion trajectory. As can be seen, if the strength of the wind is stronger, then it takes a longer time until the debris arrives the ground and impact it.

4.3. Performance Comparison for Estimation of Debris Dispersion. In this section, the performance of the proposed covariance methods for estimating the debris dispersion due to the reentry breakup is investigated in terms of the time evolution of the positional probabilistic ellipsoid. The covariance propagation methods include the

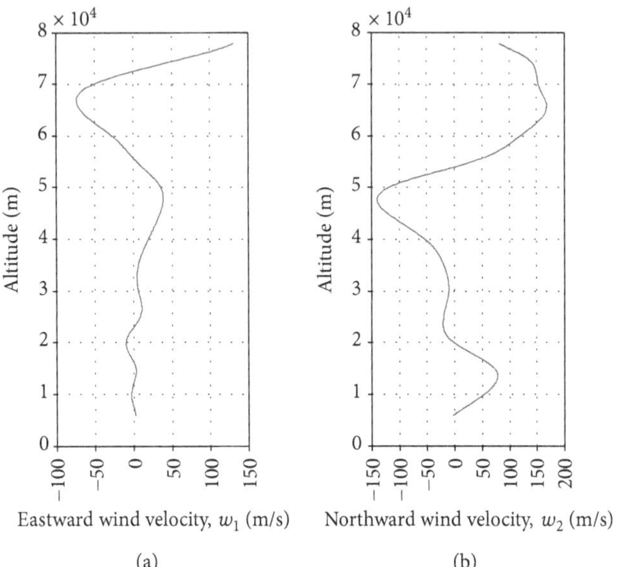

FIGURE 20: Simulated eastern and northern wind profiles for different altitudes.

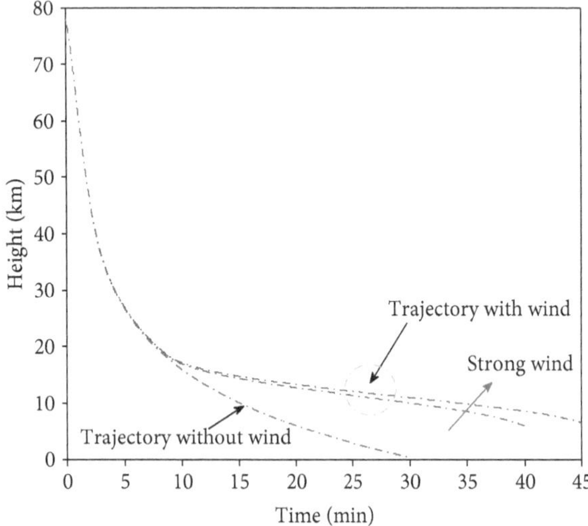

FIGURE 21: Effects of the wind to the time evolution of the decay of the debris altitude.

unscented transformation-based statistical linear regression and the Gauss-Hermite cubature-based numerical integration methods described in Section 3.

For the simulation test, it is also assumed that the breakup altitude is 78 km and the initial breakup position is located at $\mathbf{x}(t_0) = [0 \quad 0 \quad 78,000\,\text{m} \quad 709.9\,\text{m/s} \quad 0\,\text{m/s} \quad -124.0\,\text{m/s}]$. The initial breakup dispersion is modeled as a Gaussian distribution with the initial covariance ellipsoid as given in (34); the process noise covariance matrix representing modeling uncertainties is also given by $\mathbf{Q}(t) = 10^6\,\text{diag}([0 \quad 0 \quad 0 \quad 10^{-16} \quad 10^{-16} \quad 10^{-16}])$ in (35). Now, based on the initial debris dispersion covariance matrix $\mathbf{P}(t_0)$ and the process noise matrix $\mathbf{Q}(t)$, a set of sigma points

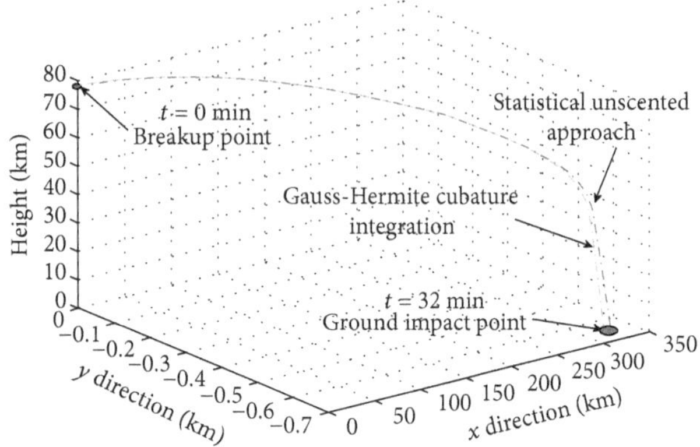

FIGURE 22: Performance comparison between unscented transformation and Gauss-Hermite cubature methods in nominal trajectory prediction and ground impact footprint.

representing a set of breakup components can be generated by using (14) for the unscented transformation (UT) method and using (29) for the Gauss-Hermite cubature (GHC) method. After the generation of the initial sigma points for the initial debris dispersion, the covariance is propagated by using (16), (17), and (18) for the UT method and (30), (31), and (32) for the GHC approach. The total simulation time used was 32 minutes.

Figure 22 depicts the estimated nominal trajectory from the breakup event to the ground impact instant, and also the footprint of the ground impact location. After the breakup event of the space debris during reentry, it is assumed that the simulation starts at $t = 0$ at the event, and then the nominal trajectory in terms of the mean value of all the debris fragments and the 3-dimensional positional ellipsoid are propagated until the debris fragments reach the ground. Based on the result in Figure 22, it is seen that the time evolution of the nominal estimates of the debris dispersion have different final locations but the overall trajectories are similar to each other. The nominal trajectory generated by the Gauss-Hermite cubature method has a little bit faster fall into the ground. The locational difference between the final footprint impact locations of two different methods was about 2 km in the x direction, and 10 m in the y direction, that is, $\delta\mathbf{x}(t_{\text{final}}) = [1.8980\,\text{km} \quad 0.00860\,\text{km} \quad 0\,\text{km}]$. It is seen from the nominal trajectory plot that the uncertainties of the motion of the debris dispersion are captured more precisely by utilizing the Gauss-Hermite cubature technique compared to the unscented transformation method, and this is because the GHC method adopts a much larger set of sigma lattice points at the initial debris dispersion which leads to a more precise estimation of the debris dispersion. In the simulation, it is also seen that the reentry motion is a nonlinear motion right after the breakup event, but from the altitude near the ground the motion has relatively small nonlinearities due to a vertical falling motion.

Figures 23 and 24 depict the profiles of the positional probability ellipsoids corresponding to a confidence interval of 99.99% for a sequence of time instants generated from

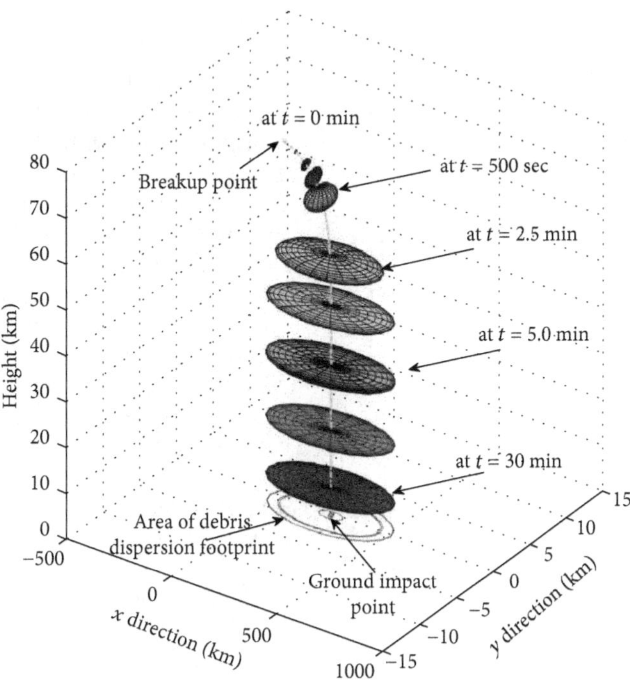

FIGURE 23: Time evolution of debris dispersion ellipsoids and impact footprint on the ground using the statistical linear regression with the unscented transformation method.

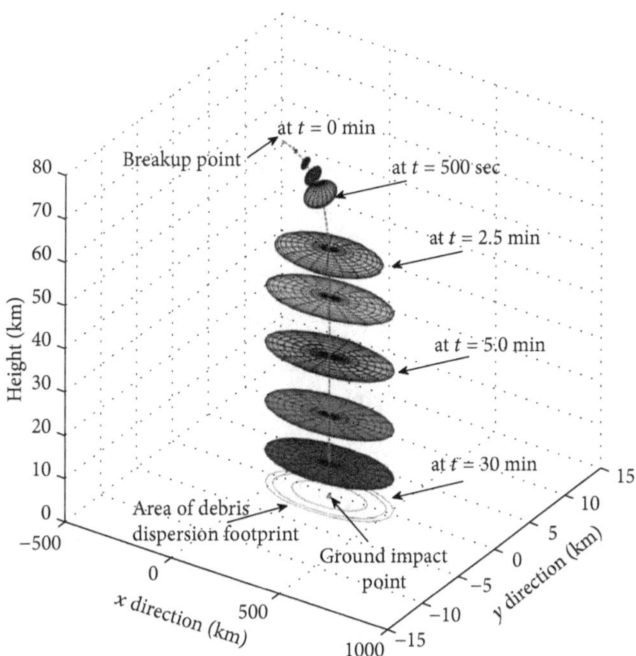

FIGURE 24: Time evolution of debris dispersion ellipsoids and impact footprint on the ground using the Gauss-Hermite cubature integration method.

the proposed covariance propagation methods, statistical linear regression, and the Gauss-Hermite moment-matching methods, respectively. In general, since the Gauss-Hermite cubature approach could include more debris breakup distribution points with $p = 3^6 = 729$ sigma points, it could generate a more accurate trajectory of the core debris as well as the covariance estimate of the breakup dispersion, while the unscented transformation method generates the initial breakup components with $p = 2 \times 6 + 1 = 13$ sigma points. In addition, the performance of the proposed approach is investigated by checking the time evolution of the dispersion ellipsoid shown in Figures 23 and 24 where the time

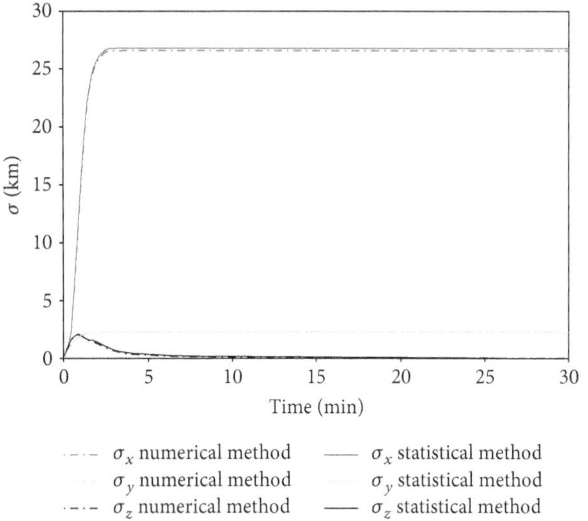

FIGURE 25: Comparison of the magnitude of positional covariances in each direction (x, y, and z).

evolution of the positional probability ellipsoids can be seen to almost match between the statistical linear regression in Figure 23 and the Gauss-Hermite cubature-based numerical integration shown in Figure 24. In fact, the close agreement of these techniques can be justified by comparing the positional covariance matrices from both of the methods in each direction. It is seen that the size of the overall debris dispersion is similar to each other, 0.6 km in the x direction and 0.01 km in the y direction. Figure 25 shows the plots of the sigma values for the positional covariance matrices for both of the methods overtime. The difference of the sigma in the x direction is large but the other directions have similar values. The Gauss-Hermite cubature method could provide a little bit more accurate estimates of the nominal trajectory and the positional ellipsoid which could affect the calculation of the risk causality, but it also requires more computational power compared to the unscented transformation method. Both methods could satisfy the computational procedures required for real-time computation in a sequential estimation, while a Monte-Carlo-based approach uses offline computational procedures.

5. Summary and Conclusion

In this paper, for advanced analysis of the time evolution of space debris dispersion due to the breakup during reentry, two effective computational approaches were proposed to estimate the statistical distribution of the debris dispersion. The estimated debris dispersion is represented by using the prediction of positional probability ellipsoids for the visualization of the results. First, the time evolution of the covariance propagation was computed by using a statistical linear regression using the unscented transformation. Second, a novel covariance estimation technique was proposed by utilizing the Gaussian moment-matching method with a Gauss-Hermite cubature-based numerical integration approach. Compared to other covariance propagation methods, the newly proposed Gauss-Hermite cubature-based covariance computation could not only represent more exact initial dispersion, but also precisely calculate the time evolution of the debris dispersion with a large set of sigma points representing debris components. The Gauss-Hermite cubature approach does not require any linearization of the nonlinear translation equations of motion of a reentering debris, which leads to precisely taking into account nonlinearities exposed on the motion of a reentering debris. Furthermore, we also carried out a parametric study in order to analyze the effects on the accuracy of the covariance propagation due to modeling uncertainties. For the detailed parametric analysis, four types of density models and drag coefficients were used in the computation of the lifetime prediction and the covariance computation. In the simulation studies, it was shown that the newly proposed statistical linear regression method and the Gauss-Hermite computational approach are in close agreement to each other, but the Gauss-Hermite covariance propagation method provides a more precise estimate of the debris dispersion of the breakup fragments.

Conflicts of Interest

The authors declare that they have no conflicts of interest.

Acknowledgments

This study was supported by the Korea Astronomy and Space Science Institute (KASI) through the research project "Study on Precise Trajectory and Trajectory Uncertainty Prediction of Space Objects for Ground Impact Risk Assessment" (Project no. 2017-1-854-01) supervised by the Ministry of Science and ICT.

References

[1] W. Ailor and L. Bonaprte, "A strategy for reducing hazards from reentry debris," in *AIAA Atmospheric Flight Mechanics Conference and Exhibit*, Keystone, Colorado, August 2006.

[2] E.-J. Choi, S. Cho, D.-J. Lee, S. Kim, and J. H. Jo, "A study on re-entry predictions of uncontrolled space objects for space situational awareness," *Journal of Astronomy and Space Sciences*, vol. 34, no. 4, pp. 289–302, 2017.

[3] L. S. Nair and R. Sharma, "Decay of satellite orbits using K-S elements in an oblate diurnally varying atmosphere with scale height dependent on altitude," *Advance in Space Research*, vol. 31, no. 8, pp. 2011–2017, 2003.

[4] R. H. Baker Jr., "Estimation of decayed satellite reentry trajectories," Air Force Institute of Technology (AFIT.EN), Wright-Peterson AFB, OH, USA, 1981, Dissertation.

[5] J. M. Tardy and C. A. Kluever, "Estimation and prediction of orbital debris reentry trajectories," *Journal of Spacecraft and Rockets*, vol. 39, no. 6, pp. 845–851, 2002.

[6] A. V. Rao, Minimum-variance estimation of reentry debris trajectories," *Journal of Spacecraft and Rockets*, vol. 37, no. 3, pp. 366–373, 2000.

[7] M. Reyhanoğlu and J. Alvarado, "Estimation of debris dispersion due to a space vehicle breakup during reentry," *Acta Astronautica*, vol. 86, pp. 211–218, 2013.

[8] A. K. Anilkumar, M. R. Ananthasayanam, and P. V. Subba Rao, "Prediction of re-entry of space debris objects: constant gain Kalman filter approach," in *AIAA Atmospheric Flight Mechanics Conference and Exhibit*, Austin, TX, USA, August 2003.

[9] D.-J. Lee, *Nonlinear Bayesian Filtering with Applications to Estimation and Navigation, [Ph.D. thesis]*, Texas A&M University, College Station, TX, USA, 2005.

[10] S. Särkkä, *Bayesian Filtering and Smoothing*, Cambridge University Press, New York, NY, USA, 2013.

[11] S. J. Julier, J. K. Uhlmann, and H. F. Durrant-Whyte, "A new method for the nonlinear transformation of means and covariances in filters and estimators," *IEEE Transactions on Automatic Control*, vol. 45, no. 3, pp. 477–482, 2000.

[12] M. Norgaard, N. K. Poulsen, and O. Ravn, "New developments in state estimation for nonlinear systems," *Automatica*, vol. 36, no. 11, pp. 1627–1638, 2000.

[13] D.-J. Lee and K. T. Alfriend, "Additive divided difference filtering for real-time spacecraft attitude estimation using modified Rodrigues parameters," *The Journal of the Astronautical Sciences*, vol. 57, no. 1-2, pp. 93–111, 2009.

[14] K. Ito and K. Xiong, "Gaussian filters for nonlinear filtering problems," *IEEE Transactions on Automatic Control*, vol. 45, no. 5, pp. 910–927, 2000.

[15] I. Arasaratnam and S. Haykin, "Cubature Kalman filters," *IEEE Transactions on Automatic Control*, vol. 54, no. 6, pp. 1254–1269, 2002.

[16] E. A. Wan and R. van der Merwe, "Chapter 7. The unscented Kalman filter," in *Kalman Filtering and Neural Networks*, S. Haykins, Ed., John Wiley & Sons, New York, NY, USA, 2001.

[17] D.-J. Lee and K. T. Alfriend, "Sigma point filtering for sequential orbit estimation and prediction," *Journal of Spacecraft and Rockets*, vol. 44, no. 2, pp. 388–398, 2007.

[18] T. Lefebvre, H. Bruyninckx, and J. D. Schutter, "Comment on "A new method for the nonlinear transformation of means and covariances in filters and estimators" (with authors' reply)," *IEEE Transactions on Automatic Control*, vol. 47, no. 8, pp. 1406–1409, 2002.

[19] A. Solin, *Cubature Integration Methods in Nonlinear Kalman Filtering and Smoothing*, School of Science and Technology, Aalto University, Espoo, Finland, 2010.

[20] B. Jia, M. Xin, and Y. Chen, "Sparse-grid quadrature nonlinear filtering," *Automatica*, vol. 48, no. 2, pp. 327–341, 2012.

[21] R. Radhakrishnan, A. K. Singh, S. Bhaumik, and N. K. Tomar, "Multiple sparse-grid Gauss-Hermite filtering," *Applied Mathematical Modelling*, vol. 40, no. 7-8, pp. 4441–4450, 2016.

[22] J. M. Picone, A. E. Hedin, D. P. Drob, and A. C. Aikin, "NRLMSISE-00 empirical model of the atmosphere: statistical comparisons and scientific issues," *Journal of Geophysical Research*, vol. 107, no. A12, pp. SIA 15-1–SIA 15-16, 2002.

[23] F. J. Regan and S. M. Anadakrishnan, *Dynamics of Atmospheric Re-entry*, American Institute of Aeronautics and Astronautics, Inc., Washington DC, USA, 1993.

[24] D. A. Vallado, *Fundamentals of Astrodynamics and Applications*, McGraw-Hill, New York, NY, USA, 1997.

[25] P. P. Rao and M. A. Woeste, "Monte Carlo analysis of satellite debris footprint dispersion," in *5th Atmospheric Flight Mechanics Conference for Future Space Systems*, Boulder, CO, USA, 1979.

[26] "System tool kit (STK)," December 2017, http://www.agi.com.

Monte Carlo Uncertainty Quantification Using Quasi-1D SRM Ballistic Model

Davide Viganò,[1] Adriano Annovazzi,[2] and Filippo Maggi[1]

[1]*Department of Aerospace Science and Technology, SPLab, Politecnico di Milano, 20156 Milan, Italy*
[2]*Space Propulsion Design Department, AVIO S.p.A., 00034 Colleferro, Italy*

Correspondence should be addressed to Filippo Maggi; filippo.maggi@polimi.it

Academic Editor: William W. Liou

Compactness, reliability, readiness, and construction simplicity of solid rocket motors make them very appealing for commercial launcher missions and embarked systems. Solid propulsion grants high thrust-to-weight ratio, high volumetric specific impulse, and a Technology Readiness Level of 9. However, solid rocket systems are missing any throttling capability at run-time, since pressure-time evolution is defined at the design phase. This lack of mission flexibility makes their missions sensitive to deviations of performance from nominal behavior. For this reason, the reliability of predictions and reproducibility of performances represent a primary goal in this field. This paper presents an analysis of SRM performance uncertainties throughout the implementation of a quasi-1D numerical model of motor internal ballistics based on Shapiro's equations. The code is coupled with a Monte Carlo algorithm to evaluate statistics and propagation of some peculiar uncertainties from design data to rocker performance parameters. The model has been set for the reproduction of a small-scale rocket motor, discussing a set of parametric investigations on uncertainty propagation across the ballistic model.

1. Introduction

In the current time frame, private space companies, supplying manned or unmanned flight services, are replacing governmental involvement in such missions. Different options are explored for what can be considered the rise of space access commercialization, spanning from the optimization of current launch options (namely, solid and liquid propulsion) to the development of advanced concepts such as hybrid propulsion, air-launched alternatives, or reusable stages. In any case, the development is targeting the reduction of costs. In this context, solid propulsion has an active role for boosting phases, initial stages, or embarked systems. In a paper describing the roadmap for solid propulsion published in 2010 the authors underlined appealing features such as the cost-effectiveness, the reliability of such technology, the capability of high thrust-to-weight ratio and the high propellant density; they also addressed well known limits, namely, reduced specific impulse and scarce flexibility [1].

In general a solid rocket motor is not throttleable, limiting the capability of in-flight corrections. Variable thrust solutions given by the actuation of a pintle-nozzle are still matter of experimentation and are not implemented in large-scale systems. The thrust profile is decided during the design process and is strictly related to the pressure-time history of the combustion chamber. Once ignited, solid rocket motors proceed till the exhaustion of all the propellant stored in the combustion chamber. Shutdown can be achieved using destructive techniques or injection of flame suppressors. Reignition is not possible anyway [2].

Deviations from the nominal behavior directly influence the mission profile and may be caused by multiple reasons. The commercially used composite solid propellant is one of the primary elements to be considered. This material is a complex mixture of oxidizer salts, metallic powders, and a binder which are compounded, cast, and cured to form the grain. The initial shape of the propellant charge evolves during combustion and affects pressure-time history [3]. Internal ballistics consists of the interaction of several details, which are not limited to the nominal propellant properties. Real performance may be altered by casting process, raw material lot variability, or environmental factors. This kind of behavior imposes strict requirements on acceptance criteria and reliability. For example, unexpected variations of the propellant properties due to casting effects have been highlighted by different experimental and modeling works and were found to be responsible for the hump behavior in pressure traces of small-scale rocket motors [4–7]. Another example is represented by rocket burnout. Ideally speaking, the pressure level should drop as the burning surface reaches the propellant liner. The actual effect can also include either a pressure spike, commonly referred to as Friedman's Curl, followed by pressure drop, or a longer burnout transient caused by grain misalignment [3, 8]. In the case of strap-on boosting units, differential thrust or burning time might be critical at stage detachment.

The analysis of internal rocket ballistics is requested for prediction and design of rocket system performance. Different degrees of complexity can be adopted, grouped into four categories: simple, engineering, full-up, and research models [9]. In the first group we find approaches based on equilibrium thermodynamics and zero-dimensional geometries and empirical models for burning rate and loss quantification. The second category mediates some simplifications with detailed description of relevant aspects, targeting a practical application. Examples of such codes are represented by the works published by Greatrix's team. These solvers do not include full solution of combustion process but can model different aspects of internal ballistics such as star-grained geometry or transient burning [10, 11]. Full-up models deal with complete physics and, in general, have commercial nature. Out of this category we mention the Rocstar code, a multiphysics multiscale computing framework used for solid rocket simulation [12]. Research codes are more focused on physics detailed investigation than on product development.

The present paper focuses on the development of POLIRocket, an engineering model for the simulation of rocket internal ballistics and performance prediction. The code implements a quasi-1D model, including evolution of grains with complex geometry, erosive burning phenomena, compressible gas dynamics, and nonuniform propellant ballistics. In the aforementioned context a tool based on Monte Carlo method was adopted for the analysis of performance uncertainties. This tool is able to treat statistically all the major sources of uncertainties for a solid rocket motor and, on this basis, their propagation towards performances.

2. Uncertainty in Rocket Motor Model

The practical interest for uncertainty analysis in rocket motors is very wide, spanning from mission analysis for a single rocket to thrust imbalance evaluation for the design of passive and active control systems in case of multiple strap-on boosters (e.g., Ariane V). In general, we refer to rocket nominal performance, such as thrust, specific impulse, or MEOP (Maximum Expected Operating Pressure). In most of applications a rocket engine never works at its nominal parameters but close to them. Statistics helps in defining the bounds of confidence for predictions which should include both the nondeterministic component of a model derived by input parameter variability or environment and the structural uncertainty introduced by the model itself and its integration. The resulting quantification supports the process of risk assessment for complex missions in mutable scenarios. Interesting approaches to uncertainty estimation and propagation for models and parameters have been proposed by Oberkampf et al. and Roy and Oberkampf [13, 14].

The propagation of uncertainty can be performed using both Taylor series (TS) and Monte Carlo (MC) methods. The former one is an analytical approach based on Taylor series expansion and requires the definition of sensitivity coefficients for each input parameter over the final result, starting from an analytical description of the problem. The second technique is a numerical statistic method for the analysis of complex models. Modern computers can run multiple instances of the same problem on the basis of probability density functions for input variables. MC methods do not need analytical differentiations and demonstrate flexibility in terms of magnitude, type of input uncertainty distributions, and nonlinearity of the model [15]. In this work MC method was implemented by treating numerically the rocket internal ballistics as a "black box." Construction parameters of the SRM were defined on the basis of Gaussian distributions of known standard deviation, obtaining the population of performance data as a result of different model runs. In order to reduce the number of iterations, latin hypersquares sampling was adopted. For m input distributions sampled with N points, this technique reduces the number of instances from N^m combinations to N calculations [16]. The construction parameters (ballistic coefficients, characteristic velocity efficiency, nozzle efficiency, propellant hump, propellant axis offset, and propellant mass) are assumed aleatory variables and are characterized by known Gaussian distributions.

3. Model for Internal Ballistics

The engineering model implemented in this work consists of a solver for internal ballistics simulation based on quasi-1D, quasi-steady, compressible, nonviscous flow equations, coupled with a zero-dimensional nozzle. This approach includes cross section propellant grain variation in both space and time. Local secondary behaviors, like vortexes and boundary layers, are not considered. Quasi-steady model can capture the evolution in time when the solid rocket motor operates under quasi-stationary condition, which happens in most cases apart from ignition and tail-off transients. Compressible fluid dynamics of the combustion chamber is described by Shapiro's ordinary differential equations. The model is specific for the solution of a flow in a duct. A simple control volume between two sections at an infinitesimal distance dx is implemented deriving a series of basic physical equations in differential logarithmic form. The resulting ordinary differential equations describe the evolution of relevant gas properties in the frame of reference. The reader is encouraged to consult Shapiro's book for details [17]. The model used in this work is represented by (1) to (4), under the assumptions of quasi-one-dimensional, steady, nonviscous, continuous variable flow and perfect gas:

$$\frac{dM^2}{M^2} = -\frac{2\left(1+((k-1)/2)M^2\right)}{1-M^2}\frac{dA}{A} + \frac{2\left(1+kM^2\right)\left(1+((k-1)/2)M^2\right)}{1-M^2}\frac{dm}{m}, \quad (1)$$

$$\frac{dV}{V} = -\frac{1}{1-M^2}\frac{dA}{A} + \frac{1+kM^2}{1-M^2}\frac{dm}{m}, \quad (2)$$

$$\frac{dT}{T} = \frac{(k-1)M^2}{1-M^2}\frac{dA}{A} - \frac{(k-1)M^2\left(1+kM^2\right)}{1-M^2}\frac{dm}{m}, \quad (3)$$

$$\frac{dp}{p} = \frac{kM^2}{1-M^2}\frac{dA}{A} - \frac{2kM^2\left(1+((k-1)/2)M^2\right)}{1-M^2}\frac{dm}{m}. \quad (4)$$

In this reduced set of equations, flow Mach number, static temperature, and pressure are dependent variables. Injected mass flow rate and cross section are independent parameters which are updated by propellant combustion modeling. The injected mass flow is proportional to the local burning rate of propellant which is expressed through Vieille's law $r_b = ap^n$. The simulation of nonconstant ballistic properties in the grain is also possible. In this respect, it is common practice to introduce a correction called hump effect. This parameter depends on r and introduces a spatial variation of propellant preexponential ballistic coefficient. In the present code the function has the shape $a_h(r) = (1 + h(r))a$. The correction factor $h(r)$ has symmetric parabolic fashion with equation $h(r) = z_1 r^2 + z_2 r + z_3$ and zero integral along the grain radius R. Parabola coefficients are reported in (5) and are uniquely defined by the grain radius R and the value of the correction factor H in the propellant midpoint:

$$z_1 = -\frac{12H}{R^2}; \quad z_2 = -\frac{12H}{R}; \quad z_3 = -2H. \quad (5)$$

The local static pressure is resolved by (4) using an iterative process, based on comparison between mass discharge from the nozzle and production by combustion. Initial guess is evaluated by means of a zero-dimensional model of the combustion chamber (whose properties are assumed constant along x-axis, at this stage). Simulations have shown that motors with low L/D ratio do not require any iteration. Once propellant is fully burnt, motor tail-off is handled by a zero-dimensional and unsteady model to compute the combustion chamber emptying. Injected mass is treated as a monophasic mixture of known features, whose thermodynamic properties (temperature, molar mass, and specific heat ratios) are tabulated as a function of pressure using a thermochemical equilibrium code [18]. The current version of the model solves Shapiro's ODEs using a fourth-order Runge-Kutta solver with variable integration step [19]. The current 1D computational domain is a simplified uniform grid, even though optimized grading might apply. Each cell contains a series of information about its geometry (position, port area, and perimeter), propellant (thickness burned both in radial and lateral directions), flow information (speed, pressure, and density), and a flag that defines if the cell represents a fixed duct, a burning propellant region, or, for special cases, a side-burning propellant. Time discretization is not uniform, due to variation of combustion velocity. Equation (6) relates Δt and combustion velocity using the coefficient c_t which is defined by the user:

$$\Delta t = c_t \frac{\Delta x}{\max(r_b)}. \quad (6)$$

For each cell, information on propellant geometry is supplied by an external module which tabulates the evolution of the initial burning surface. Burning rate can vary along the axial direction [20]. Local correction is performed by a model for erosive combustion using the Lenoir & Robillard semiempirical equation. This process is significant when high flow speed inside combustion chamber is locally registered [3, 21, 22]. The model can handle general types of grain shapes and assumes that the regression of solid phase boundary is locally normal and uniform at the considered axial position. An example of evolution for a star-shaped section is reported in Figure 1. The procedure can handle also cases of grain axis misalignment or offset. The code can handle both side-inhibited grains or lateral combustion phenomena.

Efficiency of the characteristic velocity and that of thrust coefficient are figures of merit of combustion chamber and nozzle. These values are user-defined constants in the present implementation. The code can be easily extended to account for further sources of losses such as two-dimensional nozzle exhaust, two-phase flow, or chemical kinetics using compact semiempirical formulations [23].

Code parameters have been set to reproduce the ballistic behavior of a BARIA SSRM (small-scale rocket motor), used for the experimental evaluation of propellants by *Avio S.p.A.* (Colleferro, Italy). A scheme of the rocket is reported in Figure 2. Internal grain is a simple central-perforated grain without lateral inhibition, designed to have a quasi-neutral burning surface and, consequently, quasi-constant pressure trace. Graphite convergent nozzle with different diameters is used to change the mean operating pressure. Igniter and pressure gauge are placed at the motor head-end. The propellant cartridge is 290 mm long. Inner diameter is 100 mm and propellant web thickness is 30 mm.

Validation tests were performed by comparing internal pressure trace from model results with experimental data obtained from BARIA rocket motor, finding a general good agreement for different pressure levels. Thrust is disregarded since data are not recorded during tests. One example is reported in Figure 3. Typical simulation parameters are listed in the next section.

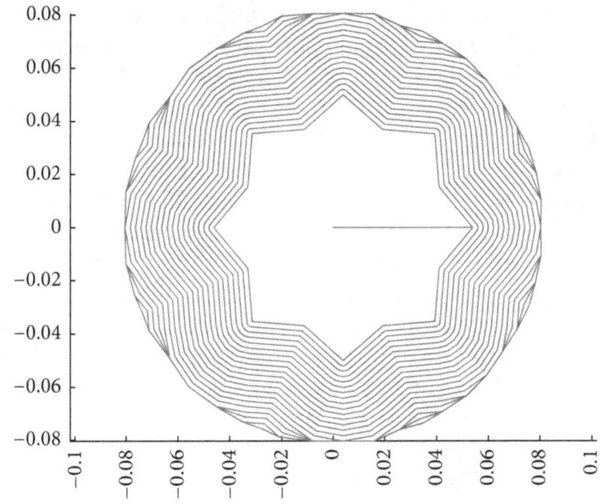

FIGURE 1: Example of central perforation evolution in time.

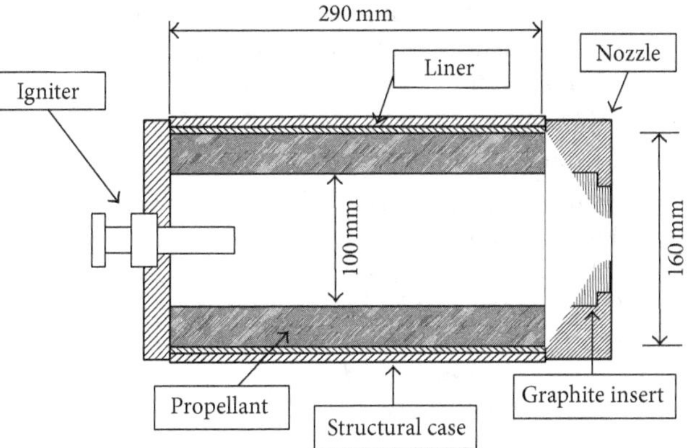

FIGURE 2: BARIA motor schematics (not in scale).

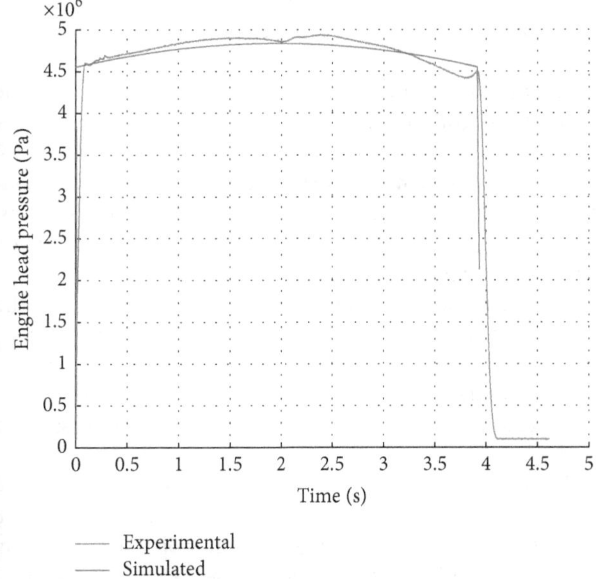

FIGURE 3: Comparison between experimental and simulated pressure traces.

4. Simulations for Uncertainty Propagation

An MC method was implemented for the analysis of uncertainty propagation towards performances, starting from construction parameters. The solver for BARIA internal ballistics was used. The aforementioned motor model was modified introducing a divergent section with area ratio of 4 to gain some sensitivity on thrust and specific impulse. Presented simulations are performed assuming a nozzle diameter of 25.25 mm, which corresponds to an expected mean operating pressure of about 45–50 bar. Gaussian distributions were assumed for Vieille's parameters a and n, c^* and c_F efficiency, hump effect, grain length (L_{grain}), and offset of axis perforation (Δr). Nominal values and relevant standard deviations of input populations, derived from comparison with available data, are reported in Table 1. Erosive combustion is not considered in the present work.

Performance parameters monitored in this work consisted in burning time, gravimetric specific impulse, total impulse, mean thrust, MEOP (Maximum Expected Operating Pressure), and mean pressure. Pressure traces were analyzed using a thickness-over-time postprocessing method [8]. Some of them (mean thrust, gravimetric specific impulse, and total impulse) are important for system-level considerations while other data gain particular interest in mission profile evaluation (burning time, mean pressure, and Maximum Expected Operative Pressure). The axial geometry was discretized with 500 evenly spaced grid points. Sensitivity study, considering up to 1000 intervals, is reported for the burning time in Figure 4.

MC method accuracy depends on the chosen population size. Generally, a bigger population will give more accurate statistics results; on the other hand the computational effort will grow accordingly. A compromise is usually requested. For this reason an analysis using sample mean value $\overline{X}$ and estimated standard deviation s was conducted to choose a proper population size. Burning time was arbitrarily chosen as monitoring variable. The convergence of the numerical method is obtained when both mean value and confidence interval become insensitive from the size of the input population. If this happens, the standard deviation of the results is generated by the sole propagation of the original uncertainty. Representative results are reported in Table 2.

We select 200 data points since the algorithm reaches a satisfactory convergence for our purposes. In the investigated range the variation of the mean value changes about 0.1% while the variation on the confidence interval is less than 3%. The choice of 200 points represents a compromise between accuracy and computational time, even though better results could be obtained for a larger population. With these parameters, the Matlab® run of the MC procedure took about 8/9 hours on a desktop PC.

Simulations were performed varying one parameter per each simulation and keeping a nominal value for the other input parameters. Populations of 200 samples were generated using Gaussian distributions for propellant ballistics (a and n), characteristic velocity c^*, nozzle efficiency η_{c_F}, hump level, propellant length (which turns into a variation on loaded mass), and propellant axis offset. Standard deviations

TABLE 1: Standard deviations for UQ propagation of BARIA rocket motor.

Input parameter	Nominal values	Relative standard deviation $(100 * \sigma/\mu)$
a	0.00410296 m/s bar^n	0.466
n	0.39396	0.721
η_{c^*}	0.95	0.500
η_{c_F}	0.95	0.500
H	1.0	0.667
L_{grain}	0.29 m	0.200
Δr	0	1.667*

*: mean value of axis offset is 0; the standard deviation of Δr refers to propellant outer radius.

TABLE 2: Sensitivity test on burning time t_b for different sizes of populations.

N	μ	σ
100	4.1007	0.0465
200	4.1042	0.0455
400	4.0988	0.0464
800	4.1017	0.0459

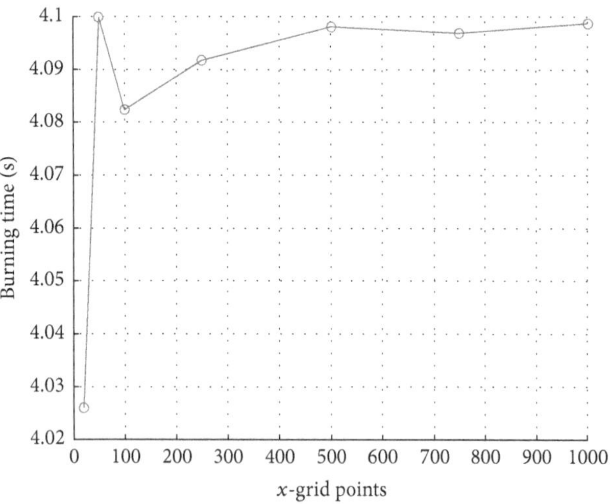

FIGURE 4: Grid convergence study for burning time.

used for the generation are reported in Table 1. Simulations were postprocessed to extract, from output populations, the relative standard deviation with respect to the mean value, in percent. Data are reported in Table 3. The last column reports results for a simulation where all input parameters contribute to uncertainty propagation.

Results allow understanding how model input parameters influence accuracy of performance predictions. The uncertainty of coefficients a and n does not propagate into gravimetric specific impulse and total impulse statistics; rather, it has substantial influence over the variability of other performance data. The preexponential coefficient amplifies its initial relative standard deviation, while the pressure exponent slightly decreases it. The uncertainty on characteristic velocity causes variability in all mapped parameters, less evident in burning time and more pronounced for thrust and pressure data. Thrust coefficient refers to nozzle expansion, so performances connected to the combustion chamber are not affected. It is important to notice that major uncertainties on impulses are caused by efficiencies. The reproducibility of the hump effect causes only a significant alteration in MEOP statistics while it does not procure other correlated effects; this is consistent since the hump is a local effect. For the tested range, variability of grain geometry does not have significant effect on gravimetric and volumetric specific impulses. The offset of grain axis causes a change of the pressure tail-off, influencing burning time and, as a consequence, total impulse computation. When tail-off becomes longer, an increased portion of the pressure curve is below the threshold value used by the TOT method to identify the burnout instant and is not considered in the relevant integral. Grain length is connected to propellant mass load as well as to geometric shape. The ballistic model amplifies the uncertainty of this parameter, propagating mainly on thrust and pressure.

The uncertainty resulting from single sources can be combined and compared with the standard deviation from the global MC simulation. According to TS method, the global standard deviation results from the formula $\sigma_{tot} = \sqrt{\sum_j \sigma_j^2}$, under the hypothesis of complete independency between uncertainty sources [16]. If not, a correlation term should be included. Table 4 reports the comparison between global standard deviation of parameters obtained from TS and MC. It appears that uncertainty contributions to most of the performance parameters are independent of each other, according to the present model. Only a minor correlation is visible for impulses. Conversely, dependence is observed in MEOP statistics and the correlation between error sources shrinks the actual uncertainty, with respect to the TS prediction.

5. Parametric Estimation of Main Input Uncertainties

The evaluation of incremented uncertainty level for some fundamental rocket parameters derived from experimentation is reported in the present paragraph. The level of accuracy for model input data is fundamental for the determination of the prediction confidence bounds of performances. The scalability of such interval with respect to input quality is of interest when experimental campaigns are planned or literature data are available. In the following paragraph the standard deviation for the population of three input parameters is varied, with respect to a reference value reported in Table 1. All sets of simulations were run with only one source of uncertainty per time. The other parameters were assumed nominal.

The first group of simulations was relevant to the ballistic parameters a and n. Resulting statistics of rocket performance are listed in Tables 5 and 6 for input populations having 50%, 100%, and 200% the respective reference standard deviation. In both cases, the input variability is reduced across the model. As already observed, there is a substantial

TABLE 3: Single source and global propagation of uncertainty. Relative standard deviations ($100 * \sigma/\mu$) are reported in columns.

Output parameters	Source of uncertainty							Global
	a	n	η_{c^*}	η_{C_F}	H	Δr	L_{grain}	
t_b	0,7597	0,6878	0,3290	0,0000	0,0347	0,1510	0,1214	1,1098
I_s	0,0000	0,0000	0,5146	0,4887	0,0091	0,0002	0,0019	0,6759
I_{tot}	0,0000	0,0000	0,5146	0,4887	0,0325	0,1448	0,1894	0,7186
T_{mean}	0,7597	0,6878	0,8441	0,4887	0,0117	0,0062	0,3108	1,4400
MEOP	0,7598	0,6977	0,8441	0,0000	1,0026	0,0024	0,3076	1,4021
P_{mean}	0,7562	0,6846	0,8453	0,0000	0,0099	0,0061	0,3099	1,3975

TABLE 4: Relative standard deviation ($100 * \sigma/\mu$) of global uncertainty model. Comparison between MC and TS methods.

	TS	MC	(TS – MC)/MC, %
t_b	1.0941	1.1098	−1.412
I_s	0.7097	0.6759	5.000
I_{tot}	0.7493	0.7186	4.277
T	1.4486	1.4400	0.594
MEOP	1.6960	1.4021	20.96
P_{mean}	1.3606	1.3975	−2.639

TABLE 5: Relative standard deviation ($100 * \sigma/\mu$) of parametric uncertainty analysis of preexponent a.

	Input variability, with respect to nominal value		
	50%	100%	200%
t_b	0.3964	0.7637	1.5773
I_s	0.0026	0.0051	0.0105
I_{tot}	0.0026	0.0051	0.0105
T	0.3988	0.7678	1.5805
MEOP	0.3992	0.7684	1.5819
P_{mean}	0.3992	0.7685	1.5819

TABLE 6: Relative standard deviation ($100 * \sigma/\mu$) of uncertainty propagation on exponent n.

	Input variability, with respect to nominal value		
	50%	100%	200%
t_b	0.3525	0.7176	1.3992
I_s	0.0055	0.0113	0.0220
I_{tot}	0.0057	0.0116	0.0226
T	0.3581	0.7304	1.4220
MEOP	0.3693	0.7531	1.4660
P_{mean}	0.3550	0.7240	1.4094

independence of impulses from ballistic coefficients. Conversely, there is a more than linear correlation between input and output uncertainty of the other parameters for the preexponential coefficient and a less than linear effect when variability of pressure exponent is considered.

A similar analysis was conducted on the efficiency of the characteristic velocity. Resulting uncertainty on performance parameters is reported in Table 7. The width of the input confidence interval decreases for burning time and for impulses, while it is amplified for pressure and thrust data. Similar results were obtained in the previous section. Variation of output standard deviations is in line with input ones. This observation is valid for all tested parameters in this model.

TABLE 7: Relative standard deviation ($100 * \sigma/\mu$) of uncertainty propagation on efficiency of c^*.

	Input variability, with respect to nominal value		
	50%	100%	200%
t_b	0.3290	0.6420	1.3345
I_s	0.5146	1.0011	2.0852
I_{tot}	0.5146	1.0011	2.0852
T	0.8441	1.6400	3.4190
MEOP	0.8441	1.6403	3.4194
P_{mean}	0.8453	1.6423	3.4235

6. Conclusion

The work presented a model for the description of rocket internal ballistics, based on Shapiro's quasi-1D formulation. The tool is a robust and simple code capable of dealing with a wide variety of phenomena that occur inside a solid rocket motor. Current implementation is based on Matlab and can easily be extended to include new features. Thanks to its capability of running fast simulations, the code is suitable for use in a MC algorithm. The numerical model made during this work proved out to be suitable for the approximation of the BARIA motor, a small-scale rocket motor used at industrial level. In the present framework, the code was used to perform an investigation on uncertainty propagation of input parameters, based on MC method.

Uncertainty on propellant ballistic coefficients were demonstrated to influence burning time, pressure, and thrust, while no effect was observed on impulses. Efficiency of characteristic velocity propagated across all mapped quantities, while thrust coefficient variability affected only quantities correlated to nozzle expansion. The only relevant effect of hump reproducibility was observed on the MEOP statistics. Variability of central perforation propagated to burning time and total impulse. The grain length was connected to propellant loaded mass and burning surface, thus influencing pressure, burning rate, thrust, and total impulse.

Future development of the POLIRocket code consists of model refinement, without losing the simplicity of the original framework. Thanks to Shapiro's equations, the inclusion of wall friction, heat exchange, chemical reactions, and phase changes is possible. The statistical framework developed so far can be applied in industrial environment by comparing predictions to resulting production uncertainty based on real firing data. Moreover, the indications obtained by the code about uncertainty propagation can be used to understand which effort for better knowledge of input data should be developed to improve global prediction accuracy. In this respect, parallel execution of Monte Carlo method will enable faster evaluations with larger populations.

Nomenclature

A: Area
a: Preexponential coefficient of Vieille law
a_h: Coefficient a after hump correction
c_t: Time coefficient
c^*: Characteristic velocity
h: Hump correction factor
H: Hump peak coefficient
I_s: Specific impulse
I_{tot}: Total impulse
I_v: Volumetric specific impulse
k: Specific heat ratio
L_{grain}: Grain length
M: Mach number
m: Mass flow rate
N: Size of population
n: Exponential coefficient of Vieille law
p: Pressure
p_{mean}: Mean pressure
r: Radial distance of consumed propellant
R: Grain radius
r_b: Burning rate
s: Estimate of the standard deviation
T: Temperature
t: Time
t_b: Burn time
V: Speed
x: Cartesian coordinate
$\overline{X}$: Sample mean for statistical analysis
z_i: Coefficients for hump correction
η: Efficiencies
μ: Population mean value
σ: Standard deviation
MC: Monte Carlo
MEOP: Maximum expected operating pressure
SRM: Solid rocket motor
TOT: Thickness over time
TS: Taylor series
c_F: Thrust coefficient.

Competing Interests

The authors declare that they have no competing interests.

References

[1] J.-F. Guery, I.-S. Chang, T. Shimada et al., "Solid propulsion for space applications: an updated roadmap," *Acta Astronautica*, vol. 66, no. 1-2, pp. 201–219, 2010.

[2] G. P. Sutton and O. Biblarz, *Rocket Propulsion Elements*, John Wiley & Sons, Hoboken, NJ, USA, 8th edition, 2010.

[3] "Solid rocket motor performance analysis and prediction," NASA Technical Report SP-8039, National Aeronautics and Space Administration, Houston, Tex, USA, 1971.

[4] P. Le Breton and D. Ribéreau, "Casting process impact on small-scale solid rocket motor ballistic performance," *Journal of Propulsion and Power*, vol. 18, no. 6, pp. 1211–1217, 2002.

[5] S. D. Heister, "Ballistics of solid rocket motors with spatial burning rate variations," *Journal of Propulsion and Power*, vol. 9, no. 4, pp. 649–651, 1993.

[6] F. Maggi, L. T. DeLuca, A. Bandera, V. S. Subith, and A. Annovazzi, "Burn-rate measurement on small-scale rocket motors," *Defence Science Journal*, vol. 56, no. 3, pp. 353–367, 2006.

[7] H. Hasegawa, M. Fukunaga, K. Kitagawa, and T. Shimada, "Burning rate anomaly of composite propellant grains," *Combustion, Explosion and Shock Waves*, vol. 49, no. 5, pp. 583–592, 2013.

[8] R. S. Fry, *Evaluation of Methods for Solid Propellant Burning Rate Measurements*, NATO/RTO Advisory Report, AVT Working Group 16, 2002.

[9] T. L. Jackson and J. M. Austin, "Solid propulsion: motor interior ballistics—modeling and design," in *Encyclopedia of Aerospace Engineering*, John Wiley & Sons, New York, NY, USA, 2010.

[10] S. Loncaric, D. R. Greatrix, and Z. Fawaz, "Star-grain rocket motor—nonsteady internal ballistics," *Aerospace Science & Technology*, vol. 8, no. 1, pp. 47–55, 2004.

[11] D. R. Greatrix, "Transient burning rate model for solid rocket motor internal ballistic simulations," *International Journal of Aerospace Engineering*, vol. 2008, Article ID 826070, 10 pages, 2008.

[12] M. D. Brandyberry, M. Campbell, B. Wasistho et al., "Rocstar simulation suite: an advanced 3-D multiphysics, multiscale computational framework for tightly coupled, fluid-structure-thermal applications," in *Proceedings of the 48th AIAA/ASME/SAE/ASEE Joint Propulsion Conference & Exhibit, Joint Propulsion Conferences*, AIAA Paper 2012-4216, August 2012.

[13] W. L. Oberkampf, S. M. DeLand, B. M. Rutherford, K. V. Diegert, and K. F. Alvin, "Error and uncertainty in modeling and simulation," *Reliability Engineering and System Safety*, vol. 75, no. 3, pp. 333–357, 2002.

[14] C. J. Roy and W. L. Oberkampf, "A comprehensive framework for verification, validation, and uncertainty quantification in scientific computing," *Computer Methods in Applied Mechanics and Engineering*, vol. 200, no. 25, pp. 2131–2144, 2011.

[15] C. E. Papadopoulos and H. Yeung, "Uncertainty estimation and Monte Carlo simulation method," *Flow Measurement and Instrumentation*, vol. 12, no. 4, pp. 291–298, 2001.

[16] H. W. Coleman and W. G. Steele Jr., "Uncertainty in a result determined from multiple variables," in *Experimentation, Validation, and Uncertainty Analysis for Engineers*, H. W. Coleman and W. G. Steele, Eds., chapter 3, pp. 61–83, John Wiley & Sons, Hoboken, NJ, USA, 3rd edition, 2009.

[17] A. H. Shapiro, *The Dynamic and Thermodynamic of Compressible Fluid Flow*, Ronald Press, New York, NY, USA, 1953.

[18] I. Glassman and R. F. Sawyer, *The Performance of Chemical Propellants*, vol. 129 of *AGARDograph*, Technivision Services, Slough, UK, 1970.

[19] A. Quarteroni, R. Sacco, F. Saleri, and P. Gervasio, *Matematica Numerica* , Springer, Milano, Italy, 3rd edition, 2008 (Italian).

[20] National Aeronautics and Space Administration (NASA), "Solid propellant grain design and internal ballistics," NASA Technical Report SP-8076, National Aeronautics and Space Administration (NASA), Houston, Tex, USA, 1972.

[21] B. A. McDonald, *The development of an erosive burning model for solid rocket motors using direct numerical simulation [Ph.D. thesis]*, Philosophy Thesis in Aerospace Engineering, Georgia Institute of Technology, Atlanta, Ga, USA, 2004.

[22] R. Badrnezhad and F. Rashidi, "Solid propellant erosive burning," in *Proceedings of the 8th World Congress of Chemical Engineering*, Paper 1111, Montreal, Canada, 2009.

[23] D. Reydellet, "Performance of rocket motors with metallized propellants," in *AGARD Advisory. Report AR-230*, AGARD PEP WG-17, AGARD, Paris, France, 1986.

7

A Novel Double Cluster and Principal Component Analysis-Based Optimization Method for the Orbit Design of Earth Observation Satellites

Yunfeng Dong,[1] Xiaona Wei,[1] Lu Tian,[2] Fengrui Liu,[1] and Guangde Xu[3]

[1] *School of Astronautics, Beihang University, Beijing 100191, China*
[2] *Department of Satellite Application System, North China Institute of Computing Technology, Beijing 100191, China*
[3] *Institute of Manned Space System Engineering, China Academy of Space Technology, Beijing 100094, China*

Correspondence should be addressed to Fengrui Liu; frliu@buaa.edu.cn

Academic Editor: Christian Circi

The weighted sum and genetic algorithm-based hybrid method (WSGA-based HM), which has been applied to multiobjective orbit optimizations, is negatively influenced by human factors through the artificial choice of the weight coefficients in weighted sum method and the slow convergence of GA. To address these two problems, a cluster and principal component analysis-based optimization method (CPC-based OM) is proposed, in which many candidate orbits are gradually randomly generated until the optimal orbit is obtained using a data mining method, that is, cluster analysis based on principal components. Then, the second cluster analysis of the orbital elements is introduced into CPC-based OM to improve the convergence, developing a novel double cluster and principal component analysis-based optimization method (DCPC-based OM). In DCPC-based OM, the cluster analysis based on principal components has the advantage of reducing the human influences, and the cluster analysis based on six orbital elements can reduce the search space to effectively accelerate convergence. The test results from a multiobjective numerical benchmark function and the orbit design results of an Earth observation satellite show that DCPC-based OM converges more efficiently than WSGA-based HM. And DCPC-based OM, to some degree, reduces the influence of human factors presented in WSGA-based HM.

1. Introduction

Earth observation satellites provide essential information on ocean, land, and atmosphere, which are very important in the environment protection and resources management. The first step of satellite mission design is usually the determination of a suitable orbit. The objective of orbit design for Earth observation satellites is to ensure that all target sites are best visited, including observation sites and ground stations. The quality of an orbit can be measured with key orbit performance indices [1]. The key orbit performance indices of an Earth observation satellite include the total coverage time, the frequency of coverage, the average time per coverage, the maximum coverage gap, the minimum coverage gap, and the average coverage gap [1, 2]. Thus, orbit design is a typical multiobjective optimization problem. Numerical methods for multiobjective orbit design optimization can be classified into three primary groups: indirect methods, direct methods, and evolutionary algorithms [3]. The last group is currently receiving research attention because of the capability of achieving global optima in very large search spaces.

In the evolutionary optimization for multiobjective orbit design, multiobjective functions are usually transformed into a single-objective function using the weighted sum method, and then a mature single-objective optimization method, such as genetic algorithm (GA), is employed to optimize the single-objective function to obtain the optimal orbit [4–9]. Abdelkhalik and Mortari [4, 5] employed GA to optimize the weighted sum of squares of the distances between each target site and the satellite at the nearest ground track point, taking five orbital elements plus all visiting times as the design variables. Abdelkhalik and Gad [6] applied a weighted function of the total number of covered sites and the ground track repetition period as fitness function and adopted GA to optimize eccentricity, inclination, space-craft's true anomaly above the first ground site, and the ground track repetition period, to design space orbits for Earth orbiting missions. Vtipil and Newman [7] and Vtipil [8] employed the sum of all time slot values of visiting as the cost function and adopted GA to conduct optimizations. The effect of population sizes was further researched. Zhang et al. [9] used a hybrid-encoding GA to optimize the sum of absolute value of velocity increment for long-duration rendezvous phasing missions. In weighted sum method [10–12], weight coefficients are utilized to transform the multiobjective function into the single-objective function. One disadvantage of the WSGA-based HM is that the artificially set values of the weight coefficients are unreasonable and subjective and depend significantly on human factors. In addition, the other disadvantage of the WSGA-based HM is the inefficient convergence of GA [13, 14].

To address these two disadvantages, this study proposes a population-based optimization method named CPC-based OM, in which candidate orbits are gradually randomly generated until the optimal orbit is obtained using a clustering via principal components based data mining method. A sufficient number of candidate orbits could ensure that the global optimal solution is obtained. In addition, the influence of the human factors from the weighted sum method is reduced in the optimization procedure because the candidate orbits are clustered based on the principal components rather than the weighted functions of the optimization objectives. Many methods have been investigated to reduce the influences of human factors of weighted sum method in multiobjective optimization [15–17]. Among them, the principal component analysis [18] is one of the most feasible methods, which transforms the original variables into a new set of variables, referred to as principal components, by using the eigenvalue-eigenvector method. The principal component analysis is thought to be the best way that explains the internal structure of the data [19, 20] and wildly applied by numerous researchers [21–26] to transform multiobjective functions for subsequent optimizations.

Methods must be introduced to accelerate convergence because the search procedure to obtain the optimal solution in CPC-based OM was a nearly exhaustive search with inefficient convergence. The methods of reducing feasible region are popular approaches [27–29]. In the methods of reducing feasible region, parts of the feasible region that do not include the optimum solution are deleted, and the subsequent optimization is accelerated because the remaining search space (feasible region) is smaller. Cluster analysis with the capability of dividing the feasible region into different regions has been used to reduce feasible regions [30–32]. Therefore, the second cluster analysis is introduced to CPC-based OM to accelerate convergence, and a novel population-based optimization method named DCPC-based OM is presented.

In this study, an orbit optimization model with constraints, six design variables, and eight optimization objectives is developed for Earth observation satellites. The process to obtain the optimal orbit using CPC-based OM is presented. To improve poor convergence of CPC-based OM, a more advanced DCPC-based OM is proposed by introducing cluster analysis based on six orbital elements. Finally, a test with numerical benchmark functions is conducted and the performances of DCPC-based OM, CPC-based OM, and WSGA-based HM on the orbit optimization of Earth observation satellites are compared.

2. Orbit Optimization Model for the Earth Observation Satellites

Abdelkhalik and Mortari [4, 5] explored the concept of developing an orbit based on target sites with no thrusters, in which design variables include five orbital elements and all the visiting times. The number of design variables increases with the increasing number of target sites. To avoid the increase of computational burden as the number of target sites increases, an optimization model with six orbital elements as design variables is employed. In addition, to deal with the increasing complexity of the observation mission, more orbit performance indices were taken into account than in prior studies [4–9].

2.1. Orbital Dynamics Model and Six Orbit Elements. For the orbit design of an Earth observation satellite without maneuvering, the relevant orbit dynamics equations in a geocentric equatorial inertial system (GEI) are as follows:

$$\begin{aligned}\frac{\mathrm{d}r_x}{\mathrm{d}t} &= v_x,\\ \frac{\mathrm{d}r_y}{\mathrm{d}t} &= v_y,\\ \frac{\mathrm{d}r_z}{\mathrm{d}t} &= v_z,\\ \frac{\mathrm{d}v_x}{\mathrm{d}t} &= \frac{F_x}{m},\\ \frac{\mathrm{d}v_y}{\mathrm{d}t} &= \frac{F_y}{m},\\ \frac{\mathrm{d}v_z}{\mathrm{d}t} &= \frac{F_z}{m},\end{aligned} \tag{1}$$

where r_x, r_y, and r_z are the components of the satellite position vector; v_x, v_y, and v_z are the components of the velocity vector; and F_x, F_y, and F_z are the components of external force, including Earth's gravity (considering Earth nonspherical shape perturbation forces), atmospheric drag perturbation forces, and solar radiation pressure as well as lunar and solar perturbations forces [1]. The positions of the satellite at each moment can be calculated using (1) and six orbital elements at the initial moment. The six orbital elements include the semimajor axis a, the eccentricity e, the inclination i, the argument of the perigee ω, the longitude of the ascending node Ω, and the true anomaly f. The key orbit performance indices of an Earth observation satellite are calculated using the positions of the satellite at each moment, the longitude and latitude data of the observation sites, the longitude and latitude data of the ground stations, and the right ascension of Greenwich at the initial moment [1].

2.2. Coverage and TT&C Performance Indices. Various key performance indices have been employed in the orbit design of the Earth observation satellites [4–9]. This paper adopts the key orbit performance indices which are systematically and comprehensively summarized by Wertz and Larson [1] and are applied by Wei et al. [2]. The key orbit performance indices of the Earth observation satellites can be separated into the coverage performance indices and the tracking telemetry and command (TT&C) performance indices. Among them, the coverage performance indices include the total coverage time (TCT), the frequency of coverage (FC), the average time per coverage (ATC), the maximum coverage gap (MCG), the minimum coverage gap (ICG), and the average coverage gap (ACG). And the TT&C performance indices include the average time interval of TT&C (ATI-TT&C) and the average time of each TT&C (AT-TT&C). The definitions of the eight orbit performance indices are as follows [1, 2].

$$\begin{aligned}
\text{TCT: } T_{\text{Cover}} &= \sum_{i=1}^{N_{\text{Cover}}} \Delta t_i \\
\text{FC: } F_{\text{Cover}} &= \frac{N_{\text{Cover}}}{T_{\text{total}}} \\
\text{ATC: } T_{\text{Average}} &= \frac{T_{\text{Cover}}}{N_{\text{Cover}}} \\
\text{MCG: } T_{\max}^{\text{Gap}} &= \max\left(\Delta t_i^{\text{Gap}}\right) \quad \left(i = 1, 2, \ldots, N_{\text{Gap}}\right) \\
\text{ICG: } T_{\min}^{\text{Gap}} &= \min\left(\Delta t_i^{\text{Gap}}\right) \quad \left(i = 1, 2, \ldots, N_{\text{Gap}}\right) \\
\text{ACG: } T_{\text{Ave}}^{\text{Gap}} &= \frac{\sum_{i=1}^{N_{\text{Gap}}} \Delta t_i^{\text{Gap}}}{N_{\text{Gap}}} \\
\text{ATI-TT\&C: } T_{\text{Ave}}^{\text{Tel}} &= \frac{\sum_{i=1}^{N_{\text{Tel}}} \Delta t_i^{\text{Tel}}}{N_{\text{Tel}}} \\
\text{AT-TT\&C: } T_{\text{Ave}}^{\text{TT\&C}} &= \frac{\sum_{i=1}^{N_{\text{TT\&C}}} \Delta t_i^{\text{TT\&C}}}{N_{\text{TT\&C}}},
\end{aligned} \tag{2}$$

where N_{Cover} is the total number of times of coverage in the simulation time T_{total}, Δt_i is the time of the ith coverage, Δt_i^{Gap} is the time of the ith coverage gap, N_{Gap} is the total number of coverage gaps, Δt_i^{Tel} is the time of the ith interval of TT&C, N_{Tel} is the total number of the intervals of TT&C, $\Delta t_i^{\text{TT\&C}}$ is the time of the ith TT&C, and $N_{\text{TT\&C}}$ is the total number of TT&C.

2.3. Orbit Optimization Model. The orbit optimization model of Earth observation satellites is shown in

$$\begin{aligned}
\text{Expected: } & T_{\text{Cover}}, F_{\text{Cover}}, T_{\text{Average}}, T_{\text{maxGap}}, T_{\text{minGap}}, T_{\text{AveGap}}, T_{\text{AveTel}}, T_{\text{AveTT\&C}} \\
\text{s.t.: } & a_{\min} < a < a_{\max} \\
\text{By find: } & a, e, i, \omega, \Omega, f.
\end{aligned} \tag{3}$$

The optimization objective is to make TCT, FC, ATC, and AT-TT&C be the maximum, the MCG, ICG, and ACG be the minimum, and the ATI-TT&C be within an expected range. The constraint is $a_{\min} < a < a_{\max}$. The six orbital elements at the initial moment are the independent variables. The orbit optimization model in (3) is a typical multidimensional and multiobjective optimization problem. The purpose of this study is to provide an optimization method for orbit decision-making for an Earth observing satellite mission. The resulting optimal orbit may need to be refined in the case of additional system designs, including orbit stability, fuel consumption in orbital maneuvering, or launch site restrictions.

3. CPC-Based OM for Orbit Optimizations

To reduce the influences of human factors in orbit optimizations [4–9], CPC-based OM is presented, and the accelerating convergence approach will be introduced in Section 4 to develop DCPC-based OM. The process flow of orbit design optimization with CPC-based OM is shown in Figure 1, including five steps.

Step 1. Randomly generate n_0 candidate orbits according to the feasible regions of the six orbital elements, and then calculate the coverage and TT&C performance indices.

Step 2. Nondimensionalize the coverage and TT&C performance indices of candidate orbits.

Step 3. Calculate the principal components of nondimensionalized orbit performance indices of candidate orbits, divide candidate orbits into classes by performing cluster analysis based on the principal components, and evaluate all class centers using the weighted sum function of key orbit performance indices to obtain the optimal class.

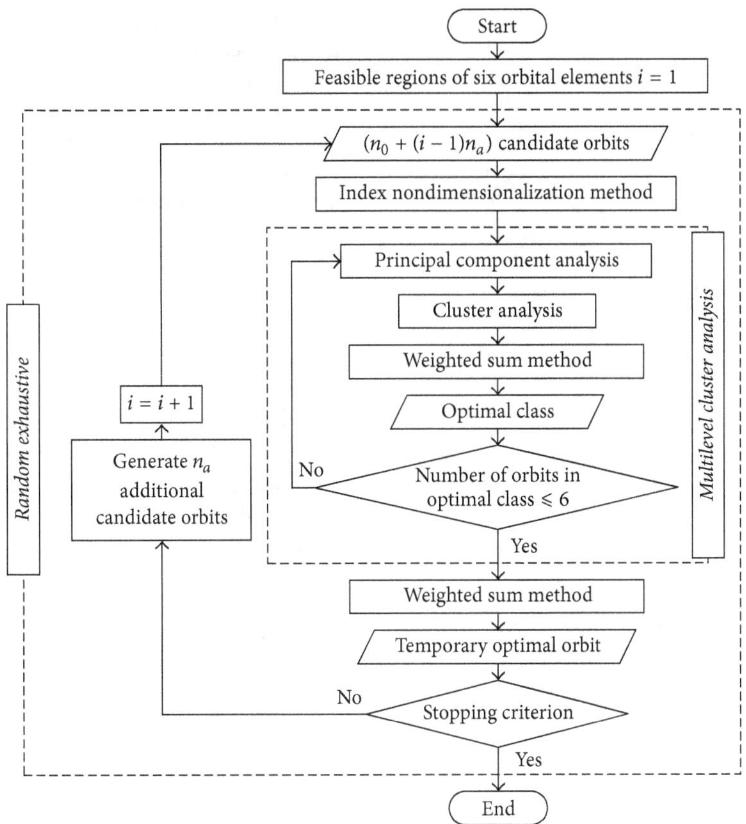

FIGURE 1: Process flow of CPC-based OM for orbit design optimizations.

Step 4. If the number of orbits in the optimal class is more than six, go to Step 3 and take orbits in the optimal class as candidate orbits in Step 3.

Step 5. If the number of orbits in the optimal class is not more than six, determine the temporary optimal orbit from the optimal class by using the weighted sum method. Randomly generate n_a additional candidate orbits and go to Step 2 with $n_0 + (i-1)n_a$ candidate orbits, until the relative improvement of the temporary optimal orbit is less than a certain bound. The applied stopping criterion refers to that used in the GA [4–9].

Note that Steps 3 and 4 constitute "multilevel cluster analysis," in which the optimal class with not more than six orbits is obtained. Steps 2~5 constitute "random exhaustive" process to obtain the optimal orbit. In addition, because the cluster analysis is based on the principal components of nondimensionalized key orbit performance indices rather than based on weighted function of the nondimensionalized key orbit performance indices in Step 3, the influence of human factors on the optimum result is reduced. The principal components, which explain the internal structure of the key orbit performance indices [19, 20], are suitable for cluster analysis. However, they do not have physical meaning and thus can not be used for evaluating the quality of orbits. Therefore, the optimal class is selected from all classes utilizing weighted sum method, rather than utilizing principal component analysis. Similarly, in Step 5, the temporary optimal orbit is obtained from the optimal class by using the weighted sum method, rather than by principal components analysis. Therefore, CPC-based OM, to some degree, can mitigate the influence of human factors and is advanced. Some details of CPC-based OM are briefly introduced below.

3.1. Index Nondimensionalization Method. When various candidate orbits are generated and orbit performance indices with different units are calculated, the performance indices are nondimensionalized using (4) [2]. If r performance indices are to be maximized, the dimensionless coefficient is equal to the performance index value divided by the maximum value in all candidate orbits. If s performance indices are to be minimized, the dimensionless coefficient is equal to the reciprocal of the performance index value divided by the minimum value in all candidate orbits. If t performance indices are to be within an expected range, the dimensionless coefficient is equal to the performance index value divided by the median of the range when the performance index is less than the median of the range and dimensionless coefficient is equal to the reciprocal of the performance index value divided by the median of the range when the performance index is greater than the median of the range.

$$(y_i)_j = \begin{cases} \dfrac{(x_i)_j}{x_i^{\max}} & i = 1, \dots, r \\ \dfrac{x_i^{\min}}{(x_i)_j} & i = 1, \dots, s \\ \dfrac{(x_i)_j}{\overline{x}_i} & (x_i)_j < \overline{x}_i,\ i = 1, \dots, t \\ \dfrac{\overline{x}_i}{(x_i)_j} & (x_i)_j > \overline{x}_i,\ i = 1, \dots, t, \end{cases} \tag{4}$$

where $(x_i)_j$ is the ith performance index of the jth orbit, $(y_i)_j$ is the dimensionless coefficient of the performance index $(x_i)_j$, $x_i^{\max}$ is the maximum value of x_i in all candidate orbits, $x_i^{\min}$ is the minimum value of x_i in all candidate orbits, and $\overline{x}$ is the median of the range.

3.2. Principal Component Analysis. Supposing the dimensionless coefficients can be represented by vector $\mathbf{Y} = (y_1, y_2, \dots, y_p)$, where p is the number of performance indices of each orbit, then $\mathbf{Y}_1, \dots, \mathbf{Y}_j, \dots, \mathbf{Y}_n$ are the dimensionless coefficients of the performance indices of n candidate orbits. In principal component analysis [2, 21], the elements u_{ij} of the covariance matrix $\mathbf{U} = (u_{ij})_{p \times p}$ are firstly calculated as follows:

$$u_{ij} = \frac{1}{n-1} \sum_{k=1}^{n} \left((y_i)_k - \overline{(y_i)} \right) \left((y_j)_k - \overline{(y_j)} \right), \tag{5}$$

where

$$\overline{(y_i)} = \frac{1}{n} \sum_{k=1}^{n} (y_i)_k, \qquad \overline{(y_j)} = \frac{1}{n} \sum_{k=1}^{n} (y_j)_k. \tag{6}$$

Next, the eigenvalues λ_i of the covariance matrix $\mathbf{U}$ are calculated. The cumulative contribution ratio of the previous m ($m \in [1, p]$) eigenvalues η_m is as follows:

$$\eta_m = \frac{\sum_{i=1}^{m} \lambda_i}{\sum_{i=1}^{p} \lambda_i}. \tag{7}$$

The number of principal components is the minimum m, which makes η_m greater than 88%. Then, by using the orthogonal normalized eigenvector $\mathbf{e}_i^T$ of the covariance matrix $\mathbf{U}$, the principal component vector $\mathbf{Z}_j$ of the jth candidate orbit is calculated as follows:

$$\mathbf{Z}_j = \left[\mathbf{e}_1^T \mathbf{Y}_j, \mathbf{e}_2^T \mathbf{Y}_j, \dots, \mathbf{e}_m^T \mathbf{Y}_j \right]^T \quad (j = 1, 2, \dots, n). \tag{8}$$

Because m is generally smaller than p, the total number of principal components to be clustered is less than the total number of performance indices.

3.3. Multilevel Cluster Analysis. The candidate orbits could be clustered according to the principal components to obtain the optimal class [33, 34]. The max–min distance method [35] is employed in this research. To avoid randomly selecting the first cluster center, the orbit most close to the zero point of principal components is selected as the first cluster center. The Euclidean distance is employed in cluster analysis, and the definition is as follows:

$$d\left(\mathbf{Z}_\zeta, \mathbf{Z}_\varsigma\right) = \sqrt{\sum_{i=1}^{m} \left(Z_\zeta^i - Z_\varsigma^i \right)^2}, \tag{9}$$

where $\mathbf{Z}_\zeta$ and $\mathbf{Z}_\varsigma$ are two arbitrary principal component vectors of candidate orbits, ζ and ς are two arbitrary natural numbers, and Z_ζ^i is the ith principal component in $\mathbf{Z}_\zeta$. A total class distance criterion is applied to determine the optimal number of classes N_c in each clustering and the candidate values of N_c include 4, 5, and 6.

The procedures of cluster analysis are as follows.

Step 1. Assume N_c is set as one of the three candidate values.

Step 2. Select initial cluster centers. Calculate Euclidean distance d_j^0 from all $\mathbf{Z}_j$ to the zero point of principal components using

$$d_j^0 = d\left(Z_j, 0\right) = \sqrt{\sum_{i=1}^{m} \left(Z_j^i \right)^2}. \tag{10}$$

$\mathbf{Z}_j$ corresponding to minimum d_j^0 is taken as the first cluster center $\mathbf{Z}_c^1$. The second cluster center $\mathbf{Z}_c^2$ will be $\mathbf{Z}_j$ which has the maximum Euclidean distance to $\mathbf{Z}_c^1$. The $(k+1)$th cluster center $\mathbf{Z}_c^{k+1}$ will be $\mathbf{Z}_j$ corresponding to the maximum $d_j^{\min c}$, where $d_j^{\min c} = \min\{d_j^{c1}, d_j^{c2}, \dots, d_j^{ci}, \dots, d_j^{ck}\}$, and d_j^{ci} is the Euclidean distance between $\mathbf{Z}_j$ and $\mathbf{Z}_c^i$. N_c cluster centers are finally selected in this way.

Step 3. Calculate Euclidean distances from each of the remaining $\mathbf{Z}_j$ to all the N_c cluster centers. Each $\mathbf{Z}_j$ will be assigned to the most close cluster center.

Step 4. Determine new cluster centers. Calculate the average value of principal components of N_c classes using

$$\overline{Z}_k^i = \left(\frac{1}{c_k} \right) \sum_{j=1}^{c_k} Z_j^i \quad (k = 1, 2, \dots, N_c), \tag{11}$$

where $\overline{Z}_k^i$ is the average value of the ith principal component in the kth class and c_k is the total number of candidate orbits in the kth class. The new cluster center of each class will be $\mathbf{Z}_j$ most close to the average value of principal components. The distance from $\mathbf{Z}_j$ to average value of principal components $\overline{\mathbf{Z}}_k$ is shown in

$$\overline{d}_j^k = d\left(\mathbf{Z}_j, \overline{\mathbf{Z}}_k\right) = \sqrt{\sum_{i=1}^{m} \left(Z_j^i - \overline{Z}_k^i \right)^2}, \tag{12}$$

where $\overline{d}_j^k$ is the distance from $\mathbf{Z}_j$ to the average value of principal components of the kth class.

Step 5. If one or more than one new cluster center is different from the corresponding last cluster center, reassign all remaining $\mathbf{Z}_j$ to the N_c cluster centers, then repeat Step 3; otherwise stop the clustering analysis corresponding to this candidate value of N_c.

Step 6. Set N_c to other candidate values, and relevant cluster analyses are carried out according to aforementioned Steps 2~5. Then, the total class distances δ in three cluster analyses corresponding to three candidates N_c are calculated [2].

$$\delta = \sum_{k=1}^{q} \delta_k,$$
$$\delta_k = \sqrt{\frac{1}{c_k}\sum_{j=1}^{c_k} d_j^2}, \tag{13}$$

where q is the candidate value of N_c in each cluster analysis ($q = 4, 5$ or 6), c_k is the number of orbits in class k, and d_j is the Euclidean distance between the class center and the other orbits in this class. The final cluster result is the cluster result corresponding to the smallest δ.

The number of candidate orbits is very large, after one cluster analysis, the number of orbits in the optimal class is usually more than six. Therefore, cluster analyses are repeatedly carried out until the number of orbits in the optimal class is less than six. This is known as multilevel cluster analysis.

3.4. Weighted Sum Method. The weighted sum method [2, 10] is adopted to determine the optimal class from all the classes and the optimal orbit from the orbits in the last optimal class. The evaluation index ξ is defined as follows:

$$\xi = \left(\sum_{i=1}^{r} (y_i) W_i + \sum_{i=1}^{s} (y_i) W_{r+i}\right) \prod_{i=1}^{t} f(y_i), \tag{14}$$

where W_i $(i = 1, 2, \ldots, r + s)$ is the weight coefficient of each performance index, in the range $[0, 1]$. The representative coefficient $f(y_i)$ is given by

$$f(y_i) = \begin{cases} 1 & \overline{x}_i^{\min} \le x_i \le \overline{x}_i^{\max} \\ 0 & x_i > \overline{x}_i^{\max} \text{ or } x_i < \overline{x}_i^{\min}, \end{cases} \tag{15}$$

where $[\overline{x}_i^{\min}, \overline{x}_i^{\max}]$ is the expected range of x_i and y_i is the dimensionless coefficient of x_i.

4. Novel DCPC-Based OM for Orbit Optimizations

4.1. The Advanced Characteristic of CPC-Based OM and Modification for Convergence Efficiency. In CPC-based OM, principal component analysis, rather than weighted sum method, is adopted to cluster orbits, and therefore the negative influence of the artificially set weight coefficients in weighted sum method is reduced. The detailed analyses are illustrated in contours resulting from principal component analysis and weighted sum method in Figure 2(a), assuming that two orbit elements are optimized and orbit performance indices are transformed into one principle component. Because artificial weight coefficients are usually different from the orthogonal normalized eigenvector (refer to Section 3.2), significantly different contours are shown in Figure 2(a). The artificial weight coefficients are unreasonable and easily influenced by human factors, while principal components are objective and therefore more reasonable. A comparison analysis of the advantage of principal component analysis is presented in Section 6.3.

In CPC-based OM, when enough candidate orbits are randomly generated, the global optimal orbit with high accuracy could be achieved. For a certain number of candidate orbits, as indicated by black stars in Figure 2(a), star A is the closest to the global optimal orbit indicated by a large red star. Therefore, star A is the optimal solution among these candidate orbits. Orbit A is not a local optimal solution but a global optimal solution with low accuracy. If the number of candidate orbits is smaller than that in Figure 2(a) and orbit A is not included in the candidate orbits, the optimal orbit obtained using CPC-based OM will be star B (as shown in Figure 2(b)); orbit B is a local optimal solution. Therefore, if there are a large number of candidate orbits, the global optimal solution can be achieved by CPC-based OM. To improve the accuracy of the optimal solution, generating more candidate orbits, is a feasible approach, as shown in Figure 2(c). When additional orbits (indicated with blue stars) are generated as candidate orbits, a better orbit marked with star C is the optimal solution. Star C is closer than star A to the global optimal orbit. Therefore, the accuracy of the global optimal solution is satisfied when sufficient candidate orbits are prepared in CPC-based OM. Additional candidate orbits are gradually generated in "random exhaustive" process in CPC-based OM as shown in Figure 1, which ensure global optimum and high accuracy in CPC-based OM. But "random exhaustive" leads to low computational efficiency for CPC-based OM.

The method of reducing the feasible region will be employed to improve the convergence. It could be known from Figure 2(c) that the additional orbits cover the entire feasible region of orbit elements. Supposing that all orbits in the green contour belong to the optimal class, the rectangular box R2 (shown in Figure 2(d)), which is envelop of green contour, is the feasible region of the optimal class, and it is far smaller than the entire feasible region R1 (shown in Figure 2(d)). If the same numbers of additional orbits as that in Figure 2(c) are generated and they cover the feasible region R2 of the optimal class, as shown in Figure 2(d), star D will be the optimal solution (better than star C). Unfortunately, some orbits (marked with a red dotted box in Figure 2(d)) in R2 do not belong to the optimal class. To avoid orbits generated in red dotted box, cluster analysis of orbits in optimal class based on the orbital elements could be introduced to divide optimal class to two parts, and the concentrated zones R3 and R4 will

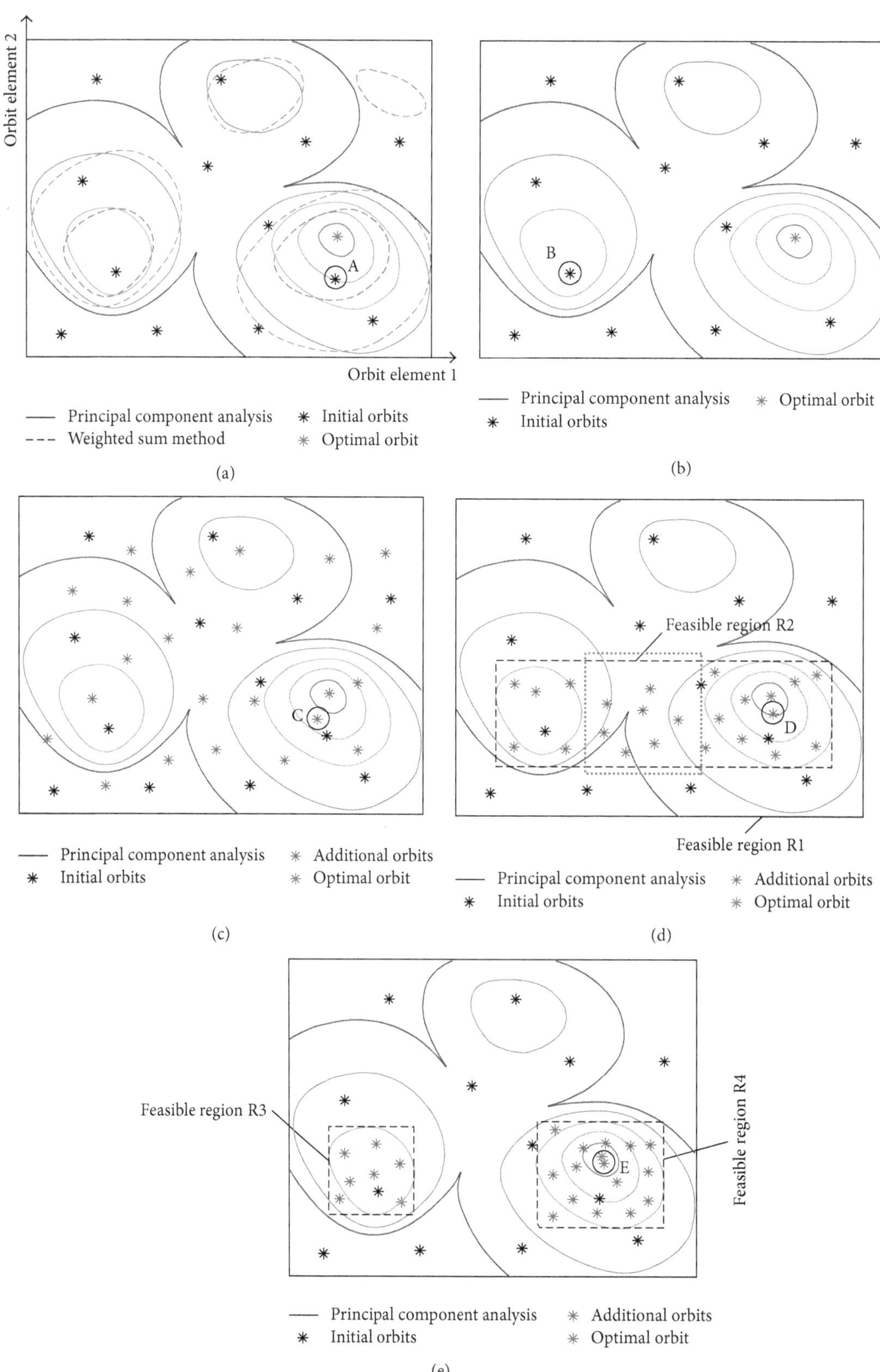

FIGURE 2: The global optimization capability and convergence efficiency of DCPC-based OM: (a) effect of principal component analysis, (b) CPC-based OM with initial orbits, (c) CPC-based OM with initial and additional orbits, (d) additional orbits generating strategy, and (e) DCPC-based OM.

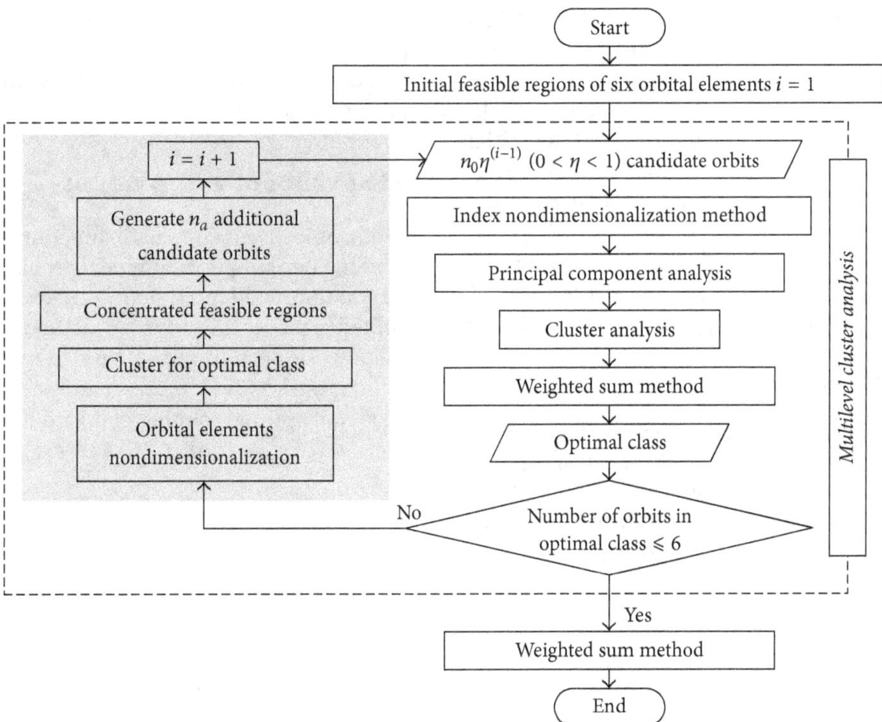

FIGURE 3: Process flow of DCPC-based OM for orbit design optimizations.

be the feasible regions of two parts of the optimal class, as shown in Figure 2(e). The areas of the new feasible regions R3 and R4 are smaller than the area of R2. If the same numbers of additional orbits in feasible regions R3 and R4 are generated, star E is the new optimal solution (shown in Figure 2(e)) and is better than star D (shown in Figure 2(d)).

4.2. Processing Flow. By introducing cluster analysis based on the six orbital elements, a novel population-based optimization method named DCPC-based OM is developed based on CPC-based OM. The processing flow is shown in Figure 3, where the gray zone indicates the cluster analysis based on the six orbital elements, including totally five steps.

Step 1. Randomly generate n_0 candidate orbits according to the initial feasible regions of the six orbital elements, and then calculate the coverage and TT&C performance indices.

Step 2. Nondimensionalize the coverage and TT&C performance indices of candidate orbits.

Step 3. Calculate the principal components of nondimensionalized orbit performance indices, divide candidate orbits into classes by performing cluster analysis based on the principal components, and evaluate class centers using the weighted sum method to obtain the optimal class.

Step 4. When the total number of candidate orbits in the optimal class is greater than six, the six orbital elements are nondimensionalized by the upper limit value of the relevant feasible region. Cluster analysis is conducted on the orbits belonging to the optimal class, by using the nondimensionalized values of the six orbital elements. The candidate number of classes N_c here is not limited to 4, 5, or 6; that is, it could change from 1 to the total number of orbits in the optimal class. After orbits with similar six orbital elements are clustered into a class, the concentrated feasible region of a class is the envelope of six orbital elements of all orbits in this class. Then n_a additional candidate orbits are generated randomly in the all concentrated feasible regions. The value of n_a is as follows:

$$n_a = n_{i+1} - n_i^o, \tag{16}$$

where n_{i+1} is the total number of candidate orbits in the $(i+1)$th cluster analysis and n_i^o is the total number of orbits in the optimal class of the ith cluster analysis. An evolutionary rate η $(0 < \eta < 1)$ is defined to ensure that the number of candidate orbits is reduced in the subsequent optimization, and the total number of candidate orbits in the ith cluster analysis is $n_i = n_0\eta^{(i-1)}$.

Step 5. Choosing the orbits in the optimal class and n_a additional orbits as candidate orbits (a total of $n_0\eta^i$ orbits), repeat Step 2, until the number of orbits in the optimal class is no more than six and then determine the optimal orbit from the optimal class by using the weighted sum method.

The characteristics of DCPC-based OM can be concluded as follows.

(1) Cluster analysis based on principal components of orbit performance indices and six orbital elements are performed, respectively. Therefore, double cluster analyses are

included in DCPC-based OM. By clustering of the orbits belonging to the optimal class using six orbital elements, the concentrated feasible regions of the optimal class are obtained. Additional candidate orbits are generated only in the concentrated feasible regions which is smaller than initial feasible region. An evolutionary rate η $(0 < \eta < 1)$ is defined to ensure that the candidate orbits is reduced in the subsequent optimization.

(2) According to the definition of evolutionary rate η in Step 4 and the flow chart in Figure 3, the total number of iterations N in DCPC-based OM is approximately as follows:

$$N \approx \log_{1/\eta} n_0 - \log_{1/\eta}(5 \times 3.5) + 1, \quad (17)$$

where 5 is the average value of the number of classes (4, 5, and 6) and 3.5 is the average number of the orbits in the last optimal class (1, 2, 3, 4, 5, and 6). Based on the hypothesis that each class has the same number of orbits in cluster analysis, the number of additional orbits in N iterations can be calculated using

$$\begin{gathered} n_0\left(\eta - \frac{1}{5}\right), \\ n_0\left(\eta - \frac{1}{5}\right)\eta, \\ n_0\left(\eta - \frac{1}{5}\right)\eta^2, \\ \vdots \\ n_0\left(\eta - \frac{1}{5}\right)\eta^{N-1}. \end{gathered} \quad (18)$$

It can be seen from (18) that the numbers of additional orbits in all iterations are in a form of geometrical sequences. Therefore the total number of generated candidate orbits n_{total} equals the number of initial orbits n_0 plus the sum of geometrical sequences in (18), as shown in

$$n_{\text{total}} \approx n_0\left[1 + \left(\eta - \frac{1}{5}\right)\frac{n_0 - (5 \times 3.5)\eta}{n_0(1-\eta)}\right]. \quad (19)$$

For orbit optimization, the calculation procedure of orbit performance indices is more time-consuming than optimization operation, because a large number of numerical computations are needed to calculate orbit performance indices. Therefore, the total optimization time approximately equals the total number of generated candidate orbits multiplied by the computation time required for the performance indices of one orbit, which means the time cost is nearly predictable. The proposed orbit design optimization method with predictable time cost is more convenient than other population-based optimization methods with unpredictable time cost for a scheduled project.

(3) In DCPC-based OM, the method of reducing the feasible region improved the computational efficiency while it might result in the reduction of capability of global optimization, because the feasible regions including the global optimum solution might be mistakenly deleted. The large value of n_0 and η can improve the capability of global optimization, and the small value of n_0 and η can improve computational efficiency but increase the risk of missing the global optimal solution.

5. Experiment and Analysis

A four-objective benchmark function (as shown in (20)) including two simple multimodal problems ($f_1(\vec{x})$ and $f_2(\vec{x})$) and two unrotated multimodal problems ($f_3(\vec{x})$ and $f_4(\vec{x})$) is adopted to testify the proposed method [36]. The Rosenbrock function $f_2(\vec{x})$ is modified to $f_2'(\vec{x})$ to ensure that global optimal solution is $[0, 0, \ldots, 0]$ and the minimum value is 0. The global optimal solution of the multiobjective benchmark function is $\vec{x}_{\text{opt}} = [0, 0, \ldots, 0]$ and $f(\vec{x}_{\text{opt}}) = 0$.

$$\begin{aligned} f(\vec{x}) &= f_1(\vec{x}) + f_2'(\vec{x}) + f_3(\vec{x}) + f_4(\vec{x}) \\ f_1(\vec{x}) &= \sum_{i=1}^{D} x_i^2 \\ f_2'(\vec{x}) &= \sum_{i=1}^{D-1}\left[100\left(x_{i+1} - x_i^2\right)^2 + x_i^2\right] \\ f_3(\vec{x}) &= -20\exp\left(-0.2\sqrt{\frac{1}{D}\sum_{i=1}^{D} x_i^2}\right) \\ &\quad - \exp\left(\frac{1}{D}\sum_{i=1}^{D}\cos(2\pi x_i)\right) + 20 + e \\ f_4(\vec{x}) &= \sum_{i=1}^{D}\left[x_i^2 - 10\cos(2\pi x_i) + 10\right]. \end{aligned} \quad (20)$$

The optimizations with dimensions D of 5 and 10 are conducted using DCPC-based OM, CPC-based OM, and WSGA-based HM [4–6]. In DCPC-based OM and CPC-based OM, the number of initial candidate elements n_0 was set as 10,000,000. With respect to CPC-based OM, the increasing number n_a was 100,000. In DCPC-based OM, the evolutionary rate η was 0.6. In WSGA-based HM, the individual number in a generation n_G was 100,000, and the fitness values were calculated using weighted sum method, as shown in Section 3.4, with weight coefficients $W_i = 1$ $(i = 1, 2, \ldots, 4)$. The initial range of $\vec{x}$ was $[-5.12, 5.12]$. At each time the clustering in DCPC-based OM was finished, the minimum values of function $f = f_1(\vec{x}) + f_2(\vec{x}) + f_3(\vec{x}) + f_4(\vec{x})$ calculated by the three methods are shown in Figure 4. Three methods started optimization process at the same time and terminated simultaneously when DCPC-based OM completed optimization. Because n_G in WSGA-based HM was far less than n_0 in DCPC-based OM and CPC-based OM, many iterations in WSGA-based HM had occurred when the first clustering in DCPC-based OM was completed. In addition, with repeat clustering, the number of candidate elements in the optimal class decreases in DCPC-based OM and the number of completed iterations of WSGA-based HM in one clustering decreases. Therefore, in each

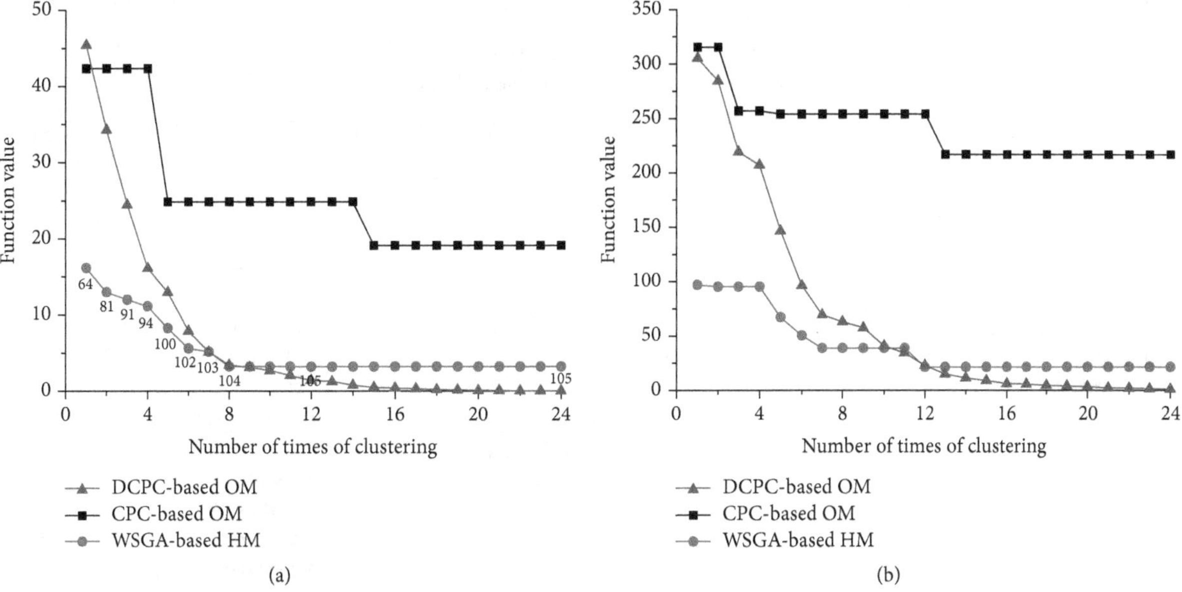

FIGURE 4: Convergence curve of test: (a) dimension is 5; (b) dimension is 10.

clustering of DCPC-based OM, the number of iterations in WSGA-based HM varies. The number of times of completed iterations in WSGA-based HM at each time when clustering in DCPC-based OM was finished is shown in Figure 4.

Figure 4(a) shows that the function values in all three methods decreased as the number of times of clustering increased. A comparative analysis between the function values of DCPC-based OM and CPC-based OM shows that, in the first clustering, the function values of DCPC-based OM were similar to that of CPC-based OM. However, in the last clustering, the function values of DCPC-based OM were far lower than that of CPC-based OM. It validated the improved convergence by the method of reducing feasible region in DCPC-based OM. When the first clustering in DCPC-based OM was completed, iterations in WSGA-based HM had been finished 64 times. Thus, the function values in WSGA-based HM were less than that in DCPC-based OM. By the 8th clustering in DCPC-based OM, 104 iterations of WSGA-based HM had been finished. And the additional 2 iterations of WSGA-BASED HM were completed by the 22nd cluster. At the end of the last clustering, the function values of DCPC-based OM were less than that in WSGA-based HM. In WSGA-based HM, the feasible region is constant, the convergence is driven by evolutionary capacity of genetic operation, but genetic operation is actually random. In a large feasible region, the global optimum is difficult to be completely randomly generated. Hence, the convergence efficiency is very low in WSGA-based HM. The candidate elements in DCPC-based OM are randomly generated. However, because reducing the feasible region is an effective method to improve optimization efficiency [27–32], DCPC-based OM achieves more efficient convergence than WSGA-based HM.

6. Orbit Design Results and Analysis

6.1. Orbit Optimization Conditions and Convergence Analysis. For a certain Earth observation satellite, five observation targets are with latitude and longitude coordinates of (25°N, 120°E), (10°N, 110°E), (40°N, 130°E), (15°N, 90°W), and (20°S, 130°E) and the same vision field angle of 25°. The minimum elevation angle, latitude, and longitude coordinates of TT&C station are 5°, 40°N, and 120°E, respectively. The expected range of ATI-TT&C is 8000 s~40000 s. The range of the semimajor axis is 400 km~600 km and $W_i = 1$ ($i = 1, 2, \ldots, 7$). The simulation time for each candidate orbit is 3 days.

The orbit design optimization is conducted using DCPC-based OM, CPC-based OM, and WSGA-based HM [4–6]. In DCPC-based OM and CPC-based OM, the number of initial candidate orbits n_0 is set as 100,000. In DCPC-based OM, the evolutionary rate η is 0.5. The number of additional orbits n_a is set as 10,000 for CPC-based OM. The number of individuals in a generation n_G is 10,000 for WSGA-based HM. The change of the evaluation indices for DCPC-based OM, CPC-based OM, and WSGA-based HM with the number of times of clustering in DCPC-based OM is shown in Figure 5.

The data in Figure 5 show that the evaluation indices for all three methods increase as the number of times of clustering increases. After the first clustering, DCPC-based OM and CPC-based OM have similar evaluation indices, which are less than that for WSGA-based HM. The optimization process for CPC-based OM and WSGA-based HM is stopped when DCPC-based OM finishes its optimization, and DCPC-based OM has the highest evaluation index of all three methods in the last clustering. The evaluation index of the orbit of DCPC-based OM is higher than that of CPC-based OM by 22.9%

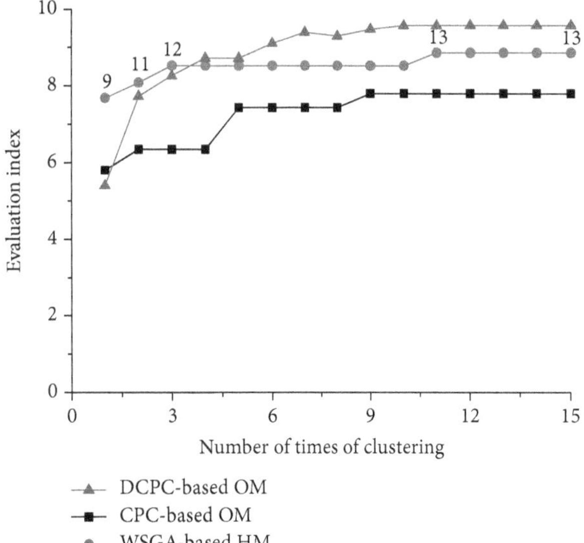

Figure 5: Evaluation indices of three optimization methods.

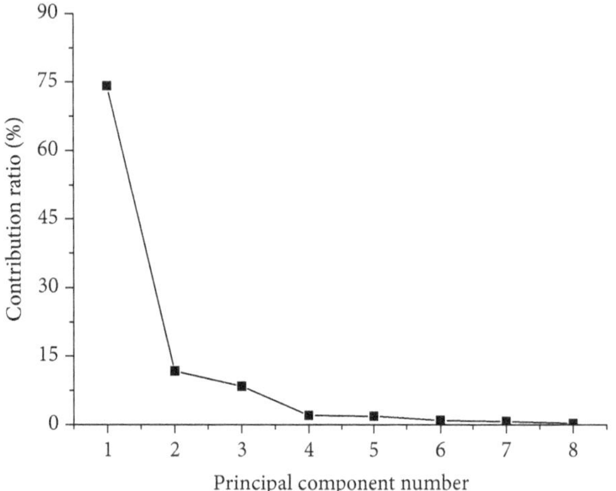

Figure 6: Principal component contribution value for the first clustering.

and higher than that of WSGA-based HM by 8.0%. In other words, DCPC-based OM has more efficient convergence than both CPC-based OM and WSGA-based HM.

6.2. Optimization Results with DCPC-Based OM. Details of orbit optimization results with DCPC-based OM are presented in this section. In the first clustering, principal component analysis was conducted after the performance indices of all candidate orbits are calculated and nondimensionalized. The contribution ratios of the first three principal components in the principal component analysis were 74.10%, 11.69%, and 8.38%, respectively, as shown in Figure 6.

The data in Figure 6 shows that the first three principal components contribute more than the others. The cumulative contribution ratio of the first three principal components is 94.17%. The number of principal components is three in all the clusters in this study, although the contribution ratios of the first three principal components vary slightly in various principal component analyses. Consistently, the multilevel cluster analysis is conducted.

Table 1: Principal components in the cluster centers for the first clusters.

Principal component	Classification				
	1	2	3	4	5
Principal component 1 (CP1)	−4.72	3.34	−1.46	3.58	−2.48
Principal component 2 (CP2)	0.46	7.38	−3.37	−2.19	−0.87
Principal component 3 (CP3)	−5.18	0.18	5.80	−0.83	4.14

According to total class distance criterion, the optimal number of classes in the first cluster was five. The principal components of the cluster centers in the five classes are shown in Table 1. The data in Table 1 illustrates the differences among the cluster centers of the five classes. Classes 4, 2, and 3 had the largest principal components 1, 2, and 3, respectively.

The data in Figure 7(a) show the distribution characteristics of the principal components in the five cluster centers. Because the principal components have no physical meaning, it is difficult to determine the optimal class according to the principal components. Therefore, the data in Figure 7(b) show the differences among the performance indices of the five cluster centers, and the weighted sum method is used to select the optimal class by evaluating the performance indices of the cluster centers. The eight dimensionless performance indices of the five cluster centers are listed in Table 2.

The data in Figure 7(b) indicate that the dimensionless orbit performance indices of Class 4 are relatively large, and the optimal class obtained by using the weighted sum method was also Class 4.

The orbits in Class 4 from the first clustering and n_a additional candidate orbits are then selected to continue with the clustering process. The principal components and dimensionless performance indices in the cluster center after clustering fifteen times using DCPC-based OM are shown in Figures 8(a) and 8(b). The analysis results using the weighted sum method indicate that Class 3 is the optimal class.

The eight dimensionless performance indices for all orbits of Class 3 are shown in Figure 9. The optimal orbit obtained by the weighted sum method is Orbit 1.

6.3. Global Optimality and Principal Component Estimation in DCPC-Based OM. The evolution process of the maximum and minimum evaluation indices of the orbits in the optimal class is shown in Figure 10. The uptrend of the two types of indices indicates that the orbits are being optimized. The phenomenon of the two sets of indices approaching each other indicates that the differences among all orbits in the optimal class are decreasing. When the two sets of indices are close enough, the optimal orbit is achieved.

A decrease in the maximum evaluation index from O1 to O2 indicates that, in the 8th clustering, the maximum evaluation index is not contained in the optimal class and

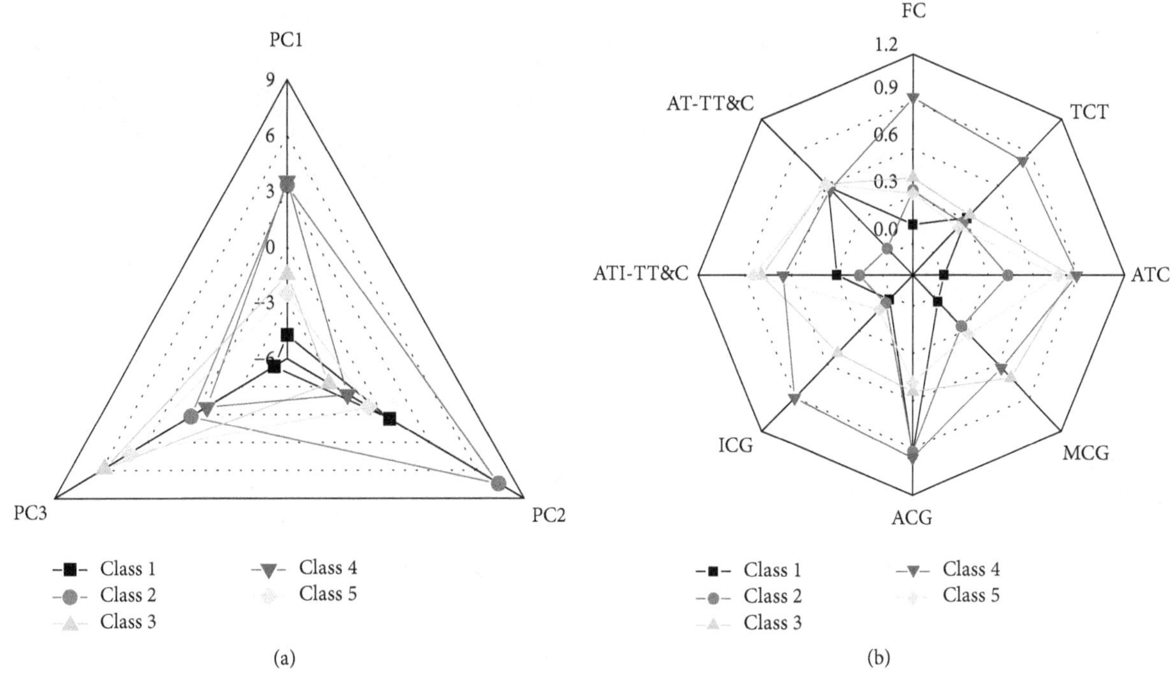

FIGURE 7: Results of the first clustering: (a) principal components of the five clustering centers and (b) dimensionless performance indices of the five clustering centers.

TABLE 2: Eight dimensionless performance indices of the five clustering centers after the first clustering.

Performance indices	Classification				
	1	2	3	4	5
Frequency of coverage (FC)	0.12	0.34	0.42	0.93	0.32
Total coverage time (TCT)	0.31	0.27	0.34	0.83	0.23
Average time per coverage (ATC)	0.01	0.43	0.86	0.89	0.77
Maximum coverage gap (MCG)	0.04	0.26	0.72	0.63	0.33
Average coverage gap (ACG)	0.94	0.92	0.54	0.96	0.48
Minimum coverage gap (ICG)	0.02	0.05	0.5	0.9	0.11
Average time interval of TT&C (ATI-TT&C)	0.30	0.15	0.79	0.65	0.84
Average time of each TT&C (AT-TT&C)	0.59	0.04	0.62	0.57	0.61

is filtered out. The coverage and TT&C performance indices of orbits O1 and O2 are listed in Table 3. Compared to orbit O2, orbit O1 exhibits better TCT and ATC, worse MCG, ICG, and AT-TT&C, and similar FC, ACG, and ATI-TT&C. Considering an obviously oversized MCG and ICG in orbit O1, orbit O2 is better than orbit O1, even though the evaluation index of orbit O2 determined by weighted sum method is less than that of orbit O1. The reason for the higher evaluation index of orbit O1 is that ATC is linearly correlated with TCT, and in the weighted sum method, both TCT and ATC are considered, that is, the coverage time is counted twice. It leads to the fact that effects of the oversized MCG and ICG are masked in weighted sum method. However, the effect of the linear correlation is eliminated in the principal component analysis, which leads to the fact that the coverage time is counted only once and the negative effects of the oversized MCG and ICG are reflected. Thus, the principal component analysis provides a more accurate estimation.

7. Conclusions

This paper proposes a population-based optimization method named DCPC-based OM, which consists of the index nondimensionalization method, principal component analysis, double cluster analysis, and the weighted sum method. Tests using numerical benchmark functions were conducted, and an example of orbit optimization for Earth observation satellites was analyzed. Both optimization results show that the proposed method, with characteristics of a predictable time cost, has the advantages of reducing the

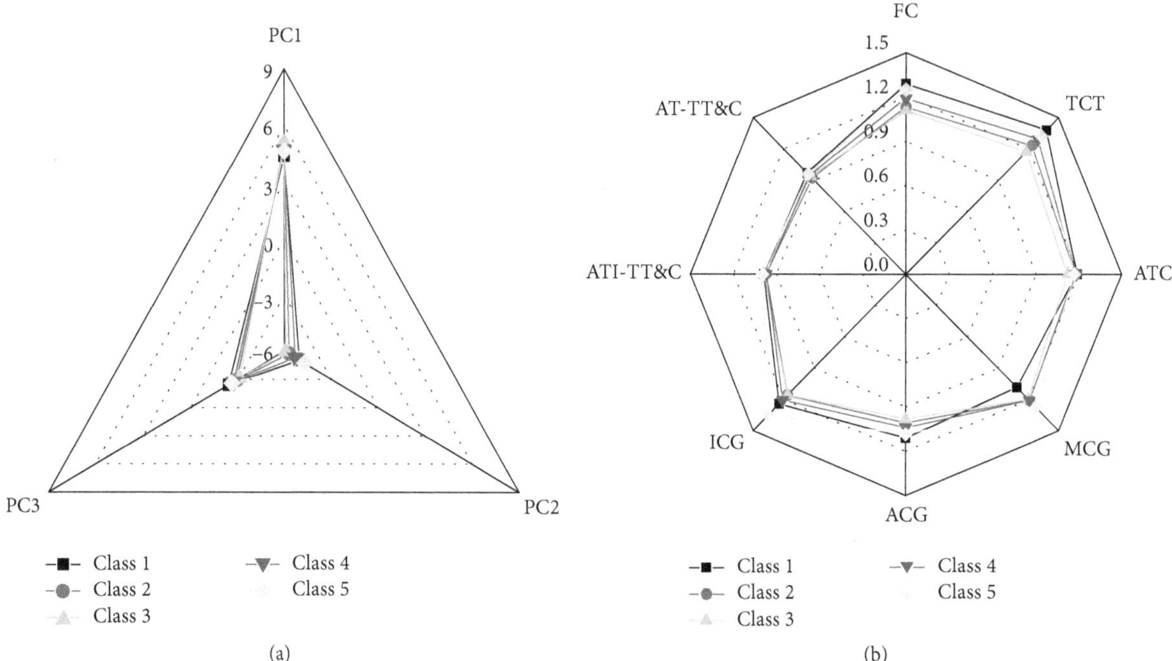

FIGURE 8: Results of the fifteenth clustering: (a) principal components of the five clustering centers and (b) dimensionless performance indices of the five clustering centers.

TABLE 3: Coverage and TT&C indices of orbits O1 and O2.

Index	Orbit O1	Orbit O2
Frequency of coverage (FC)	100	100
Total coverage time (TCT)/s	8445	5970
Average time per coverage (ATC)/s	86	60
Maximum coverage gap (MCG)/s	4213.5	2842.5
Average coverage gap (ACG)/s	1117.75	1162.5
Minimum coverage gap (ICG)/s	486.25	270.75
Average time interval of TT&C (ATI-TT&C)/s	3886	3930
Average time of each TT&C (AT-TT&C)/s	31	38

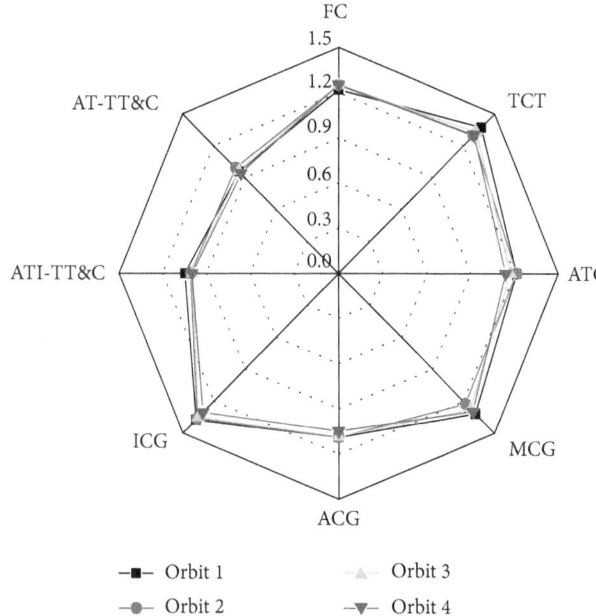

FIGURE 9: Dimensionless performance indices for all orbits in Class 3 after the fifteenth clustering.

influence of human factors that commonly exist in the weighted sum method and bring more efficient convergence than genetic algorithm.

This paper describes the results of a preliminary research study of the developed optimization method. Further study will be performed, such as determining the quantity of initial orbits to ensure the capability of global optimization, understanding how evolutionary rate influences the accuracy of the optimal solution, and determining whether a dynamic evolutionary rate is necessary. In addition, the optimal capability when the optimal solution is on the boundary of the feasible region will be further investigated.

Conflicts of Interest

The authors declare that they have no conflicts of interest.

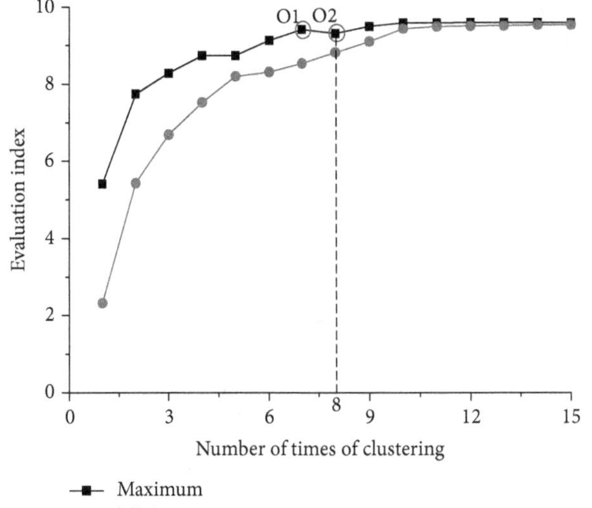

Figure 10: Evaluation index trend.

Acknowledgments

The research work is supported by the National Defense 973 Program (Grant no. 613237).

References

[1] J. R. Wertz and W. J. Larson, *Space Mission Analysis and Design*, Microcosm Press, Torrance, Calif, USA, 3rd edition, 1999.

[2] X. N. Wei, Y. F. Dong, F. R. Liu, L. Tian, Z. Hao, and H. Shi, "Principal component analysis and cluster analysis based orbit optimization for earth observation satellites," *Journal of Chongqing University English Edition*, vol. 15, no. 3, pp. 83–94, 2016.

[3] F. Simeoni, L. Casalino, A. Zavoli, and G. Colasurdo, "Indirect optimization of satellite deployment into a highly elliptic orbit," *International Journal of Aerospace Engineering*, vol. 2012, Article ID 152683, 14 pages, 2012.

[4] O. Abdelkhalik and D. Mortari, "Reconnaissance problem using genetic algorithms," in *Proceedings of the 2005 Space Flight Mechanics Meeting Conference*, AAS Paper 05-184, Copper Mountain, Colo, USA, 2005.

[5] O. Abdelkhalik and D. Mortari, "Orbit design for ground surveillance using genetic algorithms," *Journal of Guidance, Control, and Dynamics*, vol. 29, no. 5, pp. 1231–1235, 2006.

[6] O. Abdelkhalik and A. Gad, "Optimization of space orbits design for Earth orbiting missions," *Acta Astronautica*, vol. 68, no. 7-8, pp. 1307–1317, 2011.

[7] S. D. Vtipil and B. Newman, "Designing a constrained optimal orbit for earth observation satellites based on user requirements," in *Proceedings of the AIAA/AAS Astrodynamics Specialist Conference 2010*, AIAA Paper 2010-7520, Cannes, France, August 2010.

[8] S. D. Vtipil, *Constrained Optimal Orbit Design for Earth Observation [Ph.D. thesis]*, Old Dominion University, Norfolk, Va, USA, 2010.

[9] J. Zhang, G. J. Tang, and Y. Z. Luo, "Optimization of an orbital long-duration rendezvous Mission," *Aerospace Science and Technology*, vol. 58, pp. 482–489, 2016.

[10] I. Y. Kim and O. L. de Weck, "Adaptive weighted sum method for multiobjective optimization: a new method for Pareto front generation," *Structural and Multidisciplinary Optimization*, vol. 31, no. 2, pp. 105–116, 2006.

[11] L. Yang, W. C. Chen, X. M. Liu, and H. Zhou, "Steady glide dynamic modeling and trajectory optimization for high lift-to-drag ratio reentry vehicle," *International Journal of Aerospace Engineering*, vol. 2016, Article ID 3527460, 2016.

[12] S. Vtipil and J. G. Warner, "Earth observing satellite orbit design via particle swarm optimization," in *Proceedings of the AIAA/AAS Astrodynamics Specialist Conference, AIAA SPACE Forum*, AIAA Paper 2014-4428, San Diego, Calif, USA, 2014.

[13] V. Singh, S. K. Sharma, and S. Vaibhav, "Transport Aircraft Conceptual Design Optimization Using Real Coded Genetic Algorithm," *International Journal of Aerospace Engineering*, vol. 2016, Article ID 2813541, 11 pages, 2016.

[14] Y. He and R. K. Agarwal, "Shape optimization of NREL S809 airfoil for wind turbine blades using a multiobjective genetic algorithm," *International Journal of Aerospace Engineering*, vol. 2014, Article ID 864210, 13 pages, 2014.

[15] K. Deb, *Multi-Objective Optimization Using Evolutionary Algorithms*, Wiley, Chichester, UK, 2001.

[16] T. Gal and H. Leberling, "Redundant objective functions in linear vector maximum problems and their determination," *European Journal of Operational Research*, vol. 1, no. 3, pp. 176–184, 1977.

[17] M. T. Jensen, "Guiding single-objective optimization using multi-objective methods," in *Applications of Evolutionary Computing*, S. Cagnoni, C. G. Johnson, J. J. Romero Cardalda et al., Eds., pp. 268–279, 2003.

[18] M. Zhang, R. A. Kennedy, T. D. Abhayapala, and W. Zhang, "Internal structure identification of random process using principal component analysis," in *Proceedings of the 4th International Conference on Signal Processing and Communication Systems (ICSPCS '10)*, December 2010.

[19] F. Yao, J. Coquery, and K.-A. Lê Cao, "Independent Principal Component Analysis for biologically meaningful dimension reduction of large biological data sets," *BMC Bioinformatics*, vol. 13, no. 1, article 24, 2012.

[20] H. Hotelling, "Analysis of a complex of statistical variables into principal components," *Journal of Educational Psychology*, vol. 24, no. 6, pp. 417–441, 1933.

[21] K. Deb and D. K. Saxena, *On Finding Pareto-Optimal Solutions through Dimensionality Reduction for Certain Large-Dimensional Multi-Objective Optimization Problem*, Kanpur Genetic Algorithms Laboratory (KanGAL), Indian Institute of Technology Kanpur, Kanpur, India, 2005.

[22] K. Deb and D. K. Saxena, "Searching for Pareto-optimal solutions through dimensionality reduction for certain large-dimensional multi-objective optimization problem," in *Proceedings of the IEEE World Congress on Computational Intelligence*, pp. 3352–3360, Vancouver, Canada, 2006.

[23] J. H. De Freitas Gomes, A. R. S. Júnior, A. P. De Paiva, J. R. Ferreira, S. C. Da Costa, and P. P. Balestrassi, "Global Criterion Method based on principal components to the optimization of manufacturing processes with multiple responses," *Strojniski Vestnik/Journal of Mechanical Engineering*, vol. 58, no. 5, pp. 345–353, 2012.

[24] A. P. Paiva, S. C. Costa, E. J. Paiva, P. P. Balestrassi, and J. R. Ferreira, "Multi-objective optimization of pulsed gas metal arc welding process based on weighted principal component

scores," *International Journal of Advanced Manufacturing Technology*, vol. 50, no. 1–4, pp. 113–125, 2010.

[25] K. Polat and S. Günecs, "Detection of ECG arrhythmia using a differential expert system approach based on principal component analysis and least square support vector machine," *Applied Mathematics and Computation*, vol. 186, no. 1, pp. 898–906, 2007.

[26] X. Zhao, W. Lin, and Q. Zhang, "Enhanced particle swarm optimization based on principal component analysis and line search," *Applied Mathematics and Computation*, vol. 229, pp. 440–456, 2014.

[27] A. Parkinson, C. Sorensen, and N. Pourhassan, "A general approach for robust optimal design," *Journal of Mechanical Design*, vol. 115, no. 1, pp. 74–80, 1993.

[28] G. G. Wang and S. Shan, "Design space reduction for multi-objective optimization and robust design optimization problems," SAE Technical Papers 2004-10-0240, SAE International, Warrendale, Pa, USA, 2004.

[29] L. Zhu and D. Kazmer, "An extensive simplex method for mapping global feasibility," *Engineering Optimization*, vol. 35, no. 2, pp. 165–176, 2003.

[30] R. Dondo and J. Cerdá, "A cluster-based optimization approach for the multi-depot heterogeneous fleet vehicle routing problem with time windows," *European Journal of Operational Research*, vol. 176, no. 3, pp. 1478–1507, 2007.

[31] U. Halder, S. Das, and D. Maity, "A cluster-based differential evolution algorithm with external archive for optimization in dynamic environments," *IEEE Transactions on Cybernetics*, vol. 43, no. 3, pp. 881–897, 2013.

[32] M. Shams, E. Rashedi, and A. Hakimi, "Clustered-gravitational search algorithm and its application in parameter optimization of a low noise amplifier," *Applied Mathematics and Computation*, vol. 258, pp. 436–453, 2015.

[33] O. Maimon and L. Rokach, *Data Mining and Knowledge Discovery Handbook*, Springer, New York, NY, USA, 2010.

[34] S. Rajagopal, "Customer data clustering using data mining technique," *International Journal of Database Management Systems*, vol. 3, no. 4, pp. 1–11, 2011.

[35] Y. Wang and Y. Cao, "A Leukocyte image fast scanning based on maxmin distance clustering," *Journal of Innovative Optical Health Sciences*, vol. 9, no. 6, Article ID 1650022, 2016.

[36] X. Wang, Y. Shi, D. Ding, and X. Gu, "Double global optimum genetic algorithm–particle swarm optimization-based welding robot path planning," *Engineering Optimization*, vol. 48, no. 2, pp. 299–316, 2016.

8

A Shape-Based Method for Continuous Low-Thrust Trajectory Design between Circular Coplanar Orbits

Qun Fang,[1,2] Xuefeng Wang,[1,2] Chong Sun,[1,2] and Jianping Yuan[1,2]

[1]*National Key Laboratory of Aerospace Flight Dynamics, Xi'an, Shaanxi 710072, China*
[2]*School of Astronautics, Northwestern Polytechnical University, Xi'an, Shaanxi 710072, China*

Correspondence should be addressed to Qun Fang; qfang@nwpu.edu.cn

Academic Editor: Paul Williams

The shape-based method can provide suitable initial guesses for trajectory optimization, which are useful for quickly converging a more accurate trajectory. Combined with the optimal control theory, an optimized shape-based method using the finite Fourier series is proposed in this paper. Taking the flight time-fixed case and the time-free case into account, respectively, the optimized shape-based method, which considers the first-order optimal necessary conditions, can guarantee that not only an orbit designed during the preliminary phase is optimal, but also the thrust direction is not constrained to be tangential. Besides, the traditional shape-based method using the finite Fourier series, in which the thrust direction is constrained to be tangential, is developed for the time-free case in this paper. The Earth-Mars case and the LEO-GEO case are used to verify the optimized shape-based method's feasibility for time-fixed and time-free continuous low-thrust trajectory design between circular coplanar orbits, respectively. The optimized shaped-based method can design a lower cost trajectory.

1. Introduction

Recently, continuous low-thrust trajectory design and optimization are becoming increasingly popular [1, 2], although they are very challenging and time-consuming. What is particular is that the continuous low-thrust trajectory design consists of two phases: preliminary design and precise design [3]. To pursue a faster optimization and a more accurate trajectory, the preliminary design phase is expected to provide an efficient initial guess for trajectory optimizers. The shape-based (SB) method is one of the most efficient methods during this preliminary design. The SB method assumes that some functions contain a spacecraft's trajectory, and therefore boundary conditions are used to calculate the parameters of the functions, thus analytically obtaining the needed thrust during the spacecraft's flight.

Many kinds of SB methods have been proposed by researchers. For instance, Petropoulos and Longuski [4, 5] developed an exponential sinusoid (ES) method for the two-dimensional (2D) interplanetary transfer trajectory design. The ES method constrains the thrust direction to be tangential. Izzo [6] utilized this method to investigate the multirevolution Lambert's problem and simplified the interplanetary low-thrust trajectory design procedure. Cui et al. [7] proposed a new search approach algorithm for the launch window of low-thrust gravity-assist missions based on the ES method, which has fewer searching variables and is more efficient than the traditional SB methods. But the ES method cannot satisfy the full consideration conditions of circular terminal orbits unless thrusters provide impulsive propulsion; besides, the parameters of a shape cannot be solved when other constraints are introduced.

Zheng et al. [8] proposed a new trajectory shape called the logarithmic spiral-based (LS) non-Keplerian orbit. The feasibility and essential characteristics of the LS non-Keplerian orbit are analyzed. The analytical geocentric distance r expression and the phase angle θ expression about flight time subject to a tangential thrust are derived. But the LS method cannot satisfy the terminal constraints as well as the ES method.

To overcome the above-mentioned disadvantages, Wall and Conway [9, 10] developed a 2D inverse polynomial (IP) method. The fifth-order IP method can be used to design the transfer trajectory for the time-free case, while the sixth-order IP method is designed for the time-fixed case. But the thrust direction is also constrained to be tangential in the IP method, which cannot handle the thrust constraint very well. Shang et al. [11] proposed a semianalytical Lambert algorithm based on the N-degree IP method in order to improve the precision of preliminary design for an interplanetary low-thrust transfer trajectory. Considering thrust and radius constraints, Wang et al. [12] proposed a modified IP method for both the time-free transfer case and the time-fixed rendezvous case. Compared with the original IP method, the modified one can satisfy the thrust and radius constraints through optimizing the polynomial orders. To realize low-thrust trajectory design between elliptical orbits, Xie et al. [13] constructed a new shape function about two semimajor-axis parameters, which are polynomials in the polar angle. With tangential thrust, a fifth-order and sixth-order method is designed for time-free and time-fixed cases as in the IP method.

Taheri and Abdelkhalik [14] and Abdelkhalik and Taheri [15] proposed a new shape-based trajectory design method using the finite Fourier series. With the hypothesis of tangential thrust, a preliminary trajectory that satisfies the maximum thrust constraint is designed with this SB trajectory design method.

Shaping the velocity components, Gondelach et al. [16] proposed a novel low-thrust trajectory design method called hodographic-shaping (HS) method. These velocity functions are assumed to be some sets of simple base functions. Extra parameters are used to make the trajectory design and optimization more flexible.

In a word, all recent SB methods (except the HS method) design spacecraft trajectories based on the tangential thrust assumption and cannot guarantee that the designed trajectory is optimal without the first-order optimal necessary conditions. Meanwhile, almost SB methods require iterative calculations or constraint optimization to match the total flight time. However, it is difficult to determine a suitable flight time during or before the preliminary design phase.

Therefore, combined with optimal control theory, an optimized shape-based method using finite Fourier series, which can easily overcome the above-mentioned shortcomings, is proposed in this paper. Regarding spacecraft three-dimensional (3D) trajectory design, Wall [9], Novak and Vasile [17], and Taheri [18] presented their study advances. However, 3D trajectory design is not the key point in this paper, because the 2D case is enough to illustrate the idea of our method.

The paper is organized as follows. In Section 2, the spacecraft dynamics model in polar coordinate is developed. Then, the proposed method is introduced in Section 3. In the time-fixed rendezvous case and time-free transfer case, respectively, the first-order necessary conditions are derived from the Hamiltonian function. Through expanding the state variables into expressions of finite Fourier series, the optimal control problem is converted to a nonlinear programming (NLP) problem that involves Fourier series coefficients. The process of the proposed method is given in Section 4. In Section 5, two examples are used to verify the optimized shape-based method's feasibility for continuous low-thrust trajectory design between circular coplanar orbits, and the advantage that the proposed method can design a lower cost trajectory is proven by comparing it with other SB methods.

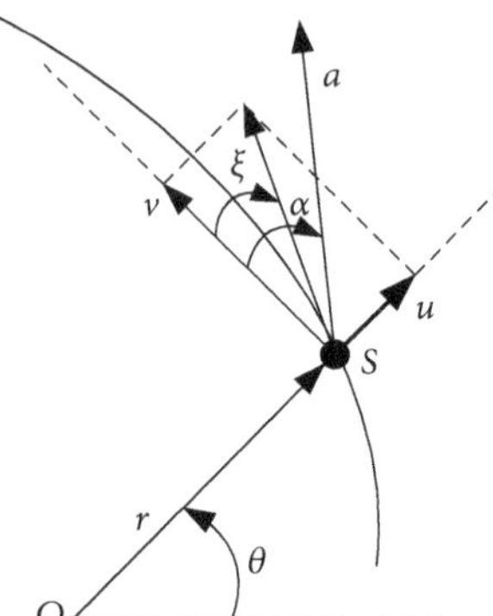

FIGURE 1: Trajectory variables.

2. Spacecraft Orbital Model

In polar coordinate, the spacecraft's orbital motion model without considering any perturbation and celestial bodies' rotation is established as

$$\begin{aligned}\dot{r} &= u,\\ \dot{\theta} &= \frac{v}{r},\\ \dot{u} &= -\frac{\mu}{r^2} + \frac{v^2}{r} + a\sin\alpha,\\ \dot{v} &= -\frac{uv}{r} + a\cos\alpha,\end{aligned} \tag{1}$$

where superscript "·" indicates a derivative with respect to time t; r is the magnitude of the position vector; θ is the polar angle; u is the magnitude of spacecraft radial velocity vector; v is the magnitude of its circumferential velocity vector; μ is the gravitational parameter; a is the thrust acceleration magnitude of the spacecraft; α is its steering angle, and ξ is its flight path angle, as shown in Figure 1.

Instead of shaping variables as a function of time, the polar angle also can be used as an independent variable. In that case, the variables have to be analytically integrable over θ to obtain the change in position.

In this paper, a new thrust acceleration parameter $\tilde{\mathbf{a}}$ is introduced, its direction is the same as $\mathbf{a}$, and its magnitude is defined as shown in

$$\tilde{a} = \frac{r}{v}a. \tag{2}$$

Instead of derivatives with respect to time d/dt, these state variables themselves are derivatives with respect to the polar angle $\mathrm{d}/\mathrm{d}\theta$. According to the second equation of (1), the spacecraft's orbital motion model can be reestablished as

$$\begin{aligned} r' &= \frac{ur}{v}, \\ u' &= -\frac{\mu}{(vr)} + v + \tilde{a}\sin\alpha, \\ v' &= -u + \tilde{a}\cos\alpha, \end{aligned} \tag{3}$$

where suffix "$'$" indicates a derivative with respect to polar angle θ.

3. The Optimized Shape-Based Method Using Fourier Series

3.1. Performance Index and Boundary Conditions. Usually the performance index for the low-thrust trajectory design is set to minimize the flight time or fuel consumption. In this paper, we consider the minimal characteristic velocity (i.e., minimal fuel consumption), as shown in

$$J = \min \Delta V = \int_{t_i}^{t_f} a \,\mathrm{d}t = \int_{\theta_i}^{\theta_f} \tilde{a} \,\mathrm{d}\theta. \tag{4}$$

To accomplish a spacecraft's transfer successfully, some constraints such as those given in (5) should be satisfied.

$$\begin{aligned} P(t_i) &= P_i, \\ V(t_i) &= V_i, \\ A(t_i) &= A_i, \\ P(t_f) &= P_f, \\ V(t_f) &= V_f, \\ A(t_f) &= A_f, \end{aligned} \tag{5}$$

where (P, V, A) represent the spacecraft's position, velocity, and acceleration conditions, respectively; (i, f) represent initial and terminal condition, respectively.

Actually, the fact that (P, V, A) mentioned in (5) are regarded as generalized boundary conditions, which may be the first-order or second-order derivative of a certain variable, is more reasonable.

3.2. The Flight Time-Fixed Case

3.2.1. Traditional Fourier Series (TFS) Method. In the time-fixed case, it is assumed that the thrust is aligned along or against the velocity vector; that is, $\alpha = \xi + n\pi$, where $n = 0, 1$. The TFS method for the time-fixed case was studied in detail in Taheri's doctoral dissertation [18]. Besides, we only consider the unconstrained version of the finite Fourier series method in this paper; because the thrust magnitude constraint is not a key point, we focus on optimizing the FS method here.

The following equation is derived from the fourth equation of (1):

$$a = \frac{r\dot{v} + uv}{r\cos\alpha}. \tag{6}$$

Substituting (6) into the third equation of (1), the following equation is derived:

$$\dot{u} + \frac{\mu}{r^2} - \frac{v^2}{r} = \frac{r\dot{v} + uv}{r}\tan\alpha, \tag{7}$$

where the tangential thrust assumption can be written as

$$\tan\alpha = \tan\xi = \frac{u}{v} = \frac{\dot{r}}{r\dot{\theta}}. \tag{8}$$

Substituting the first and second equations of (1) and the tangential thrust assumption into (7), one can be rewritten as

$$r^2\left(\ddot{r}\dot{\theta} - \dot{r}\ddot{\theta}\right) + \dot{\theta}\left(\mu - 2r\dot{r}^2\right) - \left(r\dot{\theta}\right)^3 = 0. \tag{9}$$

According to Fourier series expansion, the radius r and the polar angle θ can be approximated as follows:

$$x = \frac{a_{x0}}{2} + \sum_{i=1}^{n_x}\left\{a_{xi}\cos\left(\frac{i\pi t}{T}\right) + b_{xi}\sin\left(\frac{i\pi t}{T}\right)\right\}, \tag{10}$$

where x means (r, θ); n_x is the number of finite Fourier terms; (a_{x0}, a_{xi}, b_{xi}) are Fourier coefficients; T is the total flight time.

Substituting the state approximation (10) into (9), the differential equation is converted to a nonlinear algebraic equation, in which the only unknowns are the Fourier coefficients and the independent time variable:

$$F(a_{x0}, a_{xi}, b_{xi}; t) = 0. \tag{11}$$

This shows that, in the flight time-fixed case that considers the tangential thrust direction hypothesis, (4), (5), and (11) become an NLP problem about Fourier coefficients.

3.2.2. Optimized Fourier Series (OFS) Method. In light of (1) and (4), the Hamilton function is written as

$$\begin{aligned} H = a + \lambda_r u + \lambda_\theta \frac{v}{r} + \lambda_u\left(-\frac{\mu}{r^2} + \frac{v^2}{r} + a\sin\alpha\right) \\ + \lambda_v\left(-\frac{uv}{r} + a\cos\alpha\right). \end{aligned} \tag{12}$$

Based on the optimal control theory, costate equations and control equations are shown as

$$\begin{aligned}
\dot{\lambda}_r &= -\frac{\partial H}{\partial r} = \lambda_\theta \frac{v}{r^2} - \lambda_u \frac{2\mu}{r^3} + \lambda_u \frac{v^2}{r^2} - \lambda_v \frac{uv}{r^2}, \\
\dot{\lambda}_\theta &= -\frac{\partial H}{\partial \theta} = 0, \\
\dot{\lambda}_u &= -\frac{\partial H}{\partial u} = -\lambda_r + \lambda_v \frac{v}{r}, \\
\dot{\lambda}_v &= -\frac{\partial H}{\partial v} = -\lambda_\theta \frac{1}{r} - \lambda_u \frac{2v}{r} + \lambda_v \frac{u}{r},
\end{aligned} \tag{13}$$

$$\begin{aligned}
\frac{\partial H}{\partial a} &= 1 + \lambda_u \sin\alpha + \lambda_v \cos\alpha = 0, \\
\frac{\partial H}{\partial \alpha} &= \lambda_u a \cos\alpha - \lambda_v a \sin\alpha = 0.
\end{aligned} \tag{14}$$

Let $X = \dot{u} + \mu/r^2 - v^2/r$ and $Y = \dot{v} + uv/r$; the third and fourth equations of (1) are rewritten as

$$\begin{aligned}
X &= a \sin\alpha, \\
Y &= a \cos\alpha.
\end{aligned} \tag{15}$$

With (15), the spacecraft's thrust acceleration magnitude is solved.

$$a = \sqrt{X^2 + Y^2}. \tag{16}$$

Meanwhile, the thrust acceleration direction is defined by the results of $\sin\alpha$ and $\cos\alpha$.

The following equation is derived from (14) and (15):

$$\begin{aligned}
\lambda_u X + \lambda_v Y &= -a, \\
\lambda_u Y - \lambda_v X &= 0.
\end{aligned} \tag{17}$$

The expressions of the costate variables (λ_u, λ_v) are obtained with the solution of (17):

$$\begin{aligned}
\lambda_u &= \frac{-aX}{X^2 + Y^2}, \\
\lambda_v &= \frac{-aY}{X^2 + Y^2}.
\end{aligned} \tag{18}$$

According to the second and fourth equations of (13), the value of constant λ_θ can be represented as

$$\lambda_\theta = \left(\lambda_v u - 2\lambda_u v - \dot{\lambda}_v r\right)\Big|_i . \tag{19}$$

The following equation is rearranged from the first equation of (13):

$$r^3 \dot{\lambda}_r + rv\left(\lambda_v u - \lambda_u v - \lambda_\theta\right) + 2\mu\lambda_u = 0, \tag{20}$$

where the costate variable λ_r is solved with the third equation of (13)

$$\lambda_r = \lambda_v \frac{v}{r} - \dot{\lambda}_u. \tag{21}$$

Therefore, all the costate variables $(\lambda_r, \lambda_\theta, \lambda_u, \lambda_v)$ can be expressed as the functions of state variables (r, θ).

Substituting the state approximations of (10) into (20), the differential equation is converted to a nonlinear algebraic equation, in which the only unknowns are the Fourier coefficients and the independent time variable:

$$F_{\text{opt}}\left(a_{x0}, a_{xi}, b_{xi}; t\right) = 0. \tag{22}$$

This shows that, in the flight time-fixed with no limitation on thrust direction, (4), (5) and (22) become an NLP problem about Fourier coefficients.

3.3. *The Flight Time-Free Case*

3.3.1. TFS Method. In the flight time-free case, it is assumed that the thrust is aligned along or against the velocity, the same as what was mentioned in Section 3.2.1. This method for the flight time-free case is a new study in this paper.

The following equation is derived from the second and third equations of (3):

$$\tan\alpha = \frac{u' - v + \mu/(vr)}{v' + u}, \tag{23}$$

where the tangential thrust assumption is expressed as

$$\tan\alpha = \tan\xi = \frac{u}{v} = \frac{r'}{r}. \tag{24}$$

Substituting the first equation of (3) and (24) into (23), the following equation is derived:

$$\left(u'v - uv'\right)r - \left(u^2 + v^2\right)r + \mu = 0. \tag{25}$$

Dividing (25) by v^2, the following equation is derived:

$$r\frac{\mathrm{d}}{\mathrm{d}\theta}\left(\frac{u}{v}\right) - r\left(\frac{u}{v}\right)^2 - r + \frac{\mu}{v^2} = 0. \tag{26}$$

Considering the thrust direction assumption, (26) is rewritten as

$$r\frac{\mathrm{d}}{\mathrm{d}\theta}\left(\frac{r'}{r}\right) - r\left(\frac{r'}{r}\right)^2 - r + \frac{\mu}{v^2} = 0. \tag{27}$$

Equation (27) can be simplified as

$$r''r - 2r'^2 - r^2 + \frac{\mu r}{v^2} = 0. \tag{28}$$

According to the Fourier series expansion, the radius r and the circumferential velocity magnitude v can be approximated as follows:

$$x = \frac{a_{x0}}{2} + \sum_{i=1}^{n_x}\left\{a_{xi}\cos\left(\frac{i\pi\theta}{\Theta}\right) + b_{xi}\sin\left(\frac{i\pi\theta}{\Theta}\right)\right\}, \tag{29}$$

where x means (r, v); $\Theta = \theta_f - \theta_i + 2N_{\text{rev}}\pi$; N_{rev} is the number of revolutions around the attracting central body, as shown in Figure 2.

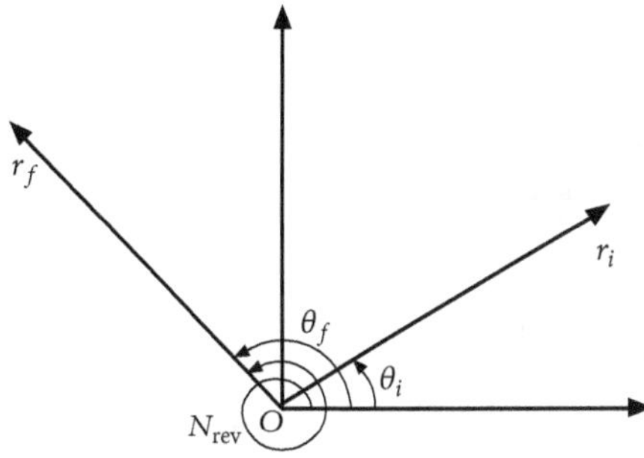

FIGURE 2: Initial and terminal position.

Substituting the state approximations of (29) into (28), the differential equation is converted to a nonlinear algebraic equation, in which the only unknowns are the Fourier coefficients and the polar angle:

$$G(a_{x0}, a_{xi}, b_{xi}; \theta) = 0. \quad (30)$$

This shows that, in the flight time-free case that considers the tangential thrust direction hypothesis, (4), (5), and (11) become an NLP problem about Fourier coefficients.

According to (3), $\tilde{a}$ is represented as

$$\tilde{a} = \left| \frac{v' + vr'/r}{\cos\alpha} \right|, \quad (31)$$

where $\cos\alpha = \pm\cos\xi$

$$\cos\alpha = \pm\frac{r\dot{\theta}}{\sqrt{\dot{r}^2 + r^2\dot{\theta}^2}} = \pm\frac{r}{\sqrt{r'^2 + r^2}}. \quad (32)$$

The spacecraft's flight time is solved with the following equation:

$$t = \int_{t_i}^{t_f} 1\,dt = \int_{\theta_i}^{\theta_f} \frac{r}{v}\,d\theta. \quad (33)$$

3.3.2. OFS Method. In light of (3) and (4), the Hamiltonian function is expressed as

$$H = \tilde{a} + \lambda_r\left(\frac{ur}{v}\right) + \lambda_u\left(-\frac{\mu}{vr} + v + \tilde{a}\sin\alpha\right) + \lambda_v(-u + \tilde{a}\cos\alpha). \quad (34)$$

Based on the optimal control theory, the costate equations and control equations are expressed as

$$\lambda_r' = -\frac{\partial H}{\partial r} = -\lambda_r\frac{u}{v} - \lambda_u\frac{\mu}{vr^2},$$
$$\lambda_u' = -\frac{\partial H}{\partial u} = -\lambda_r\frac{r}{v} + \lambda_v, \quad (35)$$
$$\lambda_v' = -\frac{\partial H}{\partial v} = \lambda_r\frac{ur}{v^2} - \lambda_u\frac{\mu}{v^2 r} - \lambda_u,$$

$$\frac{\partial H}{\partial \tilde{a}} = 1 + \lambda_u\sin\alpha + \lambda_v\cos\alpha = 0,$$
$$\frac{\partial H}{\partial \alpha} = \lambda_u\tilde{a}\cos\alpha - \lambda_v\tilde{a}\sin\alpha = 0. \quad (36)$$

With $u = vr'/r$, (35) and (3) can be rewritten as

$$\lambda_r' = -\lambda_r\frac{r'}{r} - \lambda_u\frac{\mu}{vr^2},$$
$$\lambda_u' = -\lambda_r\frac{r}{v} + \lambda_v, \quad (37)$$
$$\lambda_v' = \lambda_r\frac{r'}{v} - \lambda_u\frac{\mu}{v^2 r} - \lambda_u,$$

$$u' = -\frac{\mu}{vr} + v + \tilde{a}\sin\alpha,$$
$$v' = -\frac{vr'}{r} + \tilde{a}\cos\alpha. \quad (38)$$

Let $\widetilde{X} = u' + \mu/(vr) - v$ and $\widetilde{Y} = v' + vr'/r$; (38) is rewritten as

$$\widetilde{X} = \tilde{a}\sin\alpha,$$
$$\widetilde{Y} = \tilde{a}\cos\alpha. \quad (39)$$

The magnitude of $\tilde{\mathbf{a}}$ is solved with (39):

$$\tilde{a} = \sqrt{\widetilde{X}^2 + \widetilde{Y}^2}. \quad (40)$$

Meanwhile, the direction of $\tilde{\mathbf{a}}$ is defined by the results of $\sin\alpha$ and $\cos\alpha$.

The following equations are derived from (36) and (39):

$$\lambda_u\widetilde{X} + \lambda_v\widetilde{Y} = -\tilde{a},$$
$$\lambda_u\widetilde{Y} - \lambda_v\widetilde{X} = 0. \quad (41)$$

The expressions of the costate variables (λ_u, λ_v) are obtained with the solution of (41):

$$\lambda_u = \frac{-\tilde{a}\widetilde{X}}{\widetilde{X}^2 + \widetilde{Y}^2},$$
$$\lambda_v = \frac{-\tilde{a}\widetilde{Y}}{\widetilde{X}^2 + \widetilde{Y}^2}. \quad (42)$$

The following equation is derived from the first and third equations of (37):

$$\lambda_r' r + \lambda_v' v + 2\lambda_u\frac{\mu}{vr} + \lambda_u v = 0, \quad (43)$$

where the costate variable λ_r is solved with the second equation of (37) and (42):

$$\lambda_r = (\lambda_v - \lambda_u')\frac{v}{r}. \quad (44)$$

So all the costate variables $(\lambda_r, \lambda_u, \lambda_v)$ can be expressed as functions of state variables (r, v).

Substituting the state approximation of (29) into (43), the differential equation is converted to a nonlinear algebraic equation, in which the only unknowns are the Fourier coefficients and the polar angle:

$$G_{\text{opt}}(a_{x0}, a_{xi}, b_{xi}; \theta) = 0. \quad (45)$$

This shows that, in the flight time-free case with no limitation on thrust direction, (4), (5), and (45) become an NLP problem about Fourier coefficients.

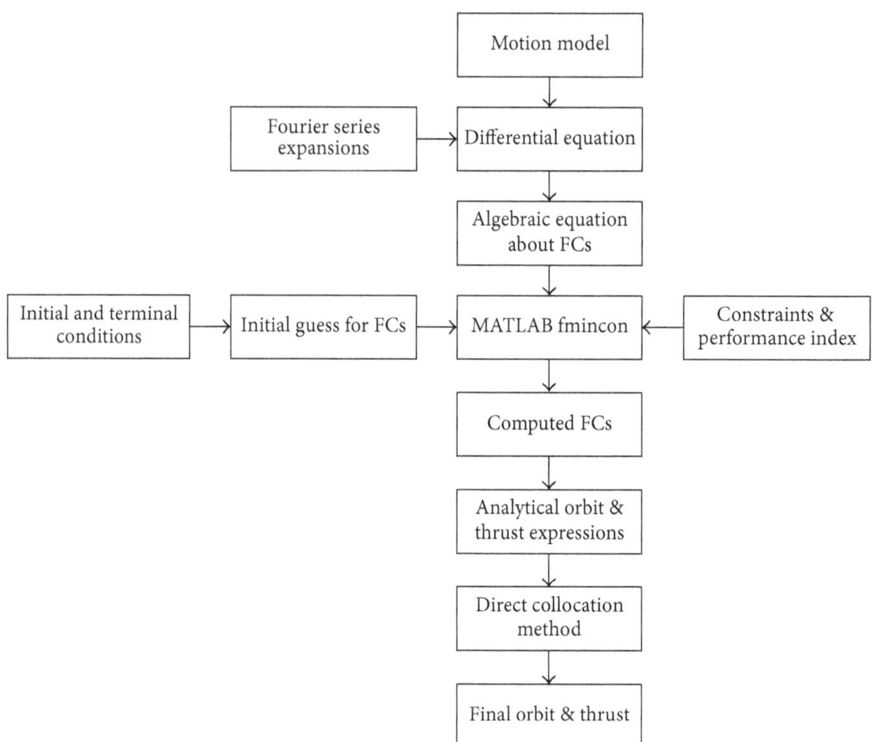

FIGURE 3: Flowchart of trajectory design using FS method.

4. Transfer Orbit Design with the FS Method

The flowchart of the spacecraft's transfer trajectory design with the FS method is shown in Figure 3, in which FCs is short for "Fourier coefficients."

The *fmincon* function is a MATLAB command used to solve multivariable, nonlinear and optimal problem with constraints. However, it requires that a set of initial guesses for the Fourier coefficients should be obtained. Reference [14] proposed some rough approaches used to gain the initial guesses. Using the FS method for designing an interplanetary low-thrust trajectory, a cubic polynomial function is used to obtain initial guesses for the Fourier coefficients. The constraint about finite Fourier series terms $n_x \geq 2$ is set to satisfy the boundary conditions. Although there is no upper limit on the number of included Fourier terms, the computational efficiency and precision are important considerations.

Based on the initial guesses, the new Fourier coefficients are calculated with the *fmincon* function. Then, the analytical trajectory and the thrust expressed with the finite Fourier coefficients are found. These expressions are used to offer initial guesses for detailed trajectory optimizers. After the preliminary design phase, trajectory optimization is required to show the advantage of the efficient initial guess from OFS method compared with the guess from TFS.

In this paper, we use the direct collocation method [19]. In the direct collocation method, the optimization model, performance index, and constraints are the same as those in other SB methods, for example, (1) or (3), (4), and (5). The optimal control problem would be converted into an NLP problem. The total flight interval is discretized into N intervals, and two endpoints of each interval are called "node." In this paper, the selection of the number of nodes is not a key point. Certainly, the simulation results will be more accurate as the number of nodes increases.

5. Simulation Examples

In the time-fixed and time-free cases, the continuous low-thrust Earth-Mars rendezvous and LEO-GEO transfer are studied, respectively. The simulation of these cases has been performed with MATLAB 2014a on an Intel Core i5 2.6 GHz computer with Windows 8.

5.1. The Earth-Mars Rendezvous. The FS method for the time-fixed case can be applied to the interplanetary exploration, asteroid deflection, rendezvous and docking missions, and so on. In the time-fixed case, we design the continuous low-thrust Earth-Mars rendezvous trajectory by using canonical units, where 1 distance unit (DU) is 1 AU and 2π time unit (TU) is 1 year. The boundary conditions and input parameters are listed in Table 1, in which N_{nod} is the number of nodes in the direct collocation method.

Figure 4 gives the spacecraft's rendezvous orbits using different FS methods. Figure 5 shows the angle relation between velocity direction and acceleration direction using the OFS method.

Different from other SB methods, which typically suppose that the acceleration direction of a spacecraft is parallel

TABLE 1: Boundary conditions and input parameters for Earth-Mars rendezvous orbit.

Boundary conditions		Input parameters	
r_i	1 DU	N_{rev}	1
θ_i	0 rad	n_r	6
r_f	1.5234 DU	n_θ	6
θ_f	3.548 rad	T	13.45 TU
$\dot{r}_i$	0 DU/TU	N_{nod}	151
$\dot{\theta}_i$	1 rad/TU		
$\dot{r}_f$	0 DU/TU		
$\dot{\theta}_f$	0.5318 rad/TU		

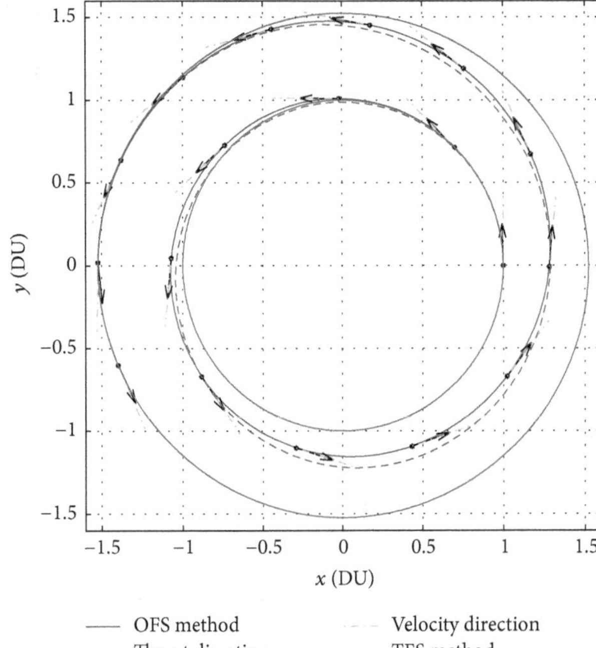

FIGURE 4: The Earth-Mars rendezvous orbits using different FS methods.

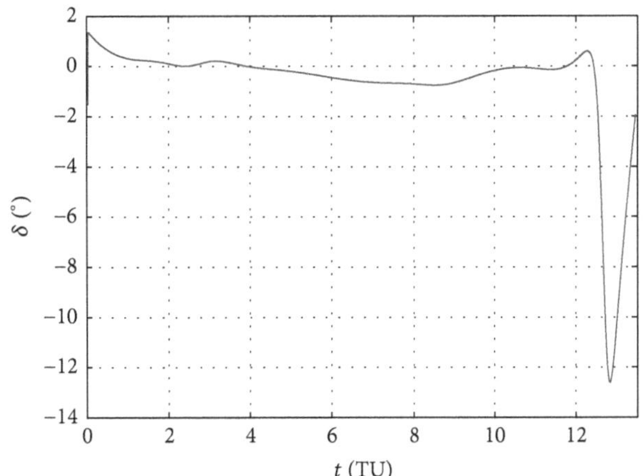

FIGURE 5: The difference δ with the OFS method for the Earth-Mars case.

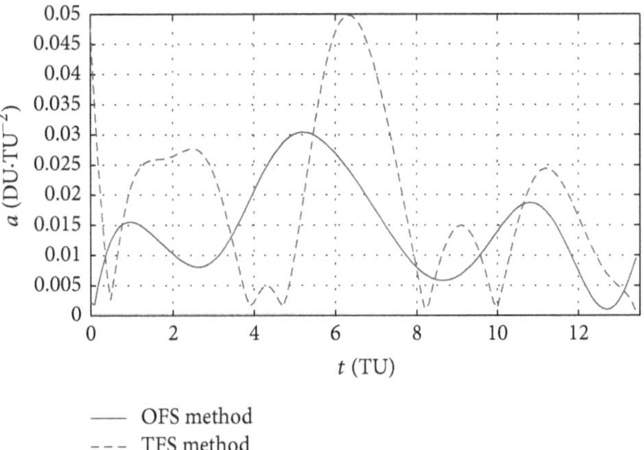

FIGURE 6: The Earth-Mars spacecraft's thrust acceleration profiles using different FS methods.

with its velocity direction, the OFS method does not need to constrain the relation between acceleration direction and velocity direction as shown in Figures 4 and 5.

In Figure 5, δ means the difference between steering angle and flight path angle; for example, $\delta = \alpha - \xi$. The figure shows that the difference fluctuates within the range of -2° to 2° during the large part of flight time, and the maximum difference is -12.59°. The acceleration direction varies sharply near the initial and terminal points, because the initial and terminal conditions result in the obvious change of the two parameters X and Y.

Figure 6 gives the spacecraft's thrust acceleration profiles using different FS methods. In this case, we can observe that the thrust acceleration profile using the OPS method is smoother than the thrust acceleration profile using the TFS method.

Regarding the thrust direction, the proposed method does not assume it to be tangential. When the obtained solution is used as initial guess in an optimizer, Figure 7 shows the thrust direction after optimization in the Earth-Mars case. The figure shows that the difference fluctuates within the range of -4° to 4° during the large part of flight time, and the maximum difference is -15°. Similar to the preliminary phase, the thrust direction after optimization varies sharply near the initial and terminal points. It justify that it is significant to relax the thrust direction or constrain it.

Table 2 gives simulation results using different initial guesses. It shows evidently that the OFS method obtains a lower cost rendezvous trajectory, and this verifies the applicability of our OFS method to continuous low-thrust trajectory design. Whether during the preliminary design phase or the precise design phase, the transfer trajectory designed with the OFS method consumes minimum fuel, but the optimization time of the OFS method is a little longer.

5.2. The LEO-GEO Transfer. The FS method for the time-free case can also be applied for a spacecraft orbit raise and maneuver mission. For the time-free case, the LEO-GEO transfer mission is considered, where 1 DU is 1 R_{earth} and

TABLE 2: Simulation results using different initial guesses in the Earth-Mars case.

	ΔV before optimization	ΔV after optimization	Optimization time
OFS method	0.1896 DU/TU	0.18787 DU/TU	12.029 s
TFS method	0.2514 DU/TU	0.18908 DU/TU	10.943 s
Hohmann		0.1877 DU/TU	

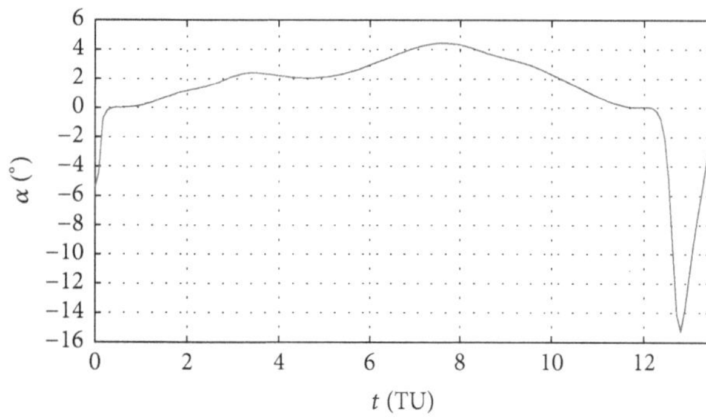

FIGURE 7: The thrust direction after optimization in the Earth-Mars case.

TABLE 3: Boundary conditions and input parameters for LEO-GEO transfer.

Boundary conditions		Input parameters	
r_i	1.0313 DU	N_{rev}	2
θ_i	0 rad	n_r	5
r_f	6.61 DU	n_θ	5
θ_f	3.1416 rad	Θ	15.708 rad
$\dot{r}_i$	0 DU/TU	N_{nod}	251
$\dot{\theta}_i$	0.9565 rad/TU		
$\dot{r}_f$	0 DU/TU		
$\dot{\theta}_f$	0.05884 rad/TU		

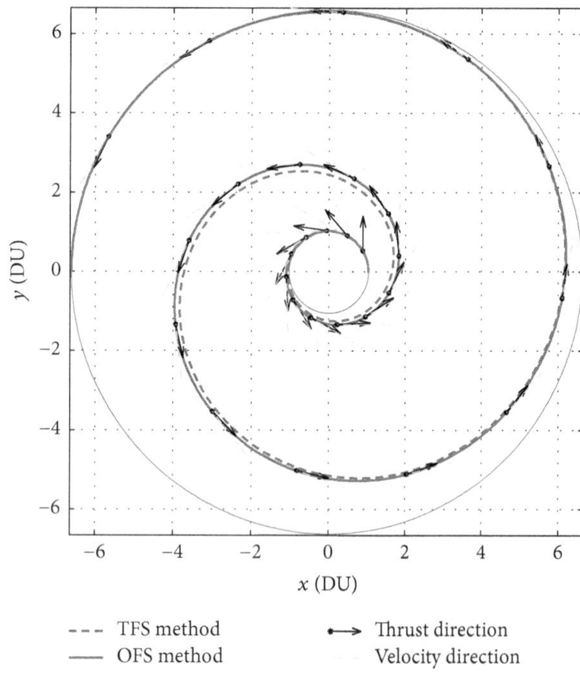

FIGURE 8: The LEO-GEO transfer orbit using different FS methods.

1 TU is 808.67 s. In this case, the flight time is unknown to us previously, but the terminal position is given. Table 3 lists the boundary conditions and input parameters.

In the time-free case, (r, v) and their first-order derivatives with respect to polar angle θ shall be calculated as new boundary conditions, as shown in

$$\begin{aligned} r &= r, \\ r' &= \frac{\dot{r}}{\dot{\theta}}, \\ v &= r\dot{\theta}, \\ v' &= -\dot{r}. \end{aligned} \tag{46}$$

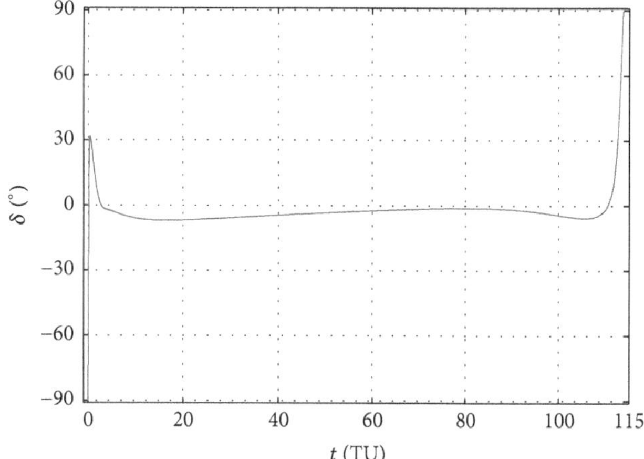

FIGURE 9: The difference δ with the OFS method for the LEO-GEO case.

Figure 8 gives the spacecraft's LEO-GEO transfer orbits using different FS methods. Figure 9 shows the angle relation between velocity direction and acceleration direction using the OFS method. Figure 10 gives the spacecraft's thrust acceleration profiles using different FS methods. Figure 11 shows the thrust direction after optimization in the LEO-GEO case.

Figures 8 and 10 prove that the shape-based trajectory design method using the finite Fourier series can be used in the flight time-free case. This paper extends the scope of

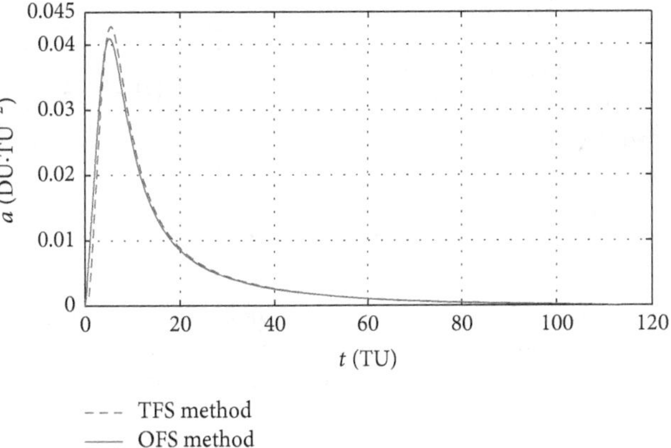

FIGURE 10: The LEO-GEO spacecraft's thrust acceleration profiles using different FS methods.

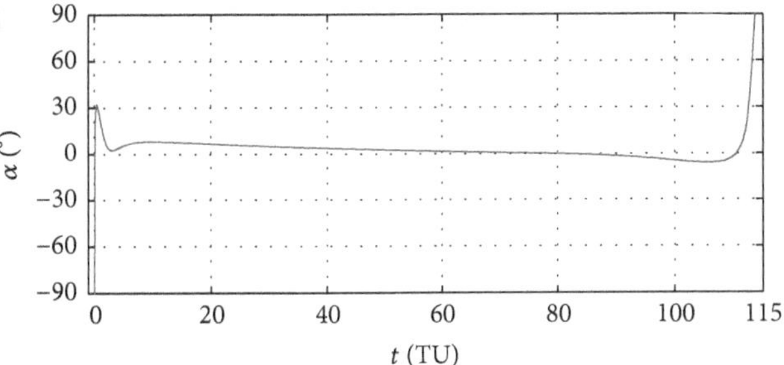

FIGURE 11: The thrust direction after optimization in the LEO-GEO case.

application of the Fourier series method to the flight time-free case. Figure 10 shows that, in the flight time-free case, different FS methods have the similar trend: during the initial phase of the flight, the thrust acceleration varies obviously and then tends to be stable.

Figures 8 and 9 also show that, similar to the Earth-Mars transfer in the flight time-fixed case, the OFS method does not need to constrain the angle relation between acceleration direction and velocity direction. As shown in Figure 9, the phenomena that the acceleration direction varies sharply near the initial and terminal points also exist, because the parameters ($\widetilde{X} = u' + \mu/(vr) - v$, $\widetilde{Y} = v' + vr'/r$) are ($\widetilde{X}_i = -0.0035$, $\widetilde{Y}_i = 0$) and ($\widetilde{X}_f = 4.5 \times 10^{-5}$, $\widetilde{Y}_f = 0$) at the initial point and terminal point, respectively, which result in $\alpha_i = -90^\circ$ and $\alpha_f = 90^\circ$. As shown in Figures 5 and 9, it is obvious to find that the phenomena where the thrust direction is not tangential near the initial and terminal points are determined by the OFS method itself, because of X and Y. As shown in Figure 11, some similar results, which are shown in Earth-Mars rendezvous case, can be found.

Table 4 presents simulation results using different initial guesses. As shown in Table 4, with the OFS method, the spacecraft has the longest flight time, and with the TFS method, its flight time is almost the same as that with the OFS method because the rough approach mentioned in Section 4, which can offer initial Fourier coefficients, leads to the approximate trajectory of the spacecraft. With the rough approach, the flight time is 112.2044 TU, which is almost the same as that with the OFS method and TFS method. Therefore, it suggests that the rough approach, which offers initial coefficients for the FS method, plays an important role.

The same as in the time-fixed case, the OFS method in the time-free case can provide better initial guesses for optimizers, and the transfer trajectory designed with the OFS method consumes minimal fuel, which is 0.5880 DU/TU before optimization and 0.5803 DU/TU after optimization. For all FS methods, the flight time after optimization becomes slightly longer. The increment with the OFS method is 0.0811 TU and TFS method is 0.2371 TU.

With regard to the optimization time, we obtain the similar simulation results that the flight time under the OFS method is a little longer than that under the TFS method, as shown in Tables 2 and 4. From the derivation in Section 3, we know that a more complicated NLP problem of the OFS method needs to be solved than the NLP problem of TFS method.

5.3. Remark. In this subsection, two important points are given to illustrate the OFS method.

With regard to the number of terms in the Fourier expansion, theoretically, the finite Fourier series expression will be very close to the real function as the number of Fourier series terms increases according to the characteristic of series theory. So the OFS method can get analytical trajectory and thrust which is very close to the optimal solution, if the number of series terms is enough. Actually, in this case, OFS looks more like a kind of hybrid optimization method, which combines direct optimization method and indirect optimization method. However, excessive series terms in OFS will lead to very time-consuming calculation. Thus, the computational efficiency is important considerations.

With regard to the sensitivity of the number of terms, we obtain a result that the selection of the number will influence the simulation obviously. Sometimes, we even cannot get the convergence solution. However, the shape-based method is proposed to get an approximate trajectory and provide an efficient initial guess for trajectory optimizers. So we only need to find a set of values which can satisfy the constraints.

6. Conclusion

A spacecraft's initial trajectory design problem is studied in this paper. On the basis of the traditional Fourier series method, this paper proposes the optimized shape-based method using the finite Fourier series.

According to these simulation examples, we can get the following conclusions: Firstly, considering the first-order optimal necessary conditions, the optimized shape-based method can design the trajectory with minimal fuel consumption and offer initial guesses for detailed optimizers; secondly, different from other shape-based methods, the thrust direction of the spacecraft is not constrained to be tangential in the optimized shape-based method; thirdly, the shape-based method using finite Fourier series could be applied in the time-free case.

TABLE 4: Simulation results using different initial guesses in the LEO-GEO case.

	Before optimization		After optimization		Optimization time
	ΔV	T	ΔV	T	
OFS method	0.5880 DU/TU	113.4903 TU	0.5803 DU/TU	113.5714 TU	44.846 s
TFS method	0.5900 DU/TU	111.1915 TU	0.5812 DU/TU	111.4286 TU	37.880 s
Hohmann			0.4974 DU/TU		

Conflicts of Interest

The authors declare that there are no conflicts of interest regarding the publication of this paper.

Acknowledgments

This research is supported by the National Natural Science Foundation of China (Grant no. 11272255).

References

[1] J. T. Betts, "Survey of numerical methods for trajectory optimization," *Journal of Guidance, Control, and Dynamics*, vol. 21, no. 2, pp. 193–207, 1998.

[2] R. J. McKay, M. MacDonald, J. Biggs, and C. McInnes, "Survey of highly-non-Keplerian orbits with low-thrust propulsion," *Journal of Guidance, Control, and Dynamics*, vol. 34, no. 3, pp. 645–666, 2011.

[3] S. Li, Y. Zhu, and Y. Wang, "Rapid design and optimization of low-thrust rendezvous/interception trajectory for asteroid deflection missions," *Advances in Space Research*, vol. 53, no. 4, pp. 696–707, 2014.

[4] A. E. Petropoulos, *Shape-based algorithm for the automated design of low-thrust, gravity assist trajectories [Ph.D. thesis]*, Purdue University, West Lafayette, Ind, USA, 2001.

[5] A. E. Petropoulos and J. M. Longuski, "Shape-based algorithm for automated design of low-thrust, gravity-assist trajectories," *Journal of Spacecraft and Rockets*, vol. 41, no. 5, pp. 787–796, 2004.

[6] D. Izzo, "Lambert's problem for exponential sinusoids," *Journal of Guidance, Control, and Dynamics*, vol. 29, no. 5, pp. 1242–1245, 2006.

[7] P.-Y. Cui, H.-B. Shang, and E.-J. Luan, "A fast search algorithm for launch window of interplanetary low-thrust exploration mission," *Journal of Astronautics*, vol. 29, no. 1, pp. 40–45, 2008 (Chinese).

[8] L.-L. Zheng, J.-P. Yuan, and Z.-X. Zhu, "Logarithmic spiral-based non-Keplerian orbit design," *Journal of Astronautics*, vol. 31, no. 9, pp. 2075–2081, 2010 (Chinese).

[9] B. J. Wall, "Shape-based approximation method for low-thrust trajectory optimization," in *Proceedings of the AIAA/AAS Astrodynamics Specialist Conference and Exhibit*, Honolulu, Hawaii, USA, August 2008.

[10] B. J. Wall and B. A. Conway, "Shape-based approach to low-thrust rendezvous trajectory design," *Journal of Guidance, Control, and Dynamics*, vol. 32, no. 1, pp. 95–102, 2009.

[11] H. Shang, P. Cui, D. Qiao, and R. Xu, "Lambert solution and application for interplanetary low-thrust trajectories," *Acta Aeronautica et Astronautica Sinica*, vol. 31, no. 9, pp. 1752–1757, 2010 (Chinese).

[12] D. Wang, G. Zhang, and X. Cao, "Modified inverse-polynomial shaping approach with thrust and radius constraints," *Proceedings of the Institution of Mechanical Engineers, Part G: Journal of Aerospace Engineering*, vol. 229, no. 13, pp. 2506–2518, 2015.

[13] C. Xie, G. Zhang, and Y. Zhang, "Simple shaping approximation for low-thrust trajectories between coplanar elliptical orbits," *Journal of Guidance, Control, and Dynamics*, vol. 38, no. 12, pp. 2448–2455, 2015.

[14] E. Taheri and O. Abdelkhalik, "Shape-based approximation of constrained low-thrust space trajectories using Fourier series," *Journal of Spacecraft and Rockets*, vol. 49, no. 3, pp. 535–545, 2012.

[15] O. Abdelkhalik and E. Taheri, "Approximate on-off low-thrust space trajectories using Fourier series," *Journal of Spacecraft and Rockets*, vol. 49, no. 5, pp. 962–965, 2012.

[16] D. J. Gondelach, R. Noomen, and D. Geller, "Hodographic-shaping method for low-thrust interplanetary trajectory design," *Journal of Spacecraft and Rockets*, vol. 52, no. 3, pp. 728–738, 2015.

[17] D. M. Novak and M. Vasile, "Improved shaping approach to the preliminary design of low-thrust trajectories," *Journal of Guidance, Control, and Dynamics*, vol. 34, no. 1, pp. 128–147, 2011.

[18] E. Taheri, *Rapid space trajectory generation using a Fourier series shape-based approach [Ph.D. thesis]*, Michigan Technological University, 2014.

[19] A. L. Herman and B. A. Conway, "Direct optimization using collocation based on high-order Gauss-Lobatto quadrature rules," *Journal of Guidance, Control, and Dynamics*, vol. 19, no. 3, pp. 592–599, 1996.

Design and Ground Verification of Space Station Manipulator Control Method for Orbital Replacement Unit Changeout

Bingshan Hu[1,2] **Feng Chen,**[3] **Liangliang Han,**[3] **Huanlong Chen,**[3] **and Hongliu Yu**[1,2]

[1]*University of Shanghai for Science and Technology, Shanghai 200093, China*
[2]*Shanghai Engineering Research Center for Assistive Devices, Shanghai 200093, China*
[3]*Shanghai Institute of Aerospace Systems Engineering, Shanghai 201109, China*

Correspondence should be addressed to Bingshan Hu; icebergh@126.com

Academic Editor: Paul Williams

Chinese space station has been in construction phase, and it will be launched around 2020. Lots of orbital replacement units (ORUs) are installed on the space station, and they need to be replaced on orbit by a manipulator. In view of above application requirements, the control method for ORU changeout is designed and verified in this paper. Based on the analysis of the ORU changeout task flow, requirements of space station manipulator's control algorithms are presented. The open loop path planning algorithm, close loop path planning algorithm based on visual feedback, and impedance control algorithm are researched. To verify the ORU changeout task flow and corresponding control algorithms, a ground experiment platform is designed, which includes a 6-DOF manipulator with a camera and a force/torque sensor, an end effector with clamp/release and screwing function, ORU module, and ORU store. At last, the task flow and control algorithms are verified on the test platform. Through the research, it is found that the ORU changeout task flow designed in this paper is reasonable and feasible, and the control method can be used to control a manipulator to complete the ORU changeout task.

1. Introduction

On orbit service is a kind of space operation, which can prolong spacecraft life and extend spacecraft task execution ability. The ORU changeout is one of the most important tasks in orbit maintenance. Human's operating ability and scope is greatly limited due to the special nature of the space environment. The space manipulator has the ability to work under an environment of microgravity, high temperature, and high radiation, so it is of great significance on economy and safety to use a space manipulator to assist or replace the astronaut to do ORU changeout mission.

In construction and service periods of the international space station, the Space Station Remote Manipulator System (SSRMS) and the Japanese Experiment Module Remote Manipulator System (JEMRMS) which are launched in the early stage are not precise enough [1], so the Special Purpose Dexterous Manipulator (SPDM) and Small Fine Arm (SFA) are launched in 2008 and 2009, respectively, and their end accuracy achieves 13 mm and 10 mm, so they can do some fine operation tasks such as small exposed load replacement, cutting, and refueling [2, 3]. When carrying out ORU replacement task, the SPDM's end effector gets close to the ORU waiting for replacing under the guidance of a hand-eye camera firstly, until the ORU's adaptor is in the end effector's capture range. Then, the SPDM's force-moment accommodation (FMA) control mode will limit the lateral force and torque in a selectable range [4].

The nature of the ORU replacement by the manipulator can be seen as pegging in hole, which is a typical problem of manipulator operation. For the ORU replacement task, Backes and Tso divide the task into several subtasks, and the control strategy of compliant control and gravity compensation is adopted for different subtasks [5]. Colombina et al. use a similar control method, and a ground test system is used to simulate the ORU replacement operation. Studies have found that the operating strategy has a smaller motion error, and the manipulator system has a tolerance as small as 5 mm/0.5° [6]. Ozaki et al. have established a ground test platform for microsatellite assembly/disassembly, and a

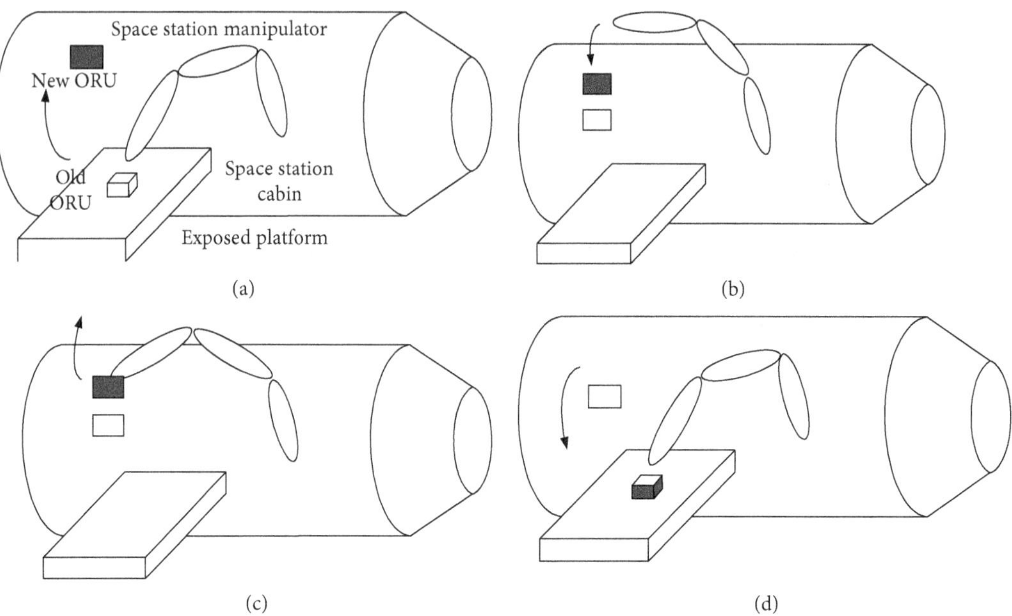

FIGURE 1: ORU replacement flow: (a) old ORU extracting, (b) old ORU inserting, (c) new ORU extracting, and (d) new ORU inserting.

heuristic search and assembly strategy is used in experiments [7]. Jiang et al. propose the MHIC (modified hybrid impedance control) strategy of the redundant robot for ORU replacements based on the ground test bed. Experimental results demonstrate that MHIC can reduce the contact force with uncertainties of constrained environments [8].

In order to ensure that the space robot implements space tasks successfully, ground experiments are required to verify the task planning and control algorithms. Because the space manipulator works in microgravity environment, the following five methods are often used to simulate the microgravity environment: air-bearing table, neutral buoyancy, airplane flying or free-falling motion, suspension system, and hardware-in-the-loop simulation system [9]. Each method has its advantages and disadvantages, but the hardware-in-the-loop simulation system is good for kinematic motion understanding and qualification of specific flight components or evaluation of integrated sensor-controller-actuator system performance [9]. The Stewart parallel robots are often used in the hardware-in-the-loop simulation system for the contact dynamics of space operation [10–12]. To verify the contact dynamics performance of SPDM performing various maintenance tasks, Dubowsky et al. developed a task verification facility. In that facility, a hydraulic robot is used to mimic the space robot performing ORU changeout, and a force sensor is installed at the base of the hydraulic manipulator to get contact force and torque [12].

Chinese space station has been in construction phase, and it will be launched into orbit around 2020. It is equipped with an exposure platform outside the cabin, and there are lots of exposed loads on the platform. During the operation phase, the exposed load should be replaced and upgraded periodically. Due to the high frequency and large quality of the ORU changeout mission, it is necessary to use a manipulator to assist [13]. Compared with the cargo transporting, astronauts assisting, and cabin inspection, ORU replacement has a higher accuracy requirement, so it is more difficult.

According to the demand of ORU replacement in Chinese space station, the second part of this paper analyzes the ORU changeout task flow and specific control algorithm requirements; the third section presents a ground verification platform to verify the task flow and control method; then corresponding algorithms are studied in Section 4; in the fifth section, the ORU replacement experiment based on the platform is introduced, and experimental results are discussed; conclusions and follow-up work are given at last.

2. ORU Changeout Mission Analysis

2.1. Mission Process Analysis. As shown in Figure 1, there are 4 stages in an ORU changeout task flow, which are old ORU extracting, old ORU inserting, new ORU extracting, and new ORU installing.

In the old ORU extracting stage, the space station manipulator grasps and pulls out an old ORU from the ORU store installed on the exposed platform. There are several subtasks. First, the manipulator gets close to the old ORU from an initial pose, and the manipulator stops above the ORU adaptor until the adaptor is in the vision range of the manipulator's hand-eye camera. Then, the manipulator approaches to the ORU adaptor in linear motion under the guidance of the camera, until the ORU adaptor begins to contact with the end effector. The manipulator continues to approach to the ORU adaptor under the guidance of guide blocks on the end effector and the arc surface on the adaptor until the end effector captures the adaptor completely. Guide blocks and the arc surface's geometry are specially designed, which ensures that the end effector can capture the adapter reliably in the presence of pose errors. After that, the end

TABLE 1: Subtasks and corresponding control algorithms of the first two stages.

	Subtask	Control algorithm
1	The manipulator gets close to the old ORU on exposed platform	Open loop path planning
2	The manipulator approaches to the ORU adaptor	Close loop path planning
3	The manipulator end effector captures the adaptor completely	Impedance control
4	The end effector locks the adaptor	—
5	The old ORU connection mechanism unlocks	—
6	The manipulator pulls out the old ORU from the ORU store on exposed platform	Impedance control
7	The manipulator with the captured old ORU moves to the top of the ORU store in the airlock cabin	Open loop path planning
8	The manipulator with the captured old ORU approaches to the ORU store in the airlock cabin	Close loop path planning
9	The manipulator inserts the old ORU	Impedance control
10	The old ORU connection mechanism locks	—
11	The end effector unlocks the adaptor of the old ORU	—
12	The manipulator withdraws to the top of the old ORU adapter	Impedance control
13	The manipulator backs to the initial pose	Open loop path planning

effector locks the old ORU and the old ORU's connection mechanism unlocks, and the electrical and mechanical connection between the ORU and the ORU store is cut off. The passive part of the connection mechanism is installed at the bottom of the ORU, and the active part is installed in the ORU store. Finally, the manipulator moves and pulls the old ORU from the ORU store installed on the exposed platform in a linear motion. The first stage named old ORU extracting is over.

In the old ORU inserting stage, the space station manipulator with the captured old ORU moves from the initial pose, which is the end pose of the first stage, to the top of the ORU store which is in the airlock cabin. Then, the manipulator approaches the ORU store under the guidance of the camera, until the ORU's edge begins to contact with the ORU store's guiding surface. Under the guidance of the specially designed guiding surface on ORU store, the manipulator inserts the old ORU into the ORU store. The active part of the connecting mechanism which is installed in the ORU store locks the passive part of the ORU connecting mechanism, and the old ORU is connected with the ORU store. At last, the end effector releases the old ORU adaptor. The second stage named old ORU inserting is over.

In the third stage, there is a new ORU extracted from the ORU store in the airlock cabin, and the manipulator's motion and control methods are the same with the first stage, except the pose of the new ORU. In the last stage, the manipulator with the captured new ORU moves from the airlock cabin to the exposed platform, and the new ORU is inserted into the ORU store on the exposed platform. The whole process is the same with the second stage.

2.2. Control Algorithm Requirements. From the above analysis, the space station manipulator has 3 working modes at each stage, and they are free motion mode, approaching mode, and contact mode individually.

In every stage, the manipulator always moves from an initial pose to the top of an ORU adaptor or the ORU store firstly, and these are called free motion modes. In this mode, the manipulator does not contact with environment and there is not a contact force. The manipulator just needs to move from an initial pose to the vicinity of the module to be replaced or the ORU store. Because the start pose and end pose are known before, an open loop path planning algorithm can be used.

The process of the manipulator's end effector approaching to the adaptor of the ORU or the manipulator with a captured ORU approaching to the ORU store is called approaching mode. At the end of this mode, the end effector must be in the adaptor's guidance scope, or the ORU's edge must be in the ORU store's guidance scope, which means that if the end effector approaches the adapter in a straight line in the presence of pose error, the effector can contact the adapter, or when the manipulator carries the ORU module to the ORU store along a linear direction in the presence of pose error, the ORU module can contact the guide surface on the ORU store. So, this mode requires a higher accuracy than the free motion mode. A close loop path planning based on visual feedback should be introduced to improve accuracy. A hand-eye camera must be installed to get the exact pose of the ORU or the ORU store.

Because an error is unavoidable when the end effector contacts with the adaptor, or the ORU edge contacts with the ORU store, some exterior force is exerted at the end of the manipulator, and these are called contact modes. To eliminate the pose error and insert the ORU into the ORU store slot precisely, a force control method must be integrated into the manipulator's control system. Also, it can prevent damages of the manipulator and ORUs. An impedance control method based on position is chosen in this paper. A 6-axis wrist force/torque sensor should be installed at the end of the manipulator between the last joint and the end effector to measure the contact force and torque.

Subtasks and corresponding control algorithms of the old ORU extracting stage and the old ORU inserting stage are illustrated in Table 1. The remaining two stages are the same.

3. Design of the Test Bed for Ground Verification

In order to verify the above task flow and corresponding control algorithms, a ground experiment platform is designed firstly as shown in Figure 2. The platform mainly consists of 6 degrees of freedom manipulator, an end effector, an adapter installed on a simulated ORU, two simulated ORU stores, robot vision system, and experiment bench. The geometry size of the real space station is big, and the space station manipulator works in a microgravity environment, but the experiment platform works under a gravity environment, so the experiment platform's size is also reduced. The simulated ORU only contains an adaptor, and there is no a connection mechanism in the platform. The real space light condition is also not simulated. Though there are some differences between the real condition and the experiment condition, these constraints do not affect verification of the ORU replacement task flow and control methods.

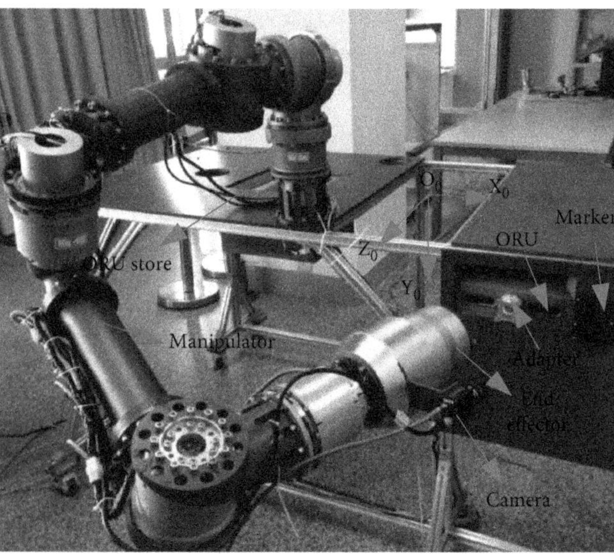

FIGURE 2: Ground experiment platform for ORU changeout.

3.1. The Manipulator. According to the analysis in Section 2, 6 degrees of freedom manipulator is enough for the ORU changeout task verification. In order to reduce the complexity of the system, this paper does not use any ground weightlessness simulator such as the air-bearing platform and the active force follow-up hanging system. All joints' output torque of the manipulator prototype in this paper is the same as the real space manipulator. In order to reduce the gravity load, joints of the manipulator are arranged in the form of shoulder yaw, shoulder roll, shoulder pitch, elbow pitch, wrist pitch, and wrist roll (Figure 3). To minimize the size of the manipulator envelope, the elbow joint adopts the bias type layout, and the other joints adopt the collinear layout. DH parameters of the manipulator are shown in Table 2. The total length of the manipulator is 1.9 m; the design value of absolute position accuracy is 5 mm/0.5 degrees, and the end load is 2 kg under the gravity environment.

To reduce weight and volume, the joint of the manipulator in this paper adopts a mechatronics integration design (Figure 4). In order to simplify the transmission mechanism, the joint uses a lightweight CSD series harmonic reducer with a large center hole. A permanent magnet synchronous motor which has a high rated torque and low rated speed is used. A speed sensor is integrated in the motor. In order to ensure the safety, there is a power off brake in the joint. A mechanical limit and an electrical safety limit are also contained in the joint. A dual-channel high-precision magnet resolver is integrated to measure the absolute position of the joint to ensure the accuracy, and the measurement accuracy of the resolver is up to $25''$. The weight of the joint is 4 kg, and the rated output torque is 150 Nm.

3.2. The Vision System. In order to measure the relative pose between the end of manipulator and the target, and to avoid collision with the surrounding objects in approaching mode at the same time, a camera is installed at the end of the manipulator. The vision system workflow is shown in Figure 5. Firstly, the pose relation between the end of the manipulator and the camera is calibrated. Secondly, images of the marker are acquired by the camera, and the image processing and pattern recognition algorithms are used to identify and extract the characteristic information of the marker. Then, the relative pose between the target marker and the camera is obtained by the relative 3D pose measurement technology. Finally, the relative pose is sent to the central controller as the input of the close loop path planning.

Referring to the marker on the exposed platform of the Japanese experimental module in space station [14], a stereo marker is presented in this paper to improve the measurement accuracy (Figure 6). On the stereo marker, four characteristic circles are painted in white color. Three of the four characteristic circles are in one plane, and the other one is on top of a rod, which is higher than the former three. The other place of the marker is painted in black. Geometric dimensions between characteristic circles are known before. In this paper, the geometric size of simulated ORU is $120 \times 20 \times 90$ mm. The geometric size of the marker is designed as 100×80 mm, and the characteristic circle diameter is 4 mm, considering requirements of the recognition distance, the measurement accuracy and the ORU's geometric size.

Because the ORU installed with a marker can be considered as a cooperative target, a monocular camera is enough. The experimental system is required to be able to monitor the area of $1\,\mathrm{m} \times 1\,\mathrm{m} \times 1\,\mathrm{m}$ when the distance between the camera and the target is 1.5 m, and the recognition accuracy is 2 mm/0.3 degrees when the distance between the camera and the marker is 200 mm. According to the above requirements, the camera GC1380H from Prosilica Company is chosen, and the lens is Computar's M1214-MP2 whose focal length is 12 mm.

3.3. The End Effector. In order to complete the ORU changeout task, the function of the end effector is to capture, lock and release. In addition, screwing screws is a necessary function additionally. According to the above functional requirements, the end effector in this paper is similar as

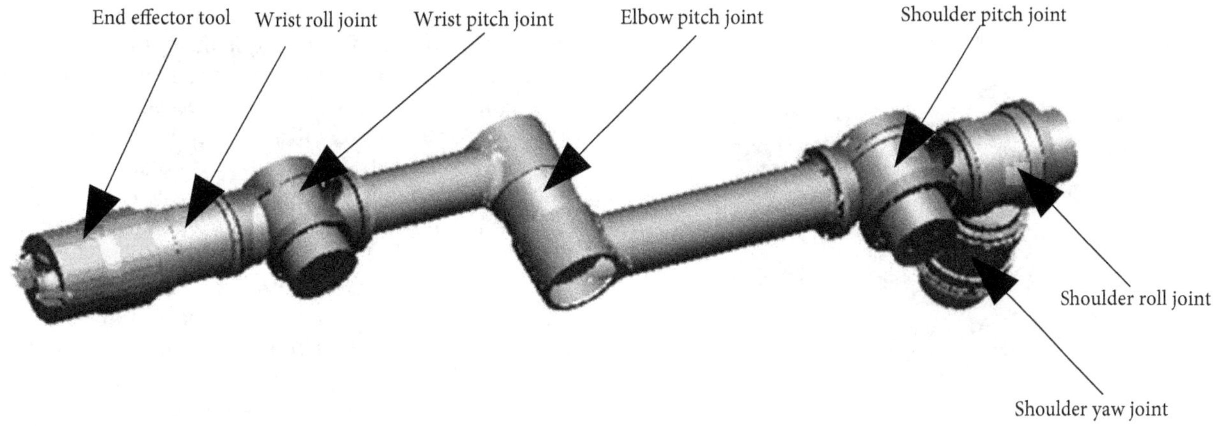

FIGURE 3: Configuration of the manipulator.

TABLE 2: DH parameters of the manipulator.

Link i	α_{i-1}	a_{i-1} (mm)	θ_i	d_i (mm)
1	0°	0	θ_1 (0°)	0
2	−90°	0	θ_2 (90°)	178.85
3	−90°	0	θ_3 (−90°)	0
4	0°	560	θ_4 (0°)	−174
5	0°	385	θ_5 (−90°)	0
6	−90°	0	θ_6 (0°)	0
7	0°	0	0	466.15

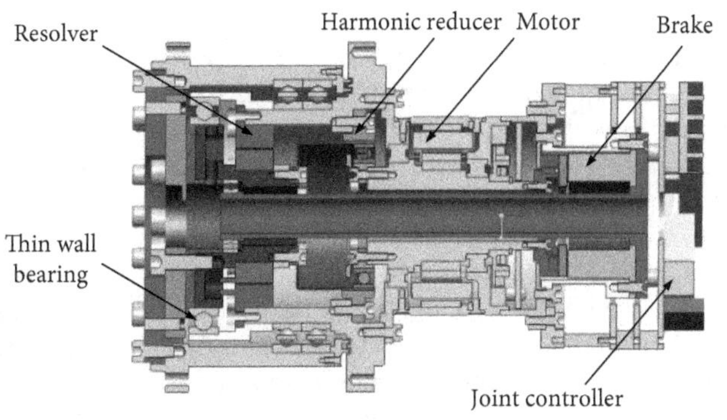

FIGURE 4: Cross section of the joint.

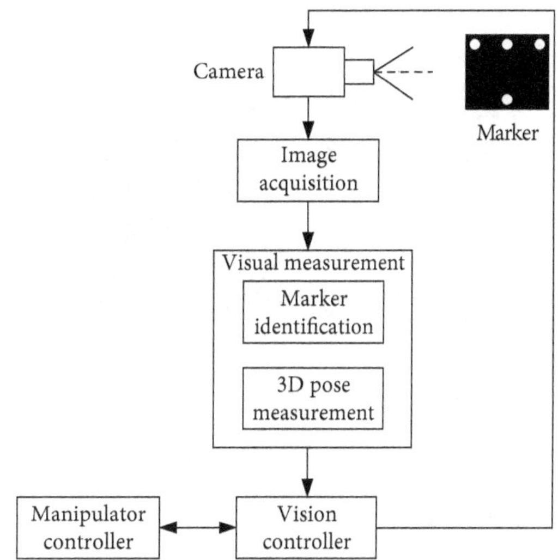

FIGURE 5: Vision system workflow of the manipulator.

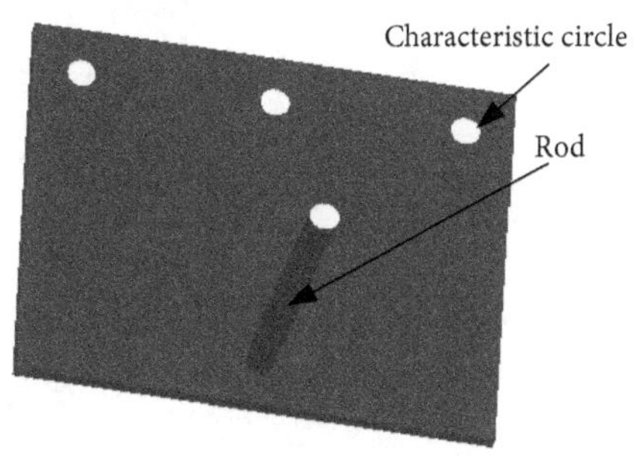

FIGURE 6: Marker of the vision system.

Nishida designed in 2005 [15]. The end effector has two degrees of freedom, which are used for clamp/release and screwing, respectively. The end effector mainly comprises a ball screw, a ball nut, torque sleeve components, clamp/release drive assembly, screwing drive assembly, two guide blocks, and clamp finger assembly (Figure 7). To measure the contact force and torque, a six-axis force sensor is installed in the end effector.

When there is a pose error, the end effector also should capture the ORU reliably, so a guide block is designed with an arc surface to obtain a tolerance larger than 10 mm as shown in Figure 8. The adaptor also has an arc surface (Figure 9). If the end effector approaches the adapter in a straight line, the guide block on the effector can contact the arc surface of the adapter when the position error is less than 10 mm. The pose bias is eliminated at last under the guidance of the guide block, and the contact force is controlled based on the feedback of the force sensor. When the end effector captures the target adapter completely as in Figure 7, the end effector begins to lock. The clamp/release drive assembly drives the ball screw to rotate, and the two clamp fingers lock the adaptor. When releasing, the clamp/release drive assembly rotates in the opposite direction.

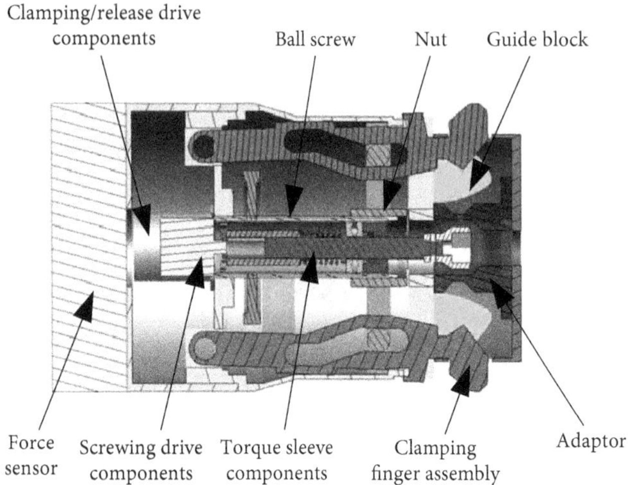

FIGURE 7: Cross section of the end effector.

FIGURE 8: The end effector prototype.

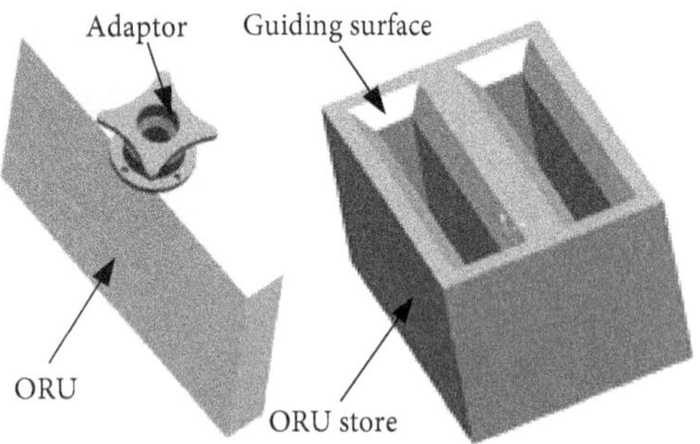

FIGURE 9: Simulated ORU and ORU store.

The opening of the ORU store also has a chamfered guide surface, and the chamfer size is 10 mm (Figure 9). The chamfer is used for adapting pose error when ORU contacts with ORU store.

3.4. Control Architecture. A distributed control system is designed to control the manipulator to complete the ORU replacement task, and the control system contains an operation console, a manipulator central controller, a visual controller, six joint controllers, an end effector controller, and a monocular camera (Figure 10).

A tel operation computer and a monitor computer are included in the operation console. A graphical user interface (GUI) is run on the tel operation computer, and the GUI can send control commands to the central controller, including single-joint control command and point-to-point movement command (Figure 11). The GUI also can display state messages of the manipulator system. A three-dimensional virtual scene which is same with the real operation scene can be shown in the GUI. The three-dimensional virtual scene is constructed by ProE and Open Inventor, and its role is to do presimulation to verify commands' correctness. The monitor computer displays the video signal which is collected by the camera at the end of the manipulator.

The central controller receives task level control commands in Cartesian space from the tel operation computer, force signals from the wrist force sensor, and pose data of the ORU from the visual controller. After running inverse kinematics, path planning, and force control algorithms, joints and the end effector's commands are sent to joints and the end effector's drive controller through communication network. Joints and the end effector's position and speed will be controlled.

Visual controller will collect the camera's images, and the vision recognition and measurement algorithm will be run to get the pose of the ORU relative to the camera waiting for operation. The pose data is sent to the central controller.

4. Design of Control Algorithms

4.1. Open Loop Path Planning. Because the space manipulator works on a floating base, the pose of the floating base (such as space station and satellite) will be changed during the manipulator's motion process, and this will affect the end pose of the manipulator. Especially when the target is moving relative to the floating base, the pose change of the floating base must be considered. In the space station ORU module replacement task, the ORU adapter, ORU store, and the manipulator base are fixed on the space station cabin. Assuming that the ORU adapter base coordinate system is Σ_1, the base coordinate system of the manipulator is Σ_2, and the space station base coordinate system is Σ_3. In the course of the ORU replacement, the relative pose between Σ_1 and Σ_3, Σ_2, and Σ_3 is not changed, so the relative pose of Σ_1 and Σ_2 is not changed, which means that when designing the space station manipulator path planning method for ORU replacement, there is no need to consider the manipulator movement's disturbance to the cabin.

In free motion mode, the manipulator just needs to move from an initial pose to the vicinity of the module to be replaced or the ORU store. The initial pose and target pose are known before because the space station manipulator works under a structured environment. It is defined that the manipulator's initial pose is $[x_{e0}, y_{e0}, z_{e0}, \alpha_{e0}, \beta_{e0}, \gamma_{e0}]$, and the target pose of free motion mode is $[x_{ef}, y_{ef}, z_{ef}, \alpha_{ef}, \beta_{ef}, \gamma_{ef}]$. To interpolate between the initial and target pose, the polynomial or spline methods are used, then the position P_e and posture ψ_e of manipulator at time t can be gotten

$$\begin{aligned} P_e &= P_e(t), \quad t_0 \le t \le t_f, \\ \psi_e &= \psi_e(t), \quad t_0 \le t \le t_f. \end{aligned} \tag{1}$$

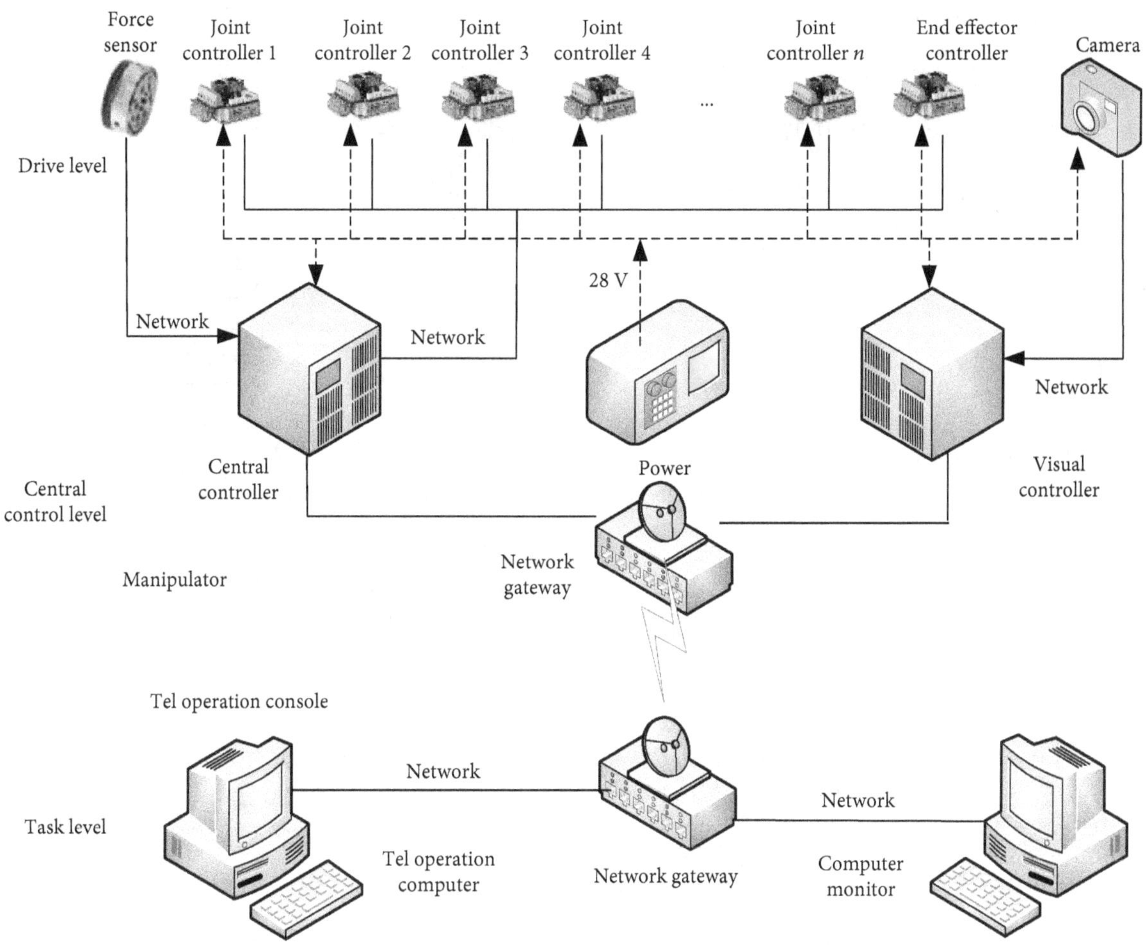

Force sensor
Joint controller 1
Joint controller 2
Joint controller 3
Joint controller 4
...
Joint controller n
End effector controller
Camera
Drive level
28 V
Network
Network
Network
Central controller
Power
Visual controller
Central control level
Manipulator
Network gateway
Tel operation console
Network
Network
Task level
Tel operation computer
Network gateway
Computer monitor

Figure 10: The distributed control system of the manipulator.

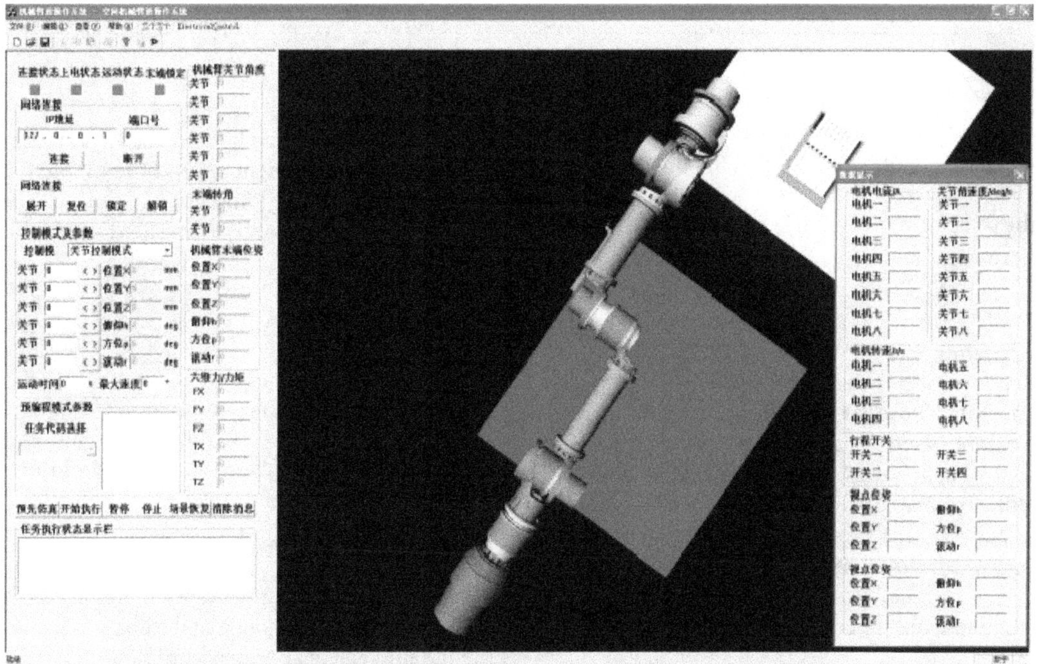

Figure 11: GUI of the tel operation computer.

The manipulator end velocity at time t is

$$\begin{bmatrix} v_e(t) \\ \omega_e(t) \end{bmatrix} = \begin{bmatrix} \dot{p}_e(t) \\ J_{\text{Euler\_}ZYX}(\psi_e(t))\dot{\psi}_e \end{bmatrix}. \quad (2)$$

In (2), v_e is the linear velocity and ω_e is the angular velocity; $J_{\text{Euler\_}ZYX}$ is a transition matrix to convert the Euler angular velocity to the end angular velocity. Every joint's velocity is calculated according to the inverse kinematics

$$\dot{\theta}(t) = J(\theta)\begin{bmatrix} v_e(t) \\ \omega_e(t) \end{bmatrix} = J(\theta)\begin{bmatrix} \dot{p}_e(t) \\ J_{\text{Euler\_}ZYX}(\psi_e(t))\dot{\psi}_e \end{bmatrix}. \quad (3)$$

In (3), $J(\theta)$ is the Jacobian matrix of the manipulator. If (3) is integrated, every joint's position $\theta(t)$ can be gotten

$$\theta(t + \Delta t) = \theta(t) + \dot{\theta}(t)\Delta t. \quad (4)$$

At last, all joints' velocity and position commands are sent to the joint controller.

4.2. Close Loop Path Planning Based on Visual Feedback. As illustrated in Section 2, a visual feedback is introduced when the manipulator is close to an ORU or to the ORU store to get a higher accuracy. The closed loop path planning algorithm flow is shown in Figure 12. First, according to the joint angles measured by the absolute position sensor, the homogeneous transformation matrix ${}^{\text{base}}T_{\text{end}}$ between the end and the mounting base of the manipulator is calculated by the manipulator's forward kinematics. The homogeneous transformation matrix ${}^{\text{end}}T_{\text{ORU}}$ between the end effector and the ORU's adapter or the ORU store is

$${}^{\text{end}}T_{\text{ORU}} = {}^{\text{end}}T_{\text{camera}}\,{}^{\text{camera}}T_{\text{marker}}\,{}^{\text{marker}}T_{\text{ORU}}. \quad (5)$$

In (5), ${}^{\text{camera}}T_{\text{marker}}$ is the transformation matrix between the camera and the marker near the ORU's adapter or the ORU store, which can be measured by the camera. The detailed process will be described below. ${}^{\text{end}}T_{\text{camera}}$ is the transformation matrix between the end effector and the camera, ${}^{\text{marker}}T_{\text{ORU}}$ is transformation matrix between the target ORU's adaptor and the ORU store and the marker. ${}^{\text{end}}T_{\text{camera}}$ and ${}^{\text{marker}}T_{\text{ORU}}$ are known before by assembly relationship.

In approaching mode, the end of the manipulator should move from an initial pose to the pose of ORU's adapter (target pose). Here, some middle poses are interpolated. Then, joint control commands gotten by inverse kinematics are sent to joint controllers. At the end of one control cycle, the tolerance between real-time pose of the manipulator and the target pose obtained by camera is computed. If the tolerance between the initial pose and the target pose is small enough, the close loop path planning finishes, vice versa the next control loop starts.

There are two steps in the marker pose's visual measurement, and they are the marker's recognition and measurement. The flow of the marker's recognition is in

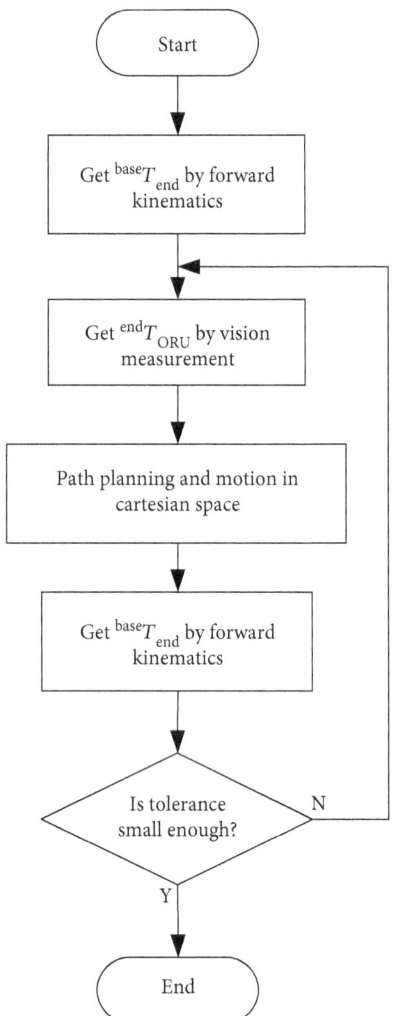

FIGURE 12: Closed loop path planning algorithm flow.

Figure 13. Here, Gaussian filter is used for denoising. A local adaptive threshold method is used for binary processing because the environment around the marker is complicated and the background luminance is not uniform. The basic model of the local adaptive threshold method is [16]

$$I_1(u, v) = \begin{cases} I(u, v), & \text{if } I(u, v) \geq T, \\ 0, & \text{if } I(u, v) < T, \end{cases} \quad (6)$$

where u and v are pixel coordinates, $I(u, v)$ is the gray value of the pixel, $I_1(u, v)$ is the gray value given after the judgment, and T is the detection threshold

$$T = \bar{B} + s\sigma. \quad (7)$$

$\bar{B}$ is the local mean value of the image background; σ is the root mean square value of the image's local random noise the parameter s is chosen according to the complexity of the image's background and the object image's grey value.

Then, Canny edge detection algorithm is used for extracting the outline of every characteristic circle. The

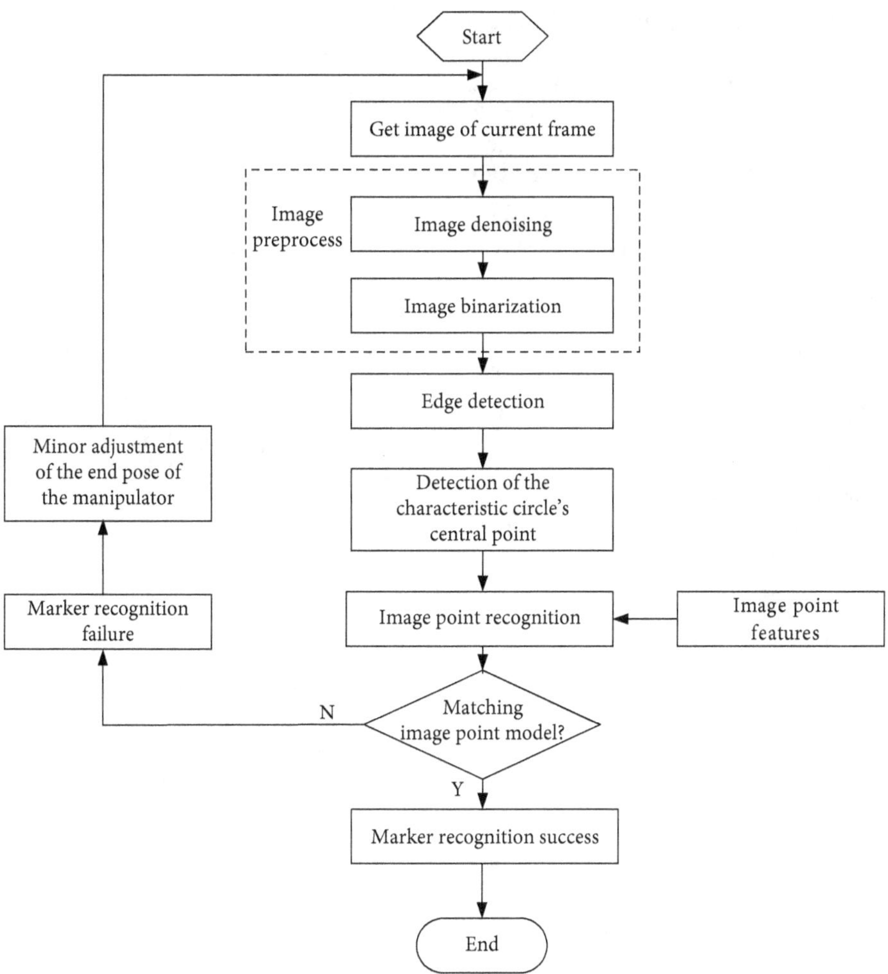

FIGURE 13: Flowchart of marker recognition.

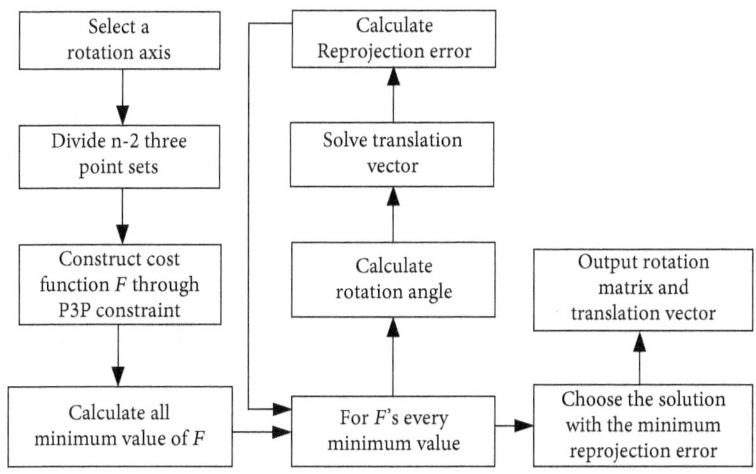

FIGURE 14: Flowchart of the RPNP algorithm.

rectangle boundary is extracted from the outline's 2D point set, and the center of a rectangular boundary is the center of the characteristic circle [17]. After the center of the characteristic circle is extracted, the center of the characteristic circle can be recognized by known geometric characteristics of the marker. The marker's pose transformation matrix ${}^{camera}T_{marker}$ relative to the camera can be measured using a three-dimensional pose solving algorithm. The RPNP algorithm is used in this paper because it has a relatively stable and more accurate solution [18]. The RPNP algorithm's flow chart is in Figure 14, and its detailed description is as follows.

(a) Select the rotation axis: if the camera's coordinate system is $O_cX_cY_cZ_c$ and the marker's coordinate system is $O_mX_mY_mZ_m$. As shown in Figure 15, a calibrated camera and n 3D reference points P_i $(i = 1, \ldots, n)$ are given. The projection of the reference point on normalized image plane is p_i. The side between any two reference points is $\{\overrightarrow{P_iP_j} \mid i \neq j, i \in \{1, \ldots, n\}, j \in \{1, \ldots, n\}\}$. If the value of $\|p_ip_j\|$ is larger, the noise's influence on the direction of $\overrightarrow{P_iP_j}$ is smaller, so the side $\overrightarrow{P_iP_j}$ corresponding to the longest projection side $\|p_ip_j\|$ is chosen as the rotation axis Z_a; the midpoint of $\overrightarrow{P_iP_j}$ is the origin O_a.

(b) Determine the least squares rotation axis: after the axis of rotation is determined, a new orthogonal coordinate system $O_aX_aY_aZ_a$ can be determined according to the cross principle. To seek the reference point P_i's coordinate in $O_cX_cY_cZ_c$, the 3D transformation matrix and translation vector from $O_aX_aY_a Z_a$ to $O_cX_cY_cZ_c$ should be determined.

To determine the direction of the Z_a axis in $O_cX_cY_cZ_c$, the depth values of the point P_{i0} and P_{j0} must be known. The solution of depth value is as follows: the N reference points are divided into n–2 3 point sets, which is $\{P_{i0}P_{j0}P_k, |k \neq i_0, k \neq j_0\}$, and each subset can get a four polynomial by P3P constraint as shown in (8), where the unknown parameter x is the square of the unknown depth of $P_{i0}.f_i(x)$, $i = 1, \ldots, n-2$ is the four-order polynomial gotten by P3P constraint; a_i, b_i, c_i, d_i, and e_i are coefficients of the four-order polynomial.

$$
\begin{aligned}
f_1(x) &= a_1x^4 + b_1x^3 + c_1x^2 + d_1x + e_1 = 0,\\
f_2(x) &= a_2x^4 + b_2x^3 + c_2x^2 + d_2x + e_2 = 0,\\
&\cdots\\
f_{n-2}(x) &= a_{n-2}x^4 + b_{n-2}x^3 + c_{n-2}x^2 + d_{n-2}x + e_{n-2} = 0.
\end{aligned} \tag{8}
$$

If the linearization method is used to solve (8), the redundancy of the equation will cause inconsistency of the solution. In this paper, the least square method is used to solve the local optimal solution of the equation [19]. First, a cost equation F is defined as

$$F = \sum_{i=1}^{n-2} f_i^{\,2}(x). \tag{9}$$

The minimum value of F can be obtained by solving its derivative F'

$$F' = \sum_{i=1}^{n-2} f_i(x) f_i'(x) = 0. \tag{10}$$

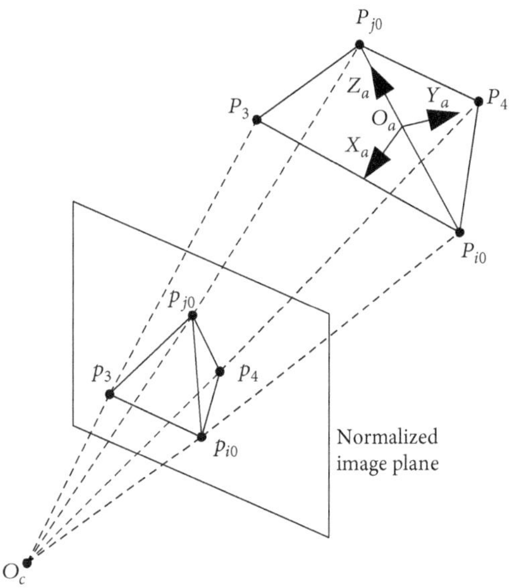

FIGURE 15: The projection of reference points.

Equation (10) is a seven equation of one variable and it can be solved by eigenvalue method. When x and the depth value of P_{i0} are determined, the depth of the other endpoint P_{j0} can be obtained by the P2P constraint. The unit vector Z_a of the coordinate system $O_aX_aY_aZ_a$ in the coordinate system $O_cX_cY_c Z_c$ is $\overrightarrow{P_{i0}P_{j0}}/\|P_{i0}P_{j0}\|$.

(c) Solve the rotation angle and translation vector: when the rotation axis Z_a is determined, the rotation matrix R between the new coordinate system $O_aX_a Y_aZ_a$ and the camera's coordinate system $O_cX_cY_cZ_c$ can be expressed as follows:

$$R = R'\mathrm{rot}(Z, \alpha) = \begin{bmatrix} r_1 & r_4 & r_7 \\ r_2 & r_5 & r_8 \\ r_3 & r_6 & r_9 \end{bmatrix} \begin{bmatrix} 1 & 0 & 0 \\ 0 & \cos(\alpha) & -\sin(\alpha) \\ 0 & \sin(\alpha) & \cos(\alpha) \end{bmatrix}. \tag{11}$$

In (11), R' is an arbitrary rotation matrix whose third column is Z_a, and it must satisfy the orthogonally constraint. $\mathrm{rot}(Z, \alpha)$ represents rotating α angle around the Z-axis. The projection from a 3D reference point to a two-dimensional normalized image plane can be expressed as follows:

$$\lambda_i \begin{bmatrix} u_i \\ v_i \\ 1 \end{bmatrix} = \begin{bmatrix} r_1 & r_4 & r_7 \\ r_2 & r_5 & r_8 \\ r_3 & r_6 & r_9 \end{bmatrix} \begin{bmatrix} 1 & 0 & 0 \\ 0 & \cos(\alpha) & -\sin(\alpha) \\ 0 & \sin(\alpha) & \cos(\alpha) \end{bmatrix} \cdot \begin{bmatrix} X_i \\ Y_i \\ Z_i \end{bmatrix} + \begin{bmatrix} t_x \\ t_y \\ t_z \end{bmatrix}. \tag{12}$$

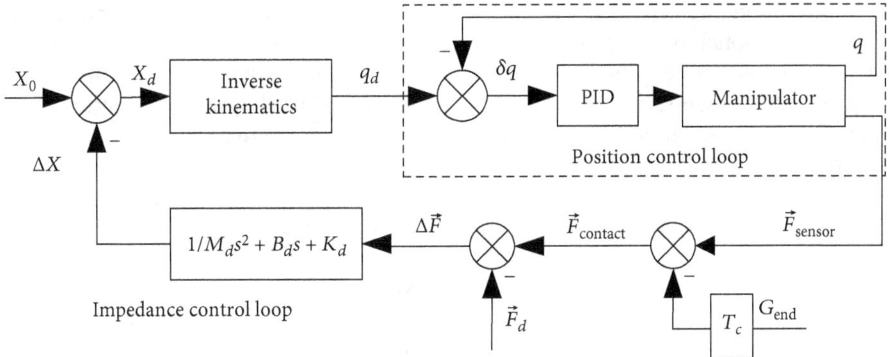

Figure 16: Flowchart of the impedance control.

In (12), $(u_i, v_i) = (u, v)/\lambda_i$ are the normalized coordinates of the image point P_i, and $t = [t_x \cdot t_y \cdot t_z]$ is the translation vector. Equation (12) can be listed as $2n \times 6$ homogeneous linear equations, and its unknown variable vector %$[\cos(\alpha)\ \sin(\alpha)\ t_x \cdot t_y \cdot t_z \cdot 1]^{\mathrm{T}}$ can be calculated. λ_i is the normalization factor. X_i, Y_i, and Z_i are three-dimensional position coordinates of space points.

(d) Calculate the relative pose ${}^{\text{camera}}T_{\text{marker}}$ between the marker and the camera: in practical application, the solution of the homogeneous linear (12) may not satisfy the constraint of the trigonometric function due to the interference of noise. Therefore, it is necessary to impose an orthogonal constraint on the rotation matrix R [20]. First, the reference point's 3D coordinates in the camera coordinate system $O_c X_c Y_c Z_c$ are estimated by unnormalized R and t, and then the standard 3D alignment method is used to solve the rotation and translation matrix. The cost function F is the sum of squares of polynomials, which contains at least 4 local minima. According to the local minimum value, the pose of the marker relative to the camera is estimated, and the solution which has the smallest inverse projection error is chosen as the optimal solution.

4.3. Impedance Control. In contact mode, there is an external force and torque at the end of the manipulator. To eliminate the pose error and insert the ORU into the ORU store slot precisely, an impedance control algorithm based on pose is used in this paper as shown in Figure 16.

In Figure 16, $M_d s^2 + B_d s + K_d$ can be seen as the desired impedance characteristic. M_d, B_d, and K_d are equivalent weight, damping, and stiffness matrix of the manipulator, respectively. q_d, q, and X_0 are joints' position command, joints' actual position, and the manipulator's pose command in Cartesian space. $\vec{F}_d$ is the desired contact force vector in the force sensor's coordinate system. $\vec{F}_{\text{sensor}}$ is the actual force vector measured by the force sensor. $\vec{F}_{\text{sensor}}$ not only contains the contact force and the gravity of the end effector, but also may contain an ORU's gravity when the ORU is captured. Because only the pure contact force vector $\vec{F}_{\text{contact}}$ is needed, the gravity terms must be compensated.

The homogeneous transformation matrix between the force sensor and the base coordinate system is

$$ {}^{\text{base}}T_{\text{sensor}} = {}^{\text{base}}T_{\text{end}}\,{}^{\text{end}}T_{\text{sensor}} = \begin{bmatrix} R_f^T & P_f \\ 0 & 1 \end{bmatrix}. \quad (13) $$

In (13), ${}^{\text{end}}T_{\text{sensor}}$ is the homogeneous transformation matrix between the force sensor and the end effector, which is known before by assembly relationship. ${}^{\text{base}}T_{\text{end}}$ can be calculated by forward kinematics. R_f^T is the posture transformation matrix between the force sensor and the manipulator's base coordinate system, and P_f is the coordinate system of the force sensor's position vector in the manipulator's base coordinate system. If the gravity field in the base coordinate system is $[0, 0, -g]^{\mathrm{T}}$, the pure contact force vector $\vec{F}_{\text{contact}}$ is

$$ \vec{F}_{\text{contact}} = \vec{F}_{\text{sensor}} - \begin{bmatrix} R_f \cdot [0, 0, -mg]^{\mathrm{T}} \\ \vec{r}_c \times R_f \cdot [0, 0, -mg]^{\mathrm{T}} \end{bmatrix}. \quad (14) $$

In (14), when there is no ORU at the end, $\vec{r}_c$ is the vector of the centroid of the end effector relative to the origin of the force/torque sensor coordinate. When there is an ORU at the manipulator's end, $\vec{r}_c$ is the vector of the centroid of the assembly composed by the end effector and the ORU relative to the origin of the force/torque sensor coordinate. With the desired impedance characteristic, if there is a contact force, the regulated pose ΔX in Cartesian coordinate system is

$$ \Delta X = \frac{\vec{F}_{\text{contact}} - \vec{F}_d}{M_d s^2 + B_d s + K}. \quad (15) $$

In order to determine the reasonable impedance parameters M_d, B_d, and K_d, the one-dimensional impedance control model is analyzed on the basis of (15), and the simplified two-order system model is expressed as

$$ m_d \ddot{e} + b_d \dot{e} + ke = f. \quad (16) $$

$\ddot{e}$, $\dot{e}$, and e are acceleration, velocity, and position adjustment value in one-dimensional space, respectively. f is the pure disturbance force. $m_d, b_d,$ and k_d are the expected inertia, damping, and stiffness, respectively. The following transfer function can be gotten by doing a Laplace transformation on (16).

$$\frac{e(s)}{f(s)} = \frac{\mu\omega_0^2}{s^2 + 2\xi\omega_0 s + \omega_0^2}. \quad (17)$$

In (17),

$$\mu = \frac{1}{k_d},$$
$$\xi = \frac{b_d}{2\sqrt{m_d k_d}}, \quad (18)$$
$$\omega_n = \sqrt{\frac{k_d}{m_d}}.$$

ξ is the damping ratio and ω_n is the natural frequency of the impedance control system. The dynamic performance of the impedance control system is determined by ξ and ω_n. k_d is related to the position deviation of the impedance control system, and it affects the stiffness of the manipulator. The larger the k_d value is, the greater the stiffness of the control system is, and the higher the tracking accuracy is. b_d is related to the velocity deviation. A reasonable b_d value can reduce the overshoot and oscillation, and it can also shorten the response time. m_d is related to the acceleration deviation of the impedance control system. When the expected inertia coefficient m_d is larger, it will cause a larger impact on the environment, resulting in an increase in trajectory tracking error and slow response of the system; therefore, in an impedance control system, m_d usually takes a very small value. In this paper, the speed and acceleration changes of the space station manipulator are very slow when performing the ORU module replacement task, which means that $m_d\ddot{e}$ can be ignored, so the value of M_d in this paper is set to 0.

When the impedance parameter satisfies the following formula, the transition from free movement to restricted space can reach stability [21].

$$\alpha = \frac{k_e}{k_d} \gg 1,$$
$$\xi \geq \frac{\sqrt{1+\alpha} - 1}{2}. \quad (19)$$

In (19), k_e is the environment stiffness, assuming that the ORU inserting and extracting direction is the Z-axis in the manipulator base coordinate system, and the Z-axis is also perpendicular to the ORU's mounting surface. At the final stage of approaching mode, target poses (the ORU adapter or the ORU store slot's pose) that the manipulator will reach are obtained by the hand-eye camera, and it can be expressed as $(x, y, z, \alpha, \beta, \text{and } \gamma)$, but the Cartesian space control command sent by the central controller is $(x, y, z + \Delta z, \alpha, \beta, \gamma)$, which means that the end effector is right above the ORU's adaptor or above the ORU store at the end of the approaching mode. In contact mode, take inserting ORU as an example, the manipulator carries the ORU along the Z-axis and approaches the ORU store slot to insert the ORU. In the presence of pose error, the bottom edge of the ORU will contact with the guide surface of the ORU store slot. If the impedance control method is used in the Z direction, and the expected stiffness and damping are k_{pz} and b_{pz}, respectively, the manipulator will stop moving but in a balance state in the Z-axis when the position error reaches $F_z/(k_{pz} + b_{pz}s)$. F_z is the contact force in the Z-axis. In order to ensure that the ORU module can be inserted into the ORU slot in the Z-axis successfully, the manipulator requires a high stiffness in the Z-axis. Therefore, the manipulator adopts the position control mode in this direction. In order to prevent the excessive contact force in the Z direction, the contact force threshold F_{tz} is set in this direction. When the contact force is greater than the threshold, the manipulator stops moving or carries the ORU exit along the Z direction and prevents the contact force from being too large. In the other directions $(x, y, \alpha, \beta, \text{and } \gamma)$, the impedance control is adopted to correct pose deviation under the guidance of the guide surface of the ORU store slot. In summary, the manipulator is in position control mode in Z direction, and it moves a distance of Δstep for every control cycle ΔT until the contact force is bigger than a threshold of F_{tz}. In other directions, the manipulator is in impedance control mode using the control method in (15). The motion adjustment value ΔX is

$$\Delta X = \begin{bmatrix} \Delta p_x \\ \Delta p_y \\ \Delta p_z \\ \Delta\alpha \\ \Delta\beta \\ \Delta\gamma \end{bmatrix} = \begin{bmatrix} \frac{1}{k_{px} + b_{px}s} & 0 & 0 & 0 & 0 & 0 \\ 0 & \frac{1}{k_{py} + b_{py}s} & 0 & 0 & 0 & 0 \\ 0 & 0 & 1 & 0 & 0 & 0 \\ 0 & 0 & 0 & \frac{1}{k_\alpha + b_\alpha s} & 0 & 0 \\ 0 & 0 & 0 & 0 & \frac{1}{k_\beta + b_\beta s} & 0 \\ 0 & 0 & 0 & 0 & 0 & \frac{1}{k_\gamma + b_\gamma s} \end{bmatrix} \begin{bmatrix} \Delta F_x \\ \Delta F_y \\ \Delta p_z \\ \Delta T_x \\ \Delta T_y \\ \Delta T_z \end{bmatrix}$$
$$\Delta P_z(t + \Delta T) = \Delta P_z(t) + \Delta\text{step}. \quad (20)$$

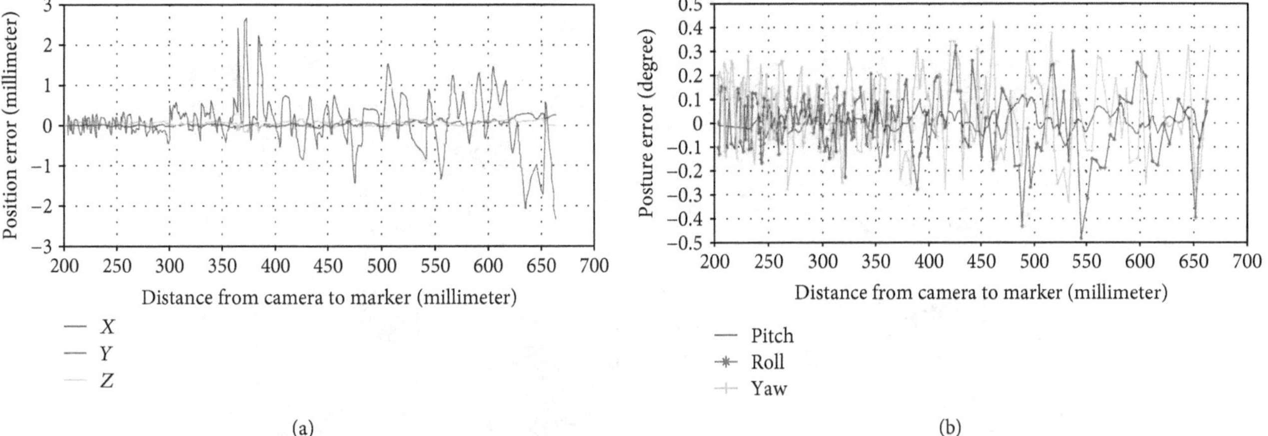

FIGURE 17: Pose error of vision measurement: (a) position error and (b) posture error.

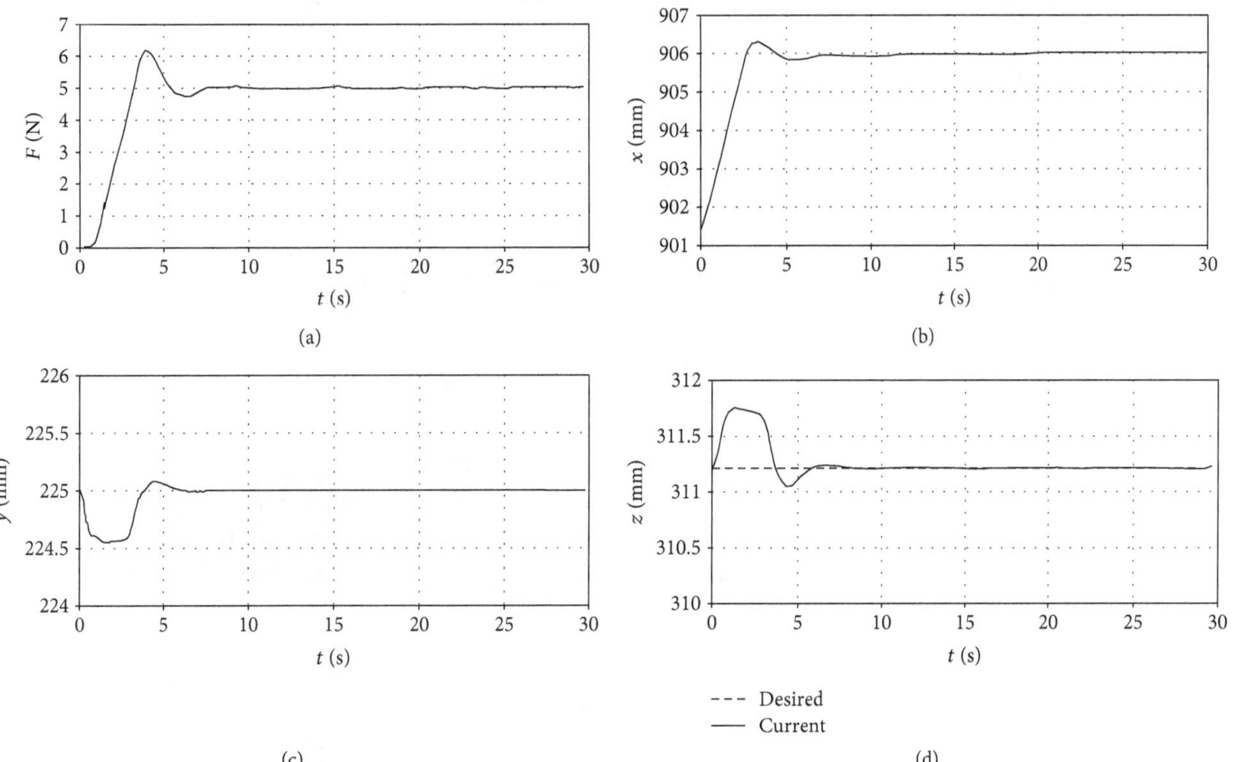

FIGURE 18: Force and position response curve under impedance control: (a) force applied in x direction, (b) position response in x direction, (c) position response in y direction, and (d) position response in z direction.

The pose command sent to the manipulator is

$$X_d = X_0 - \Delta X. \tag{21}$$

5. Experiments and Discussion

5.1. Experiment Introduction. In Section 2.1, the process of ORU changeout is divided into old ORU extracting, old ORU inserting, new ORU extracting, and new ORU installing. Because the control method and operation flow of a new or an old ORU are the same, so only one ORU's extracting and inserting experiment is described in this paper. As shown in Figure 2, the reachable operating space of the manipulator for extracting ORU in the inertial coordinate system is

$$\begin{aligned} 480\,\text{mm} &\le x \le 680\,\text{mm}, \\ 0\,\text{mm} &\le y \le 200\,\text{mm}, \\ 375\,\text{mm} &\le z \le 625\,\text{mm}. \end{aligned} \tag{22}$$

FIGURE 19: ORU changeout experiments: (a) initial state, (b) free motion, (c) adapter approaching, (d) ORU capturing and locking, (e) ORU extracting, (f) free motion, (g) ORU inserting, and (h) ORU inserting finished.

The reachable operating space for inserting ORU in the inertial coordinate system is

$$\begin{aligned} -400\,\text{mm} &\le x \le -200\,\text{mm}, \\ -250\,\text{mm} &\le y \le 0\,\text{mm}, \\ -100\,\text{mm} &\le z \le 100\,\text{mm}. \end{aligned} \tag{23}$$

The inertial coordinate system origin O_0 is the midpoint of the experiment platform. Axis direction of the inertial coordinate system is shown in Figure 2. Axis O_0X_0 and axis O_0Z_0 are parallel to the platform's two sides, respectively.

5.2. Results and Discussion. In this section, control algorithms in Section 4 are verified individually. First, an operator gives a target point to the central controller of the manipulator by the GUI. Using the open loop path planning algorithm, the manipulator moves from the initial pose to the target pose. Experiment results show that joints' motion is continuous, and the absolute pose accuracy is 3.2 mm/0.1° which is measured by a laser tracker.

When verifying the visual recognition and measurement algorithm, the end of the manipulator moves close to the marker, and the relative pose of the camera to the marker is measured in real time. At the same time, a laser tracker is used to measure the exact pose of the camera to the marker. Compared with visual measurement results, the measurement error curve at different distance is gotten as shown in Figure 17. It can be found that the error decreases as decreasing of the distance between the camera and the marker, and the measurement error is less than 1 mm/0.3° when the distance is below 300 mm. Experiment results show that the

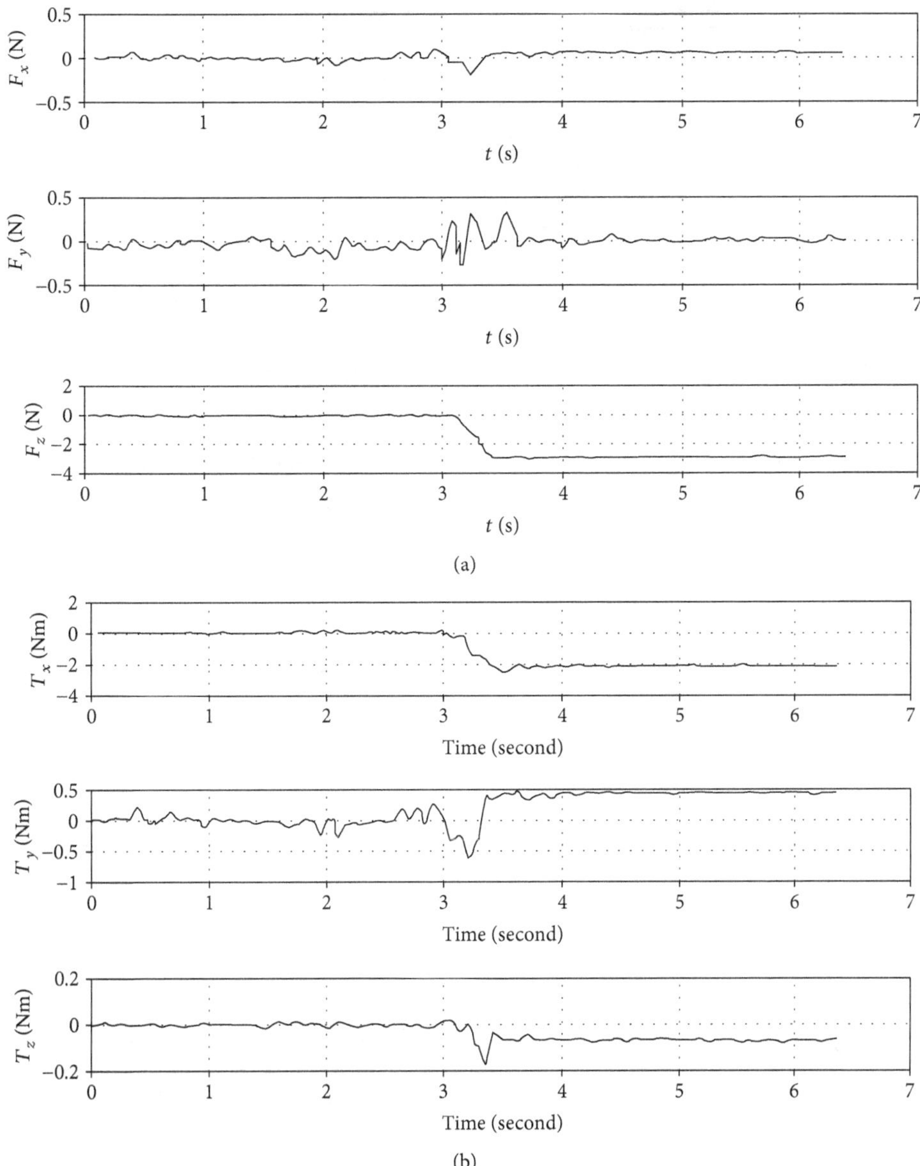

FIGURE 20: End force and moment under impedance control: (a) force and (b) moment.

accuracy can meet the tolerance requirement when the manipulator enters into the contact mode.

Because the principle of impedance control method in any one of the 6 directions of the terminal position and posture is the same as that in the other 5 directions. Therefore, the impedance control is selected in the x direction in this section just for verifying the impedance control algorithm, and position control mode is set in the y and z directions. The initial contact force in the x-axis F_{dx} is 0 N. The initial positions in x, y, and z directions are 901 mm, 225 mm, and 311 mm. The desired stiffness is set to 1 N/mm, and desired damping is 50 Ns/m. Figure 18 is the force and position response curve. From 0 s, the external force is applied, so the manipulator moves along the x direction. After about 5 s, the external force is increased and maintained at 5 N. The position along the x direction is changed quickly with the change of the external force, and it maintained at about 906 mm at last. The position change caused by the 5 N external force is about 5 mm, and this proves that the stiffness in the x direction is 1 N/mm. The experimental results verify the correctness of the impedance control method. The position error in y and z directions is lower than 0.5 mm, which means that the manipulator nearly keeps still in y and z directions.

The whole process of ORU extracting and inserting experiment is shown in Figure 19, and all working modes and control algorithms are contained in the experiment. Figure 19(a) is the initial state. After receiving the target pose, the manipulator moves above the ORU adapter in free motion mode (Figure 19(b)). Then, the manipulator approaches to the adapter under the guide of the camera (Figure 19(c)). In Figure 19(d), the manipulator continues to get close to the adapter, until the end effector captures the adapter completely. In Figure 19(e), the end effector

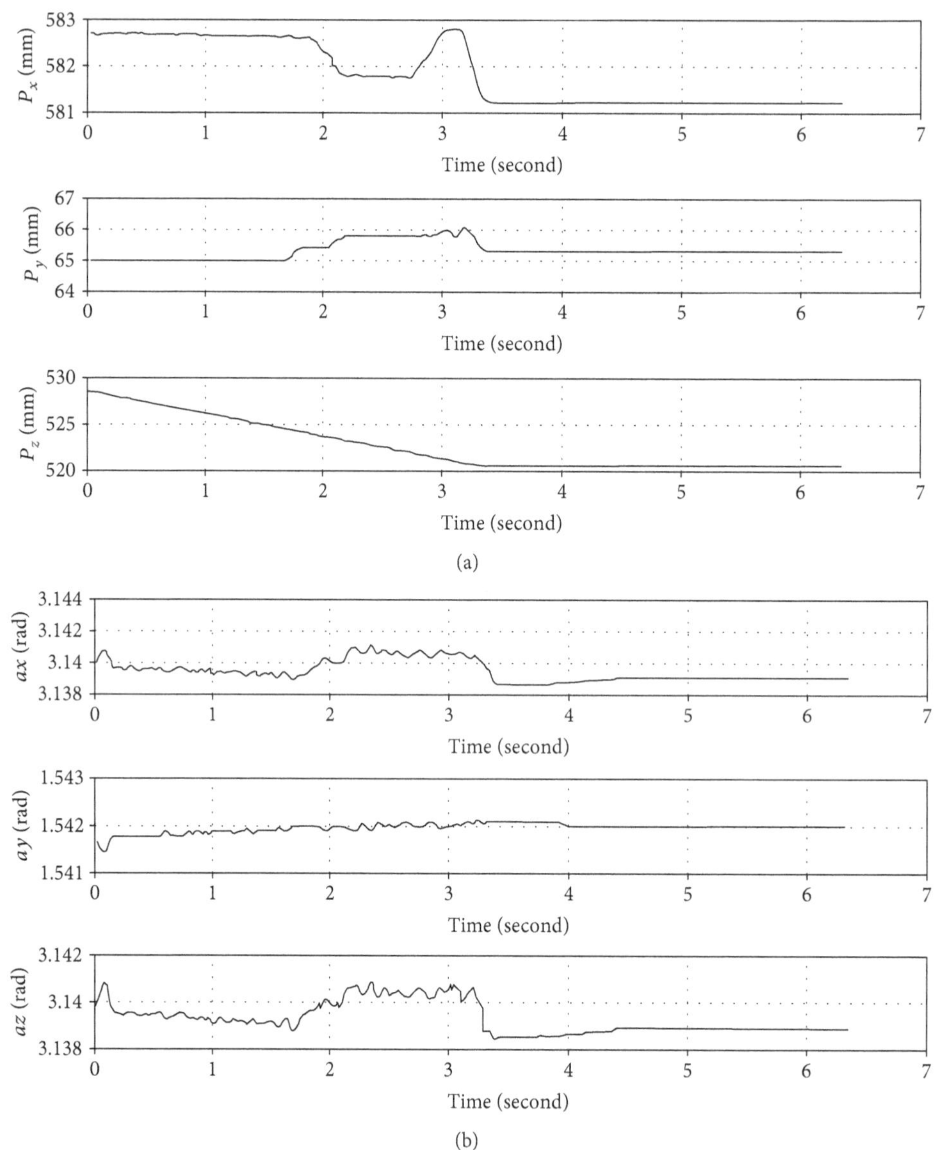

FIGURE 21: The end poses under impedance control: (a) position and (b) posture.

locks the adapter and extracts the ORU. Then, the manipulator sends the ORU to the ORU store (Figure 19(f)). In Figure 19(g), the deviation of the ORU relative to the ORU store slot is completely corrected by the guide surface on the ORU store using impedance control. At last, the ORU is inserted into the ORU store slot completely (Figure 19(h)). In this experiment, the impedance parameters are determined as $B_d = \text{diag}\ ([500, 500, 500, 10, 10, 10])$ (Ns/m, Nms/rad) and $K_d = \text{diag}\ ([1000, 1000, 1000, 20, 20, 20])$ (N/m, Nm/rad).

In Figure 19(d), the manipulator is in contact mode and the impedance control method in Section 4.3 is used. Figure 20 is the end force and moment in that process and its reference coordinate system is the force sensor's coordinate system. Figure 21 is the manipulator's end pose of that process, and it is in inertial coordinate system. The sampling period of the sensor is 20 ms, while the impedance control cycle is also 20 ms. It is defined that F_{tz} is 2.5 N. From 0 s to 3 s, the contact force and moment does not change obviously because of impedance control. The position in Z direction reduces linearly because the manipulator is in position control mode in that direction. The position in other directions changes slightly according to contact force or moment. At 3 s, the contact force in Z direction begins to increase gradually. When the force in direction Z is equal to F_{tz} at last the manipulator stops to move in Z direction, which means that the ORU has been inserted completely. These experiment data verify the correctness of impedance control method in Section 4.3. After $t = 3$ s, the ORU module began to contact with the ORU store slot, so the force/torque in other directions also changed besides the Z axis. In structural design, there is a small gap between ORU store slot and ORU. Because the manipulator is in position control mode in the direction of Z, and the manipulator in this article has a certain flexibility, so when the bottom of the ORU module begins to contact with the slot, the high stiffness in Z

direction and the gap mentioned before will cause slight deformation of the manipulator, and the result is that the other surface of the ORU will contact the module store slot, which leads to changes of the contact force/torque in the other directions.

6. Conclusions and Future Work

Replacing of orbital replacement units by a manipulator has great significances in space station. In this paper, the full process of ORU changeout is analyzed first, and it is divided into old ORU extracting, old ORU inserting, new ORU extracting, and new ORU installing. For every stage, there are 3 working modes, which are free motion mode, approaching mode, and contact mode. Referring to different working modes, the open loop path planning algorithm, the vision-based close loop path planning algorithm, and the impedance control algorithm are researched individually. To verify the ORU changeout task flow and corresponding control algorithms, a ground experiment platform is designed, which includes a 6-DOF manipulator and an end effector with clamp/release and screwing function. The manipulator has a distributed control system and a monocular camera. Then, some experiments are done first to verify every algorithm singly. Results show that the absolute pose accuracy of the manipulator controlled by the open loop path planning algorithm is 3.2 mm/0.1°. The visual measurement accuracy is less than 1 mm/0.3° when the distance between the camera and the marker is below 300 millimeters. Impedance control experiment shows that the force control algorithm based on position and end force feedback is correct. At last, the whole process of ORU extracting and inserting experiments is done. Experiments results show that the control architecture and control algorithms in this paper are correct. In the future, experiments will be done on a more real environment as in a space station.

Conflicts of Interest

The authors declare that there are no conflicts of interest regarding the publication of this paper.

Acknowledgments

This research is partially supported by a government grant from the Shanghai Engineering Research Center for Assistive Devices, Shanghai, China.

References

[1] A. Flores-Abad, O. Ma, K. Pham, and S. Ulrich, "A review of space robotics technologies for on-orbit servicing," *Progress in Aerospace Sciences*, vol. 68, pp. 1–26, 2014.

[2] O. Ma, J. Wang, S. Misra, and M. Liu, "On the validation of SPDM task verification facility," *Journal of Field Robotics*, vol. 21, no. 5, pp. 219–235, 2004.

[3] Y. Fukazu, N. Hara, Y. Kanamiya, and D. Sato, "Reactionless resolved acceleration control with vibration suppression capability for JEMRMS/SFA," in *2008 IEEE International Conference on Robotics and Biomimetics*, pp. 1359–1364, Bangko, Thailand, February 2009.

[4] R. Mukherji, D. A. Ray, M. Stieber, and J. Lymer, "Special purpose dexterous manipulator (SPDM) advanced control features and development test results," in *Proceedings of the International Symposium on Artificial Intelligence, Robotics and Automation in Space*, Montreal, Canada, 2001.

[5] P. G. Backes and K. S. Tso, "Autonomous single arm ORU changeout—strategies, control issues, and implementation," *Robotics and Autonomous Systems*, vol. 6, no. 3, pp. 221–241, 1990.

[6] G. Colombina, F. Didot, G. Magnani, and A. Rusconi, "Automation and robotics technology testbed for external servicing," in *Proceedings of IEEE/RSJ International Conference on Intelligent Robots and Systems (IROS'94)*, vol. 3, pp. 1954–1963, Munich, Germany, September 1994.

[7] F. Ozaki, K. Machida, J. Oaki, and T. Iwata, "Robot control strategy for in-orbit assembly of a micro satellite," *Advanced Robotics*, vol. 18, no. 2, pp. 199–222, 2004.

[8] L. Jiang, X. Huo, Y. Liu, and H. Liu, "Optimization-based compliance control strategy of redundant robot for ORU replacements," in *2016 12th World Congress on Intelligent Control and Automation (WCICA)*, pp. 2192–2197, Guilin, China, June 2016.

[9] W. Xu, B. Liang, and Y. Xu, "Survey of modeling, planning, and ground verification of space robotic systems," *Acta Astronautica*, vol. 68, no. 11-12, pp. 1629–1649, 2011.

[10] C. Qi, A. Ren, F. Gao, X. Zhao, Q. Wang, and Q. Sun, "Compensation of velocity divergence caused by dynamic response for hardware-in-the-loop docking simulator," *IEEE/ASME Transactions on Mechatronics*, vol. 22, no. 1, pp. 422–432, 2017.

[11] C. Qi, F. Gao, X. Zhao, A. Ren, and Q. Wang, "A force compensation approach toward divergence of hardware-in-the-loop contact simulation system for damped elastic contact," *IEEE Transactions on Industrial Electronics*, vol. 64, no. 4, pp. 2933–2943, 2017.

[12] S. Dubowsky, W. Durfee, T. Corrigan, A. Kuklinski, and U. Muller, "A laboratory test bed for space robotics: the VES II," in *Proceedings of IEEE/RSJ International Conference on Intelligent Robots and Systems (IROS'94)*, vol. 3, pp. 1562–1569, Munich, Germany, September 1994.

[13] H. Liu, "An overview of the space robotics progress in China," in *Proceedings of the International Symposium on Artificial Intelligence Robotics and Automation in Space*, Montreal, Canada, 2014.

[14] Z. Wen, Y. Wang, J. Luo, A. Kuijper, N. Di, and M. Jin, "Robust, fast and accurate vision-based localization of a cooperative target used for space robotic arm," *Acta Astronautica*, vol. 136, pp. 101–114, 2017.

[15] S. I. Nishida, H. Hirabayashi, and T. Yoshikawa, *A New Space Robot End-Effector for on-Orbit Reflector Assembly*, Japan Aerospace Exploration Agency (Chofu) Advanced Space Technology Research Group, 2005.

[16] H. Wu, X. Zheng, Y. Zhao, and N. Li, "A new thresholding method applied to motion detection," in *2008 IEEE Pacific-Asia Workshop on Computational Intelligence and Industrial Application*, vol. 1, pp. 119–122, Wuhan, China, December 2008.

[17] C. X. Deng, G. B. Wang, and X. R. Yang, "Image edge detection algorithm based on improved canny operator," in *2013*

International Conference on Wavelet Analysis and Pattern Recognition, pp. 168–172, Tianjin, China, July 2013.

[18] S. Li, C. Xu, and M. Xie, "A robust O(n) solution to the perspective-n-point problem," *IEEE Transactions on Pattern Analysis and Machine Intelligence*, vol. 34, no. 7, pp. 1444–1450, 2012.

[19] L. Quan and Z. Lan, "Linear n-point camera pose determination," *IEEE Transactions on Pattern Analysis and Machine Intelligence*, vol. 21, no. 8, pp. 774–780, 1999.

[20] S. Umeyama, "Least-squares estimation of transformation parameters between two point patterns," *IEEE Transactions on Pattern Analysis and Machine Intelligence*, vol. 13, no. 4, pp. 376–380, 1991.

[21] D. Surdilovic, "Contact stability issues in position based impedance control: theory and experiments," in *Proceedings of IEEE International Conference on Robotics and Automation*, vol. 2, pp. 1675–1680, Minneapolis, MN USA, April 1996.

10

Characteristics of Orbit Determination with Short-Arc Observation by an Optical Tracking Network, OWL-Net

Jin Choi,[1,2] Jung Hyun Jo,[1,2] Kyung-Min Roh,[1] Hong-Suh Yim,[1] Eun-Jung Choi,[1] and Sungki Cho[1]

[1]*Korea Astronomy and Space Science Institute, Daejeon 34055, Republic of Korea*
[2]*University of Science and Technology, Daejeon 34113, Republic of Korea*

Correspondence should be addressed to Jung Hyun Jo; jhjo39@kasi.re.kr

Academic Editor: Yue Wang

An optical tracking network, the Optical Wide-field patroL Network (OWL-Net), has been developed to maintain the orbital ephemeris of 11 domestic low Earth orbit satellites. The schedule overlapped events were occurred in the scheduling of the OWL-Net with reduction of the optical observation chances. A short-arc observation strategy for the OWL-Net was tested to reduce schedule overlapped events with the optical observation simulation and the orbit determination. In the full-scale optical observation simulation from January 2014 to December 2016, the most frequent overlapped events were occurred 127, 132, and 116 times in the 4th, 34th, and 18th weeks of 2014, 2015, and 2016, respectively. The average number of overlapped event for three years was over 10% for the whole observation chances of five stations. Consequently, the short-arc observation strategy reduced the schedule overlapped events for every observation target of the OWL-Net. In case of the 5 s and 10 s cases, the most schedule overlapped events were removed. The test results of the orbit determination results show that the most maximum orbit prediction errors after seven days are maintained at <10 km in the in-track direction for the short-arc observation simulations. The results demonstrate that the short-arc optical observation strategy is more optimal to maintaining the accuracy of orbital ephemeris with more observation chances.

1. Introduction

Space objects can be tracked with various types of equipment [1], and optical tracking devices have been used to collect observation data for space objects from low Earth orbit (LEO) to the geosynchronous orbit (GSO). Among the 43,269 known space objects, 13,393 objects were registered as low Earth orbit (LEO) objects as of 6 April 2018. LEO is defined as a volume of space that has the altitude from 0 to 2000 km by the Inter-Agency Space Debris Coordination Committee (IADC) Space Debris Mitigation Guidelines [2]. The Baker–Nunn camera was an early type of optical tracking network installed at 11 sites around the world. The primary goal of the Baker–Nunn camera was to track launched vehicles and satellites with a wide field of view [3]. Ground-Based Electro-Optical Deep Space Surveillance (GEODSS) was another optical tracking network. GEODSS replaced the Baker–Nunn camera in 1979 [4]. The United States operates these facilities and also provides a catalogue of unclassified space objects publicly via the website Space-Track.

The Korean Astronomy and Space Science Institute (KASI) has been developing an optical tracking network called the Optical Wide-field patroL Network (OWL-Net) since 2010. KASI plans to install five optical tracking systems in Mongolia, Morocco, Israel (at the Wise observatory), the United States (at the Mt. Lemmon Astronomical Observatory (LOAO)), and South Korea (at the Bohyun Astronomical Observatory (BOAO)). The main observation targets of the OWL-Net are 11 South Korean LEO satellites shown in Figure 1. Monitoring the South Korean geostationary Earth orbit (GEO) region near 128 degrees of longitude is another goal of the OWL-Net. The candidate sites were selected in the light of results for an optical tracking

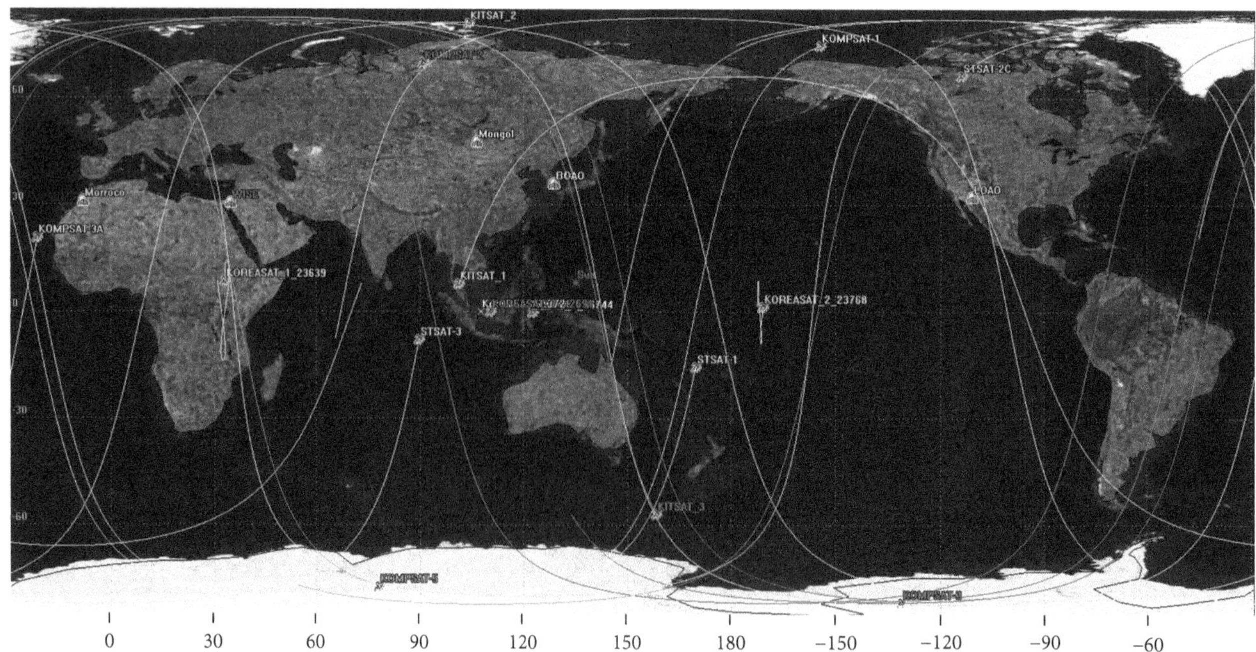

FIGURE 1: OWL-Net site locations and observation target, 11 LEO satellites and 6 GEO satellites (4 GEOs are located near 128 E of longitude Koreasat_1 and 2 are drifting eastward).

simulation with 11 observation targets of LEO satellites [5]. The OWL-Net can make an observation with a chopper to obtain dozens of measurements per shot [6, 7]. The authors have performed test observations and orbit determination research for various satellites [8–11]. OWL-Net headquarters create daily schedules [based on two-line element (TLE) data from the Joint Space Operations Center (JSpOC)] for automatic observation of scheduled targets.

Occasionally, two or more targets will be visible to the OWL-Net at the same time (overlapped tracks). The optical tracking equipment is a passive observation system, and the observation time of LEO observation targets is limited to terminator conditions at dawn and dusk (when the sky is dark but the target remain sun-illuminated). When the target visibility of multiple targets is overlapped, only one target can be tracked at a time. When this happens, the number of observations for nonselected targets is reduced. However, too sparse or irregular observation data can affect the quality of orbit determination [12]. We designated this problem as an optical observation overlapping problem.

Observation conditions such as frequency of track and arc length can affect the orbit estimation accuracy. Vallado and Carter showed that highly accurate orbit determination is possible with dense observation data for 120 s (a few hundred observations per satellite pass) [13]. Bennett et al. showed that short-arc observations for 5 s from two passes are sufficient to maintain the orbital elements for the catalogue [14]. Park et al. analysed the minimum amount of data for observed LEO satellites by using the OWL-Net [9]. In this analysis, the orbit determination for the various observation scenarios with the same arc-length were analysed to find the optimal observation condition. In the case of GEO objects, the relation between the orbit determination accuracy and arc length was analysed theoretically. Analysis was verified again with real observation data [15]. Tombasco and Axelrad also showed the relation between orbit determination accuracy and observation frequency for the Galaxy 15 GEO satellite with tracking and data relay satellite (TDRS) data [16]. The arc length was the dominant factor for orbit determination accuracy of GEO objects. However, more frequently observed data also improved orbit determination accuracy. Orbit determination accuracy with the angle-only data case was also analysed for sparse, sporadic, and dense cases using a single optical telescope [11]. The orbit determination results showed that the arc length was the dominant factor for angle-only orbit determination of GEO satellites. However, even in GEO satellite orbit determination tests, multiarc observation cases have shown much more accurate orbit determination results than the single arc observation case.

In this paper, we analyse the effectiveness of a short-arc optical observation strategy for orbital ephemeris maintenance using the OWL-Net. Firstly, the orbital characteristics of 11 LEO satellites and observation simulation statistics of the OWL-Net were analysed using consecutive TLEs. Secondly, we simulated weekly overlapped observations for 2014, 2015, and 2016. Thirdly, an orbit determination test was done with short-arc observation strategy. Generally, a short-arc optical observation strategy can allow for more frequent observation opportunities. We have attempted to verify various short-arc strategies for effective observation scheduling. Finally, the orbit estimation test for the short-arc optical observation strategy was performed using the real observation data from the OWL-Net.

Table 1: Orbital characteristics of 11 Korean LEO satellites on 6 April 2018.

NORAD ID	Satellite name	International designation	Period	Inclination	Apogee	Perigee	Designed Orbit	Operational
22077	OSCAR 23 (KITSAT-1)	1992-052B	111.93	66.08	1319	1312	LEO	no
22825	KITSAT-B	1993-061C	100.7	98.78	797	786	LEO	no
25756	KITSAT-3	1999-029A	99.06	98.41	723	705	LEO	no
26032	KOMPSAT-1	1999-070A	98.03	97.84	667	662	SSO	no
27945	STSAT-1	2003-042G	98.31	97.91	686	670	SSO	no
29268	KOMPSAT-2	2006-031A	98.48	98.12	699	673	SSO	no
38338	KOMPSAT-3	2012-025B	98.53	98.17	695	682	SSO	yes
39068	STSAT-2C	2013-003A	95.94	80.23	852	276	SSO	no
39227	KOMPSAT-5	2013-042A	95.71	97.6	554	552	SSO	yes
39422	STSAT-3	2013-066G	96.69	97.69	615	585	SSO	yes
40536	KOMPSAT-3A	2015-014A	95.22	97.57	537	521	SSO	yes

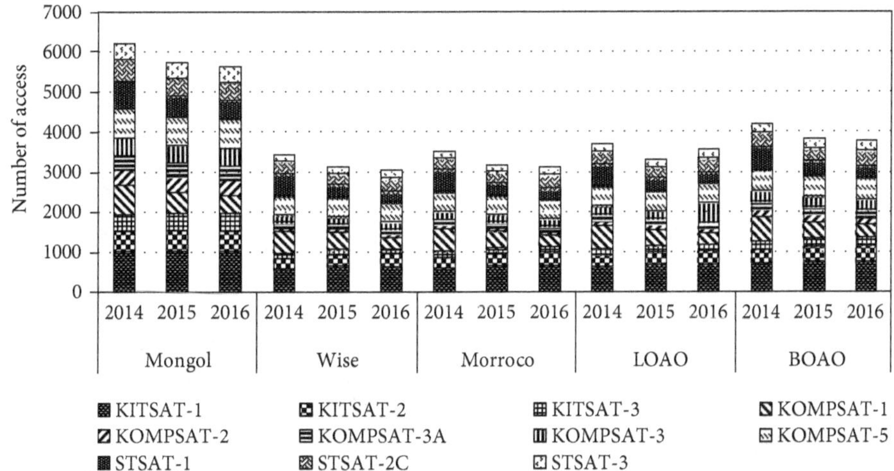

Figure 2: Number of accesses for the 11 LEO observation targets at OWL-Net from 2014 to 2016.

2. Optical Observation Schedule Overlapping Problem

2.1. OWL-Net Observation Target and Observation Simulation. The primary observation target of the OWL-Net is 11 Korean LEO satellites. Only four LEO satellites were in operation on 6 April 2018. Table 1 shows identification (ID) and orbital characteristics for the 11 LEO satellites. All of LEO satellites (except for STSAT-2C) have circular orbits, with eight in Sun-synchronous orbit (SSO). The SSO satellites revisit the same local position at the same local time with the fixed line of nodes of orbit to the sun [17]. Because optical tracking systems can only make observations near dawn and dusk for LEO satellites, the observation opportunities are very limited for these SSO satellites.

We analysed the orbital characteristics of these satellites to look for the dominant parameters of optical observation overlap events for the ground-based observation. Firstly, the orbital shape can be a dominant parameter; we note that, except for STSAT-2C, the remaining 10 domestic LEO satellites have similar eccentricities. The orbital plane location can be another dominant parameter. The right ascension of the ascending node (RAAN) and the inclination determine the rotation of the orbit from the vernal equinox and degree of tilt of the orbital plane, respectively. Except for KITSAT-1 and STSAT-2C, the inclinations of domestic LEO satellites are similar to each other. If the orbital shape and the inclination are similar and the orbital planes are close longitudinally, the possibility of an optical observation overlap event is increased as these satellites can pass above the observing station at a similar local time. If a ground station is positioned near 30 degrees of latitude, LEO satellites from 500 to 800 km can be observed at the same time from a single ground station within the orbital plane's maximum longitudinal difference of 40 degrees.

OWL-Net observations were simulated from 2014 to 2016 with optical observation constraints. Figure 2 shows the number of visible passes for the 11 LEOs with the OWL-Net from 2014 to 2016 (from Mongolia, Wise, Morocco, LOAO, and BOAO). Almost all LEOs have similar numbers of observable passes for each site over the three years. The Mongolian site has the most observation opportunities for polar orbiting satellites due to its higher latitude. The BOAO site has the second most observable passes. KITSAT-1 has the largest number of observation opportunities among the 11 satellites. KOMPSAT-1 and KOMPSAT-5

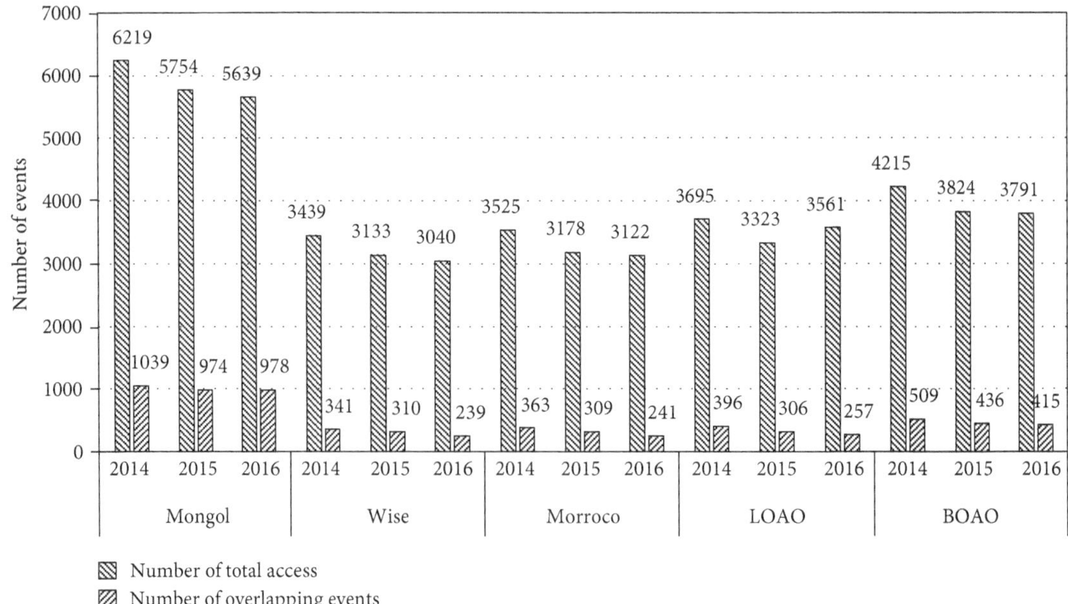

FIGURE 3: Number of total accesses and overlapped events for the 11 LEO observation targets at the OWL-Net from 2014 to 2016.

also have a larger number of observation opportunities than the others. KITSAT-3, KOMPSAT-2, and KOMPSAT-3A have fewer observation opportunities than the others, with <200 observation opportunities every year for four sites (excepting the Mongolian site). Weather conditions and the possibility of system anomalies were not considered in this simulation.

2.2. Statistics of Overlapped Event Simulation of the OWL-Net. The OWL-Net system was designed to operate an orbit determination process every week. We analysed weekly based overlap events for 2014, 2015, and 2016. TLEs from JSpOC were used for this simulation. We used a general optical tracking condition with −6 degrees as the sun elevation limit and 5 degrees as the site elevation limit.

The number of overlap events for the OWL-Net stations is presented in Figure 3. The number of overlap events schedule comprised almost 10% of the total number of observable passes for every site during the three years. In 2015, the number of overlap events at the Mongolian site was the highest, with 974 overlapped events. This is ~40% of the total number of overlapped events for five OWL-Net stations in 2015. For the other four stations, the numbers of overlap events in 2015 were 310 at Wise, 309 at Morocco, 306 at LOAO, and 436 at BOAO. The Mongolian station has a higher latitude than the other four stations. We could not discern any annual variation for the total overlap events for the 11 LEOs at any of the OWL-Net sites. Figure 4 shows the variation of weekly number of overlap events at each OWL-Net site for 2014, 2015, and 2016. This weekly base analysis was undertaken because the OWL-Net system was designed to implement weekly base orbit determination. The Mongolian site still had the most overlap events for the three years as well as the most total available passes. Among the 52 weeks of 2015, overlap events occur most frequently in the 34th week (140 events). The 4th and 18th weeks experienced the most overlapping events of 2014 and 2016, respectively. However, we also could not discern a weekly tendency of change in the number of overlapped events.

When the overlapped events were occurred, the one observation schedule should be selected among some observation chances. Therefore, the other observation chances were cancelled. Furthermore, the optical observation chance can be reduced by the weather condition and system condition in reality. These too sparse observation conditions can result in failure or bad quality of the orbit determination.

As the number of domestic LEO satellites increases, the frequency of overlapped events will also increase. The overlap problem can be solved either by using more ground stations for covering more observation time or with an effective observation strategy for orbit determination accuracy. However, more ground stations are not an option due to the higher costs. The short-arc optical observation strategy could provide a more acceptable and better solution for the optical observation overlap problem.

3. Orbit Determination Test for Short-Arc Optical Observation Strategy

3.1. Analysis of the Number of Overlapped Event for Short-Arc Optical Observation Strategy. Various short-arc optical observation strategies were devised with the arc length of each pass. Firstly, we considered Bennett et al.'s work [14]. In this research, the arc length was set as 5 seconds for two passes. We fixed the arc length as 5 s, but the number of passes was not limited. Because the OWL-Net system was designed for weekly orbit determination. Bennett's work focused on the follow-up observation. Therefore, some satellites can have more optical observation opportunities than the others in this study. Secondly, we selected 120 s from Vallado and Carter's work [13]. Furthermore, we selected

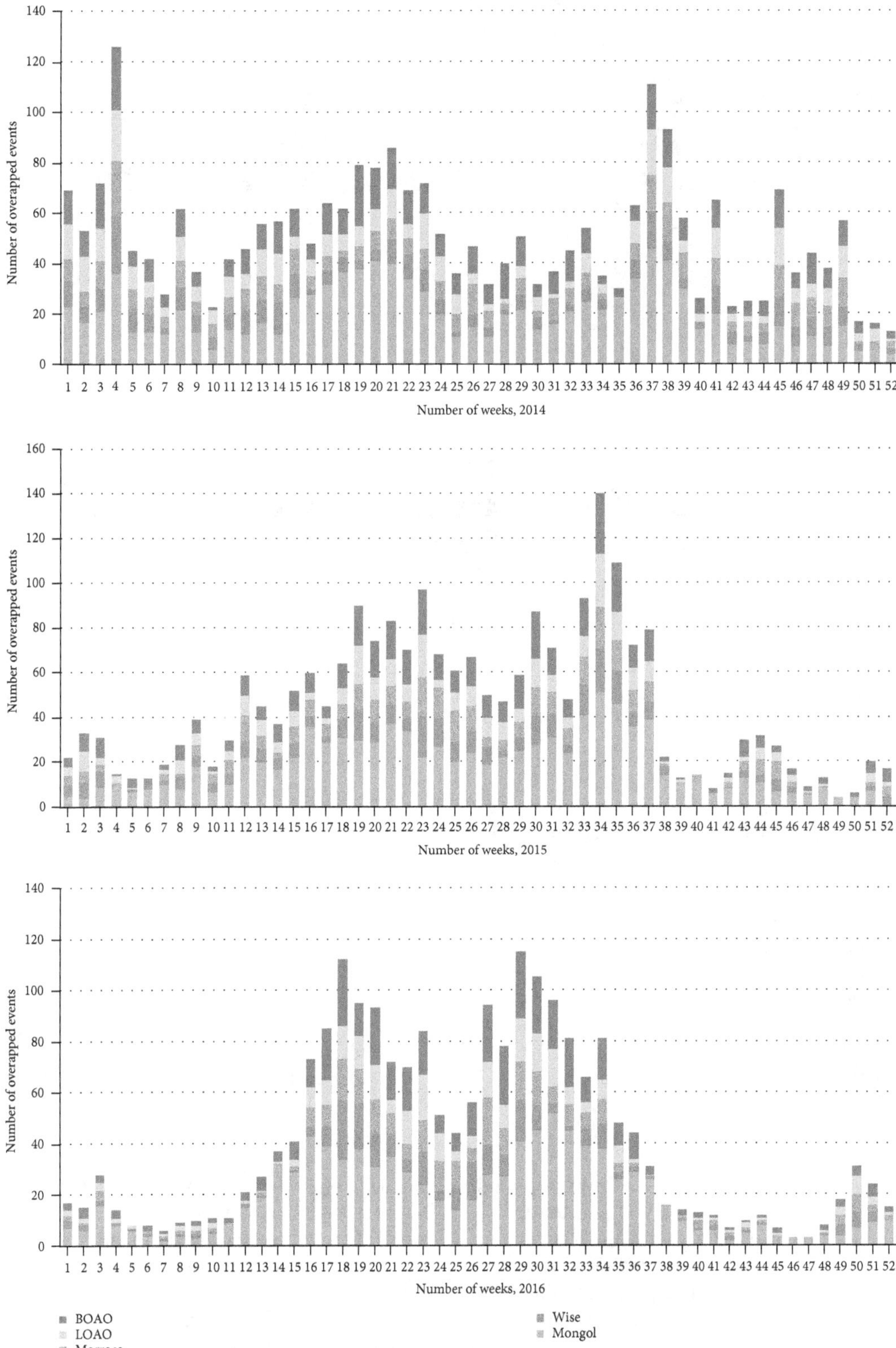

FIGURE 4: Weekly variation of number of optical observation overlapped events for 11 LEO observation targets at the five OWL-Net stations from 2014 to 2016.

250 s as a test case. The average observation duration for the 11 domestic LEO satellites at the OWL-Net stations in 2015 was ~250 s. We also selected three other cases (10 s, 60 s, and full arc) to investigate the number of overlap events. The full arc indicated the maximum observation time duration under the optical observation constraints.

The number of optical observation overlap events is presented in Figure 5. Most overlap events occurred at the

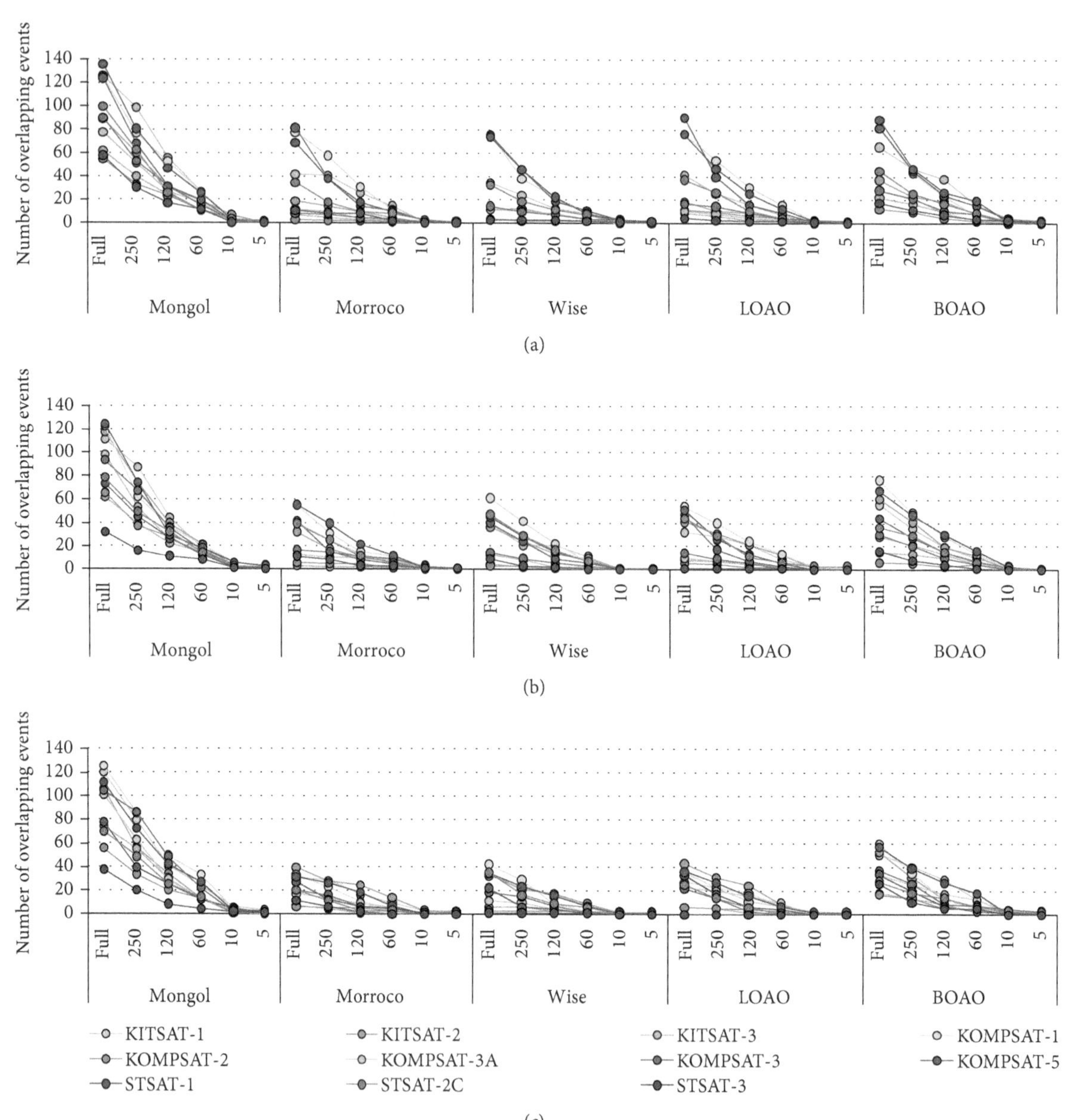

FIGURE 5: Variation of number of overlapped events for the 11 LEO observation targets for normal cases (full) and five short-arc strategies from 2014 to 2016 ((a) 2014, (b) 2015, and (c) 2016).

Mongolian site for all satellites, because it was supposed that the number of overlapped events was affected by the latitude of site and designed orbit of satellites. Except for the Mongolian site, the rest of the sites showed a similar number of overlap events. In the overall trend, almost all overlap events in a ~10 s short-arc strategy were eliminated.

The short-arc observation strategy for the 3 selected weeks with the most overlapped events were analysed with an orbit determination test. In a short-arc observation strategy, only the arc length is considered with the same observation frequency in each observation span. The TLEs from JSpOC were used as a priori for 11 LEOs. Generally, TLE information is good to fit for the orbit of target within a few days in forward propagation [18]. However, there are sometimes abnormal TLEs for target of orbits. Therefore, we made one reference orbit for 2 weeks considering orbital information from consecutive TLEs for each observation target. This reference orbit information was used to simulate the optical observation overlapped event and to test orbit determination accuracy.

We selected five short-arc optical observation cases to compare the orbit determination effectiveness by arc length: full arc, 120, 10, 60, and 5 s. The number of overlapped events was almost 70% against the full-arc case. Table 2 shows the number of overlapped events of five cases for 11 LEOs in the selected weeks of 2014, 2015, and 2016. These selected weeks experienced the most overlapping events for each year as we mentioned in Section 2.2. In the case of the selected week in 2014, there were no observation opportunities for six LEOs. However, there were the most overlapping events

Table 2: Number of overlapped events for 11 domestic LEO satellites at the five OWL-Net stations during the selected week for orbit determination test by the number of weekly optical observation overlapped events (2014: 4th week; 2015: 34th week; and 2016: 18th week).

Satellite	Year	Total number of accesses	Number of overlapped events for short-arc optical observation strategy					
			Full-arc	250	120	60	10	5
KITSAT-1	2014	160	18	12	6	2	0	0
	2015	114	6	9	6	1	0	0
	2016	106	16	12	3	0	0	0
KITSAT-2	2014	0	0	0	0	0	0	0
	2015	63	11	7	4	1	0	0
	2016	64	17	8	5	1	0	0
KITSAT-3	2014	0	0	0	0	0	0	0
	2015	20	6	2	2	2	0	0
	2016	42	2	2	1	1	0	0
KOMPSAT-1	2014	100	39	25	9	4	1	0
	2015	50	19	4	1	1	0	0
	2016	56	14	7	5	3	0	0
KOMPSAT-2	2014	0	0	0	0	0	0	0
	2015	21	5	2	2	1	0	0
	2016	36	2	1	0	0	0	0
KOMPSAT-3	2014	0	0	0	0	0	0	0
	2015	37	5	4	3	1	0	0
	2016	56	18	13	6	3	0	0
KOMPSAT-3A	2014	0	0	0	0	0	0	0
	2015	15	4	3	2	2	0	0
	2016	33	9	8	4	3	1	1
KOMPSAT-5	2014	104	22	14	7	4	1	1
	2015	32	8	4	2	0	0	0
	2016	20	3	3	2	2	0	0
STSAT-1	2014	100	42	25	13	6	0	0
	2015	62	24	12	3	2	0	0
	2016	60	18	10	8	3	0	0
STSAT-2C	2014	24	6	5	1	1	0	0
	2015	64	35	27	8	3	0	0
	2016	50	17	8	6	4	1	0
STSAT-3	2014	0	0	0	0	0	0	0
	2015	36	9	5	3	1	0	0
	2016	53	0	0	0	0	0	0

in the fourth week in 2014, with figures similar to those in 2015 and 2016. As we confirmed earlier in Figure 5, there were a few overlapped events for 10 s and 5 s in those weeks. In this most overlapped event week, almost all overlapped events in a <10 s short-arc optical observation strategy were also eliminated like we have shown in yearly analyses plotted in Figure 5.

3.2. Orbit Determination Result for Five Short-Arc Optical Observation Strategies. The orbit determination test with the short-arc observation simulation was implemented. TLE was used as the a priori of the reference orbit as we described in the previous section. Simulated optical observation data with various arc lengths were used with assumed seeing errors of 5 arc-seconds for the right ascension and declination, respectively. In case of the tracking for LEO satellites, the time synchronization error is the dominant source of error. We assumed that the overall errors were included in that 5 arc-seconds. In the system requirements of the OWL-Net, the maximum astrometric error was assumed as 5 arc-seconds. We considered error modelling for the atmospheric density, solar radiation pressure, measurement biases, process noise, and measurement white noise. Although OWL-Net telescope can make tens of observation

Table 3: Orbit estimator setup and used perturbations.

Perturbation	Gravity	EGM2008 (70 × 70) Solid tide
	Third body	Sun and moon, planets
	Drag	JB2008, spherical shape
	SRP	Spherical shape
Estimator	Integrator	RKF7(8)
	Filter	Kalman filter
	Initial uncertainty	Range: 50 m in-track: 100 m cross-track: 20 m

points per second by using the chopper system, we deliberately simulated one point per second interval observations to check the effect of the arc length and the number of arcs. The estimator setup is summarised in Table 3. The commercial software System Tool Kit and Orbit Determination Tool Kit by Analytical Graphics Inc. were used for simulation and orbit estimation. The initial uncertainty in Table 3 meant initial covariance for sequential filtering.

The orbit determination results with observation data for seven days were compared with the propagated orbit for seven days after the end of the epoch of filtering. The results were analysed in the radial, in-track, and cross-track frame (RIC frame). Among those, radial and cross-track position uncertainty and differences from reference orbit showed a small and similar tendency for all cases. Therefore, we compared only in-track position uncertainty and difference. Figure 6 shows one-sigma position uncertainty in meters from the orbit estimation results versus the number of arcs for selected weeks in 2014, 2015, and 2016. The position uncertainty was displayed versus the total number of arcs without distinction of targets or year. Understandably, the position uncertainty was inversely proportional to the arc length for each satellite. However, the position uncertainties of 5 s and 10 s were similar to each other for all cases of each satellite. It indicates that frequent short-arc observation is more effective in maintaining the precision of the orbit than sparse longer-arc observation is.

We summarise the in-track direction orbit determination comparison results for seven days after the orbit determination in Figure 7. The estimated orbit was propagated seven days and compared with the reference orbit. The goal of the OWL-Net was maintaining orbit ephemeris with seven days to support follow-up tracking. Therefore, the maximum in-track difference is as shown in Figure 7. Even if the arc length was reduced to 5 s for each pass, the orbit estimation results show similar results to those of the original, full-arc case for the next seven days. In the case of STSAT-2C, however, the estimated orbit showed large differences (tens of kilometers or more). Due to its elliptical orbit, STSAT-2C has more geometrical limitations on observation opportunities. Therefore, the observation chances were more frequent when the satellite is near the apogee. The number of observation chances was similar with other satellites; the orbit determination of STSAT-2C failed in this simulation. Use of single- or double-station observations can also lead to greater

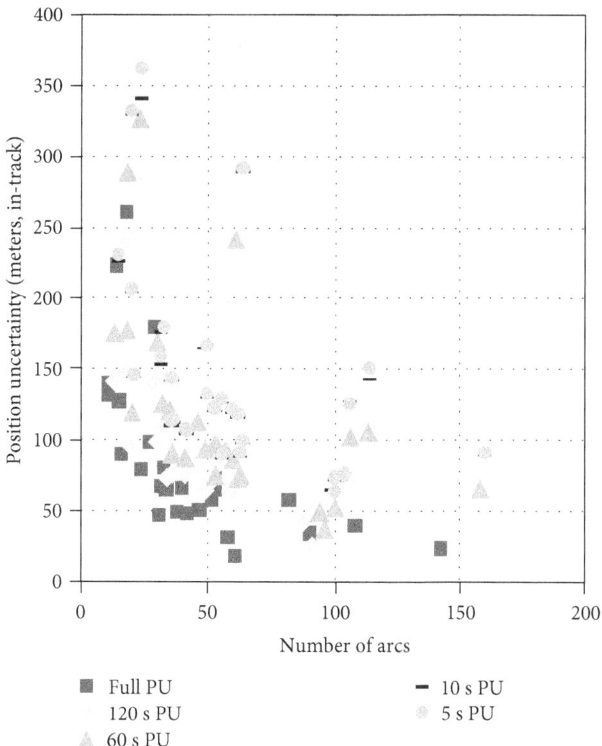

Figure 6: Final in-track direction position uncertainty versus the number of arcs from 7-day estimation for 11 LEOs on the most overlapped event week.

differences for the orbit estimation. We could not find any tendency of the orbit estimation accuracy between the arc length and in-track difference. This means short-arc optical observation strategy can make the orbit estimation of similar accuracy against full-arc optical observation. It can also provide more observation opportunities to many observation targets.

We confirmed that the short-arc optical observation strategy was more effective than the full-arc optical observation strategy with similar orbit accuracy and more observation opportunities. In the case of real observing conditions, the short-arc observation strategy provides a higher possibility of successful observation when weather conditions or other observation conditions like hardware maintenance are taken into account. Using short arcs, it is sometimes possible to get one or two successful shots for multishots. When using a CCD (charge-coupled device), the short-arc observing strategy also allows for a signal read-out and telescope positioning to the target. Therefore, we propose the short-arc strategy with 60 s arc-length as most effective optical observation tasking for the OWL-Net.

3.3. Orbit Determination Result for Five Short-Arc Optical Observation Strategies Using the Real OWL-Net Observation Data. We validated the short-arc optical observation strategy with the real observation data from the OWL-Net. The observation condition for LEO target varies with path direction, range, and track duration. We could not consider the

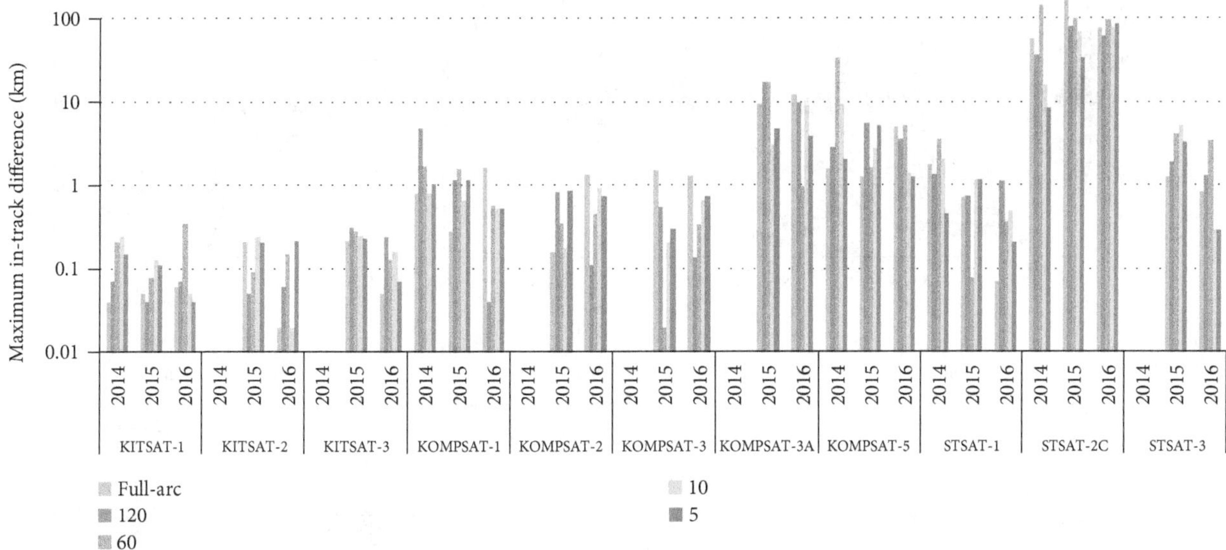

Figure 7: Maximum in-track direction difference in kilometers between the estimated orbit and the reference orbit during seven days after the end of orbit estimation with 5 optical observation test cases for seven days.

Table 4: Observation summary of KOMPSAT-1 from OWL-Net.

Site	Date	Shot	Number of points	Arc-length (seconds)
Wise	2017-06-23	7	566	145
Wise	2017-06-25	5	301	170
Wise	2017-06-26	7	500	130
Wise	2017-06-29	8	729	220

same observation condition with the previous simulation for multi-LEO targets. Therefore, we performed the orbit determination test with the real observation data for one target with various arc-lengths.

The target satellite was KOMPSAT-1. KOMPSAT-1 is a nonoperational satellite with the altitude of 660 km above the ground. We observed KOMPSAT-1 for four days from 23rd to 29th of June of 2017 at the Wise observatory of the OWL-Net. The observation summary is described in Table 4. We collected almost 2000 points with a high-speed chopping system [6]. The orbit estimation test was done for full-arc, 120 s, 60 s, and 5 s cases. Figure 8 shows the comparison results of the estimated orbit and consecutive TLEs. The propagated orbit for seven days was also compared with TLEs. The differences of the four cases were very similar to each other except the 5 s case. Radial and cross-track differences were smaller and stable than the in-track direction fluctuation. The radial and cross-track differences were maintained within 1 km for duration of the estimation and the prediction. The differences for the radial and cross-track directions in root mean square (RMS) were maintained within 0.1 km and 0.2 km for 4 cases, respectively. The in-track direction fluctuation within 3 km came from uncertainty of TLE data. The 5 s case only showed a slightly bigger difference, but still the maximum difference did not exceed 2 km for 4 cases. In case of the differences for the in-track direction in RMS, the full-arc, 120 sec, and 60 sec case errors were under 0.66 m. And the difference of the in-track direction in RMS for the 5 sec case was 0.67 m. We confirmed that a 60 s short-arc observation is enough to maintain the orbit information stably in real observation cases.

4. Summary and Discussion

We analysed the effectiveness of a short-arc optical observation strategy for orbit determination using the OWL-Net. In the case of the LEO satellites, the optical observation time is limited to dawn and dusk. 11 South Korean LEO satellites are the main target of the observation of the OWL-Net. The OWL-Net is aimed at maintaining the orbital elements with weekly based orbit determination. However, the observation schedule for each target can overlap occasionally. The schedule overlapped event can result in bad quality of orbit determination with limited optical observation chances. Therefore, an effective observation strategy is required for an optical tracking system with multiple targets.

The number of weekly base optical observation overlapping events was analysed for the observation targets of the OWL-Net in 2014, 2015, and 2016. In 2015, the number of optical observation overlap events peaked in the 34th week. The 4th and 18th weeks experienced the most overlapping events of 2014 and 2016, respectively. The optical observation overlap events occurred during more than 10% of total observation opportunities for the five sites but did not show any obvious weekly trend. We simulated observation data during these 3 selected weeks in 2014, 2015, and 2016. Short-arc optical observation data were simulated for 250, 120, 10, and 5 s of observation pass duration. In case of 10 s and 5 s, only a few overlapping events remained.

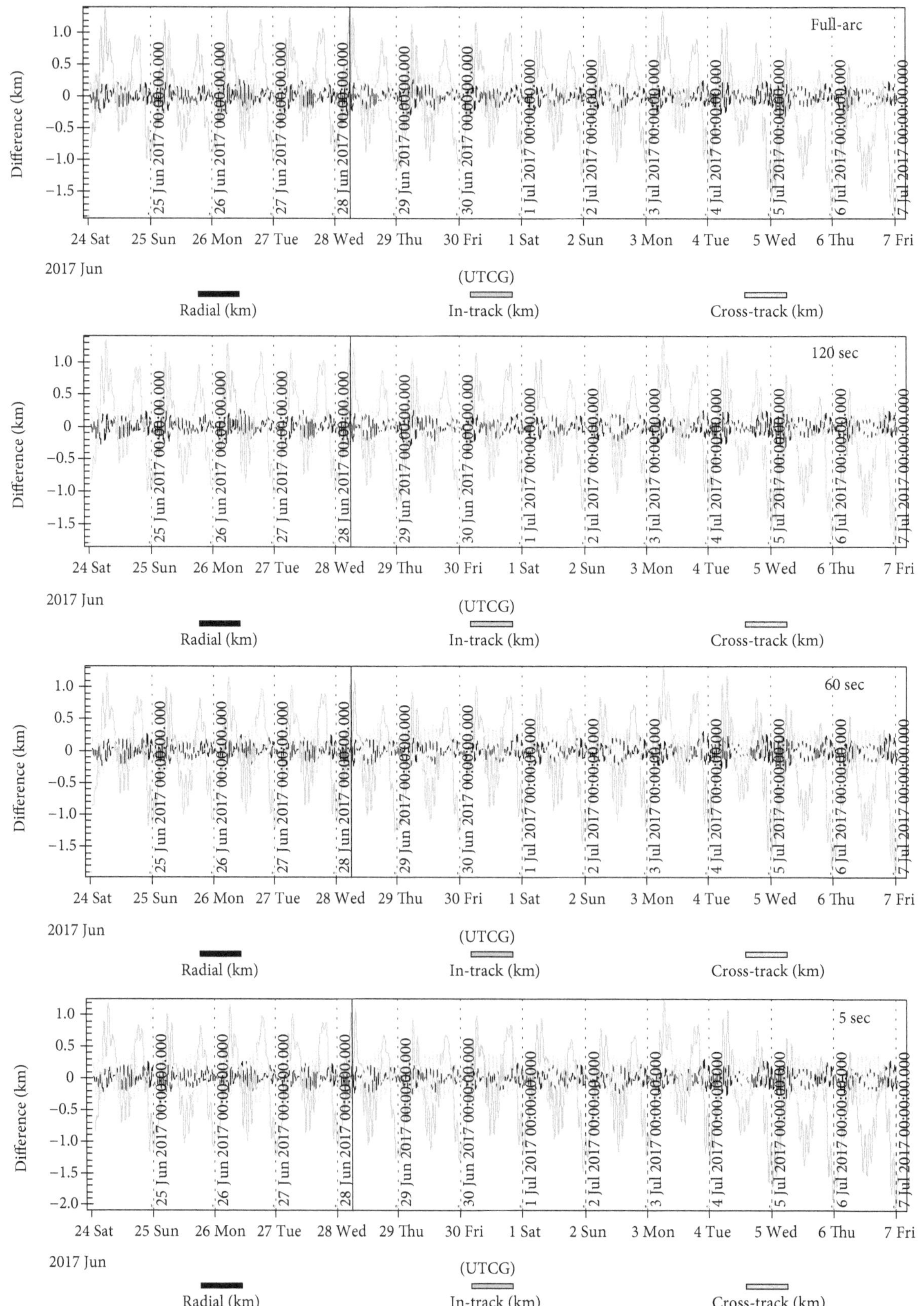

Figure 8: Comparison on the RIC frame with estimated orbit and consecutive TLEs for full-arc, 120 s, 60 s, and 5 s cases (first seven days: estimated orbit, last seven days: propagated orbit).

The short-arc optical observation cases show similar orbit determination results with more observation opportunities. The estimated orbits were compared with reference orbits generated by using TLE for the given target. The in-track direction orbital differences were similar to each other without any clear trends. In the case of STSAT-2C, the orbit determination test failed due to the geometrical condition of the optical observation. This result from simulation was also confirmed with orbit estimation results for KOMPSAT-1 using the real observation data from OWL-Net. We proposed 60 s arc-length short-arc observation strategy for scheduling strategy of real observation as seen at Section 3.3.

In this study, we have not incorporated weather conditions or the possibility of system anomalies. In a real situation, the observation opportunities can be reduced by these environmental conditions. Furthermore, the number of target satellites tracked by the OWL-Net system will increase consistently in the years to come, and growth in the number of observation targets will increase the number of overlap events. As we demonstrated in our work, the short-arc optical observation strategy can offer greater observation opportunities with similar orbit determination accuracy. Therefore, the short-arc observation strategy should be taken into account for the effective optical tracking operation of the OWL-Net.

Conflicts of Interest

The authors declare that there is no conflict of interest regarding the publication of this paper.

Acknowledgments

This study was supported by the Korea Astronomy and Space Science Institute (KASI) with the research project "Development of Space Situational Awareness Technology." The authors express their sincere appreciation to colleagues for the observation of the OWL-Net and support for this research: Myung-Jin Kim, Dong-Goo Roh, Sooyoung Kim, Aeri Kim, and Jang-Hyun Park.

References

[1] G. Veis, "Optical tracking of artificial satellites," *Space Science Reviews*, vol. 2, no. 2, pp. 250–296, 1963.

[2] "IADC Space Debris Mitigation Guidelines," IADC-02-01 Revision 1, 2007, http://www.iadc-online.org/Documents/.

[3] A. G. Massevitch and A. M. Losinsky, "Photographic tracking of artificial satellites," *Space Science Reviews*, vol. 11, no. 2-3, pp. 308–340, 1970.

[4] M. Shoemaker and L. Shroyer, "Historical trends in ground-based optical space surveillance system design," in *Proceedings of the Advanced Maui Optical and Space Surveillance Technologies Conference*, Wailea, HI, USA, 2007.

[5] J.-H. Kim, J.-H. Jo, J. Choi et al., "Visibility analysis of domestic satellites on proposed ground sites for optical surveillance," *Journal of Astronomy and Space Sciences*, vol. 28, no. 4, pp. 319–332, 2011.

[6] S.-Y. Park, K.-H. Keum, S.-W. Lee et al., "Development of a data reduction algorithm for Optical Wide Field Patrol," *Journal of Astronomy and Space Sciences*, vol. 30, no. 3, pp. 193–206, 2013.

[7] S.-Y. Park, J. Choi, D.-G. Roh et al., "Development of a data reduction algorithm for optical wide field patrol (OWL) II: improving measurement of lengths of detected streaks," *Journal of Astronomy and Space Sciences*, vol. 33, no. 3, pp. 221–227, 2016.

[8] J. H. Jo, J. Choi, M. Park et al., "Early CAL/VAL process for an optical tracking system by Korea," in *Proceedings of the Advanced Maui Optical and Space Surveillance Technologies Conference*, Wailea, HI, USA, 2015.

[9] M. Park, J. H. Jo, S. Cho et al., "Minimum number of observation points for LEO satellite orbit estimation by OWL network," *Journal of Astronomy and Space Sciences*, vol. 32, no. 4, pp. 357–366, 2015.

[10] J. Y. Son, J. H. Jo, J. Choi et al., "Optical orbit determination of a geosynchronous earth orbit satellite effected by baseline distances between various ground-based tracking stations II: COMS case with analysis of actual observation data," *Journal of Astronomy and Space Sciences*, vol. 32, no. 3, pp. 229–235, 2015.

[11] J. Choi, J. H. Jo, M.-J. Kim et al., "Determining the rotation periods of an inactive LEO satellite and the first Korean space debris on GEO, KOREASAT 1," *Journal of Astronomy and Space Sciences*, vol. 33, no. 2, pp. 127–135, 2016.

[12] A. Herz and F. Stoner, "SSA sensor tasking approach for improved orbit determination accuracies and more efficient use of ground assets," in *Proceedings of the Advanced Maui Optical and Space Surveillance Technologies Conference*, Wailea, HI, USA, 2013.

[13] D. A. Vallado and S. S. Carter, "Accurate orbit determination from short-arc dense observational data," *Advances in the Astronautical Sciences*, vol. 97, pp. 1587–1602, 1998.

[14] J. C. Bennett, J. Sang, C. Smith, and K. Zhang, "An analysis of very short-arc orbit determination for low-earth objects using sparse optical and laser tracking data," *Advances in Space Research*, vol. 55, no. 2, pp. 617–629, 2015.

[15] S. Kawase, "Orbit determination accuracy for optically tracked near-synchronous objects," in *Proceedings of the International Symposium on Space Dynamics*, Biarritz, France, 2000.

[16] J. Tombasco and P. Axelrad, "A study of the achievable geosynchronous angles-only orbit estimation accuracy," *The Journal of the Astronautical Sciences*, vol. 58, no. 2, pp. 275–290, 2011.

[17] D. A. Vallado and W. D. McClain, *Fundamentals of Astrodynamics and Applications*, Kluwer Academic Publisher, Boston, MA, USA, 2nd edition, 2001.

[18] D. A. Vallado, B. B. Virgili, and T. Flohrer, "Improved SSA through orbit determination of two-line element sets," in *6th European Conference on Space Debris*, Darmstadt, Germany, 2013.

11

Optimization of Allocation and Launch Conditions of Multiple Missiles for Three-Dimensional Collaborative Interception of Ballistic Targets

Burak Yuksek and N. Kemal Ure

Aerospace Research Center, Department of Aeronautical Engineering, Istanbul Technical University, 34469 Istanbul, Turkey

Correspondence should be addressed to N. Kemal Ure; ure@itu.edu.tr

Academic Editor: Mahmut Reyhanoglu

We consider the integrated problem of allocation and control of surface-to-air-missiles for interception of ballistic targets. Previous work shows that using multiple missile and utilizing collaborative estimation and control laws for target interception can significantly decrease the mean miss distance. However, most of these methods are highly sensitive to initial launch conditions, such as the initial pitch and heading angles. In this work we develop a methodology for optimizing selection of multiple missiles to launch among a collection of missiles with prespecified launch coordinates, along with their launch conditions. For the interception we use 3-DoF models for missiles and the ballistic target. The trajectory of the missiles is controlled using three-dimensional extensions of existing algorithms for planar collaborative control and estimation laws. Because the dynamics of the missiles and nature of the allocation problem is highly nonlinear and involves both discrete and continuous variables, the optimization problem is cast as a mixed integer nonlinear programming problem (MINP). The main contribution of this work is the development of a novel probabilistic search algorithm for efficiently solving the missile allocation problem. We verify the algorithm by performing extensive Monte-Carlo simulations on different interception scenarios and show that the developed approach yields significantly less average miss distance and more efficient use of resources compared to alternative methods.

1. Introduction

Missile systems form an integral part of a nation's defensive capability. In particular, high precision control of surface-to-air-missiles (SAMs) is a key technology for intercepting maneuvering targets, such as hostile aircraft and ballistic missiles [1]. Successful mid-air interception of such targets might prevent significant losses and enable preparing for counter-attacks. Hence, there has been a significant amount of previous work that focuses on development of guidance, control, and estimation laws for interception of maneuvering targets by a single missile in the past years [2].

Intercepting ballistic targets is generally more challenging compared to interception of static targets or targets executing simple maneuvers (such as an aircraft in level flight), due to lack of precise dynamic models and high reentry speed of the target. Hence, even the smallest perturbations in the guidance errors might translate into large miss distances for the interception of ballistic targets. In order to reduce the miss distance, there has been a growing interest in using multiple collaborating missiles for the interception of a single target. There are two main advantages for using multiple missiles; (1) probability to hit increases due to increased number of missiles targeted at the hostile; (2) missiles can fuse their measurements to improve the estimation of the target's state, which results in increased guidance performance and hence reduced miss distance.

That being said, the initial conditions of the launched missiles still have a significant impact on their overall performance. Ballistic target's total speed is usually greater than the interceptors speed; hence the launching coordinates of the SAMs directly affect the miss distance. In addition, launching pitch and heading angles of the SAMs are also critical for the guidance performance. Even though errors in these angles are compensated by the missile autopilots in later phases of the flight, due to high kinetic energy of the ballistic target, having 1-2 degrees off in the initial pitch and heading angles of the SAMs might lead to significant deviations from the desired performance, even in the presence of restoring actions of the autopilot. Hence, the selection of the launching coordinates of the SAMs along with their initial launching angles should be adjusted carefully to gain optimal collaborative missile interception performance. The main contribution of this paper is the development of an optimization algorithm that drives the initial conditions of collaborative guidance and estimation laws. In particular, the missile launching parameters and the number of launched missiles are selected using a probabilistic search algorithm, which attempts to optimize an objective function that favors minimum miss distance and maximum efficiency in use of resources.

1.1. Previous Work on Collaborative Estimation and Control for Missiles. There has been a significant amount of previous work on control and estimation of multiple missiles. Chen and Speyery [3] formulated the multiple missile coordination problem as a Linear Exponential Gaussian Differential game and applied their algorithm to interception of ballistic missiles in terminal and boost phase. Wang and Fu [4] formulated the multiple missile interception problem as a multiplayer pursuit and evasion game and studied the interception of a ballistic target in three dimensions. Jeon et al. [5] developed a cooperative proportional navigation (PN) law that enables multiple missiles to close simultaneously on a stationary target. Similarly, Daughtery and Qu [6] also developed an algorithm for multiple missiles attacking simultaneously a target, and they also showed that their algorithm is robust to communication losses between the missiles.

Shaferman and Oshman [7] developed an extended Kalman filter (EKF) algorithm that fuses information gathered from multiple interceptors. They were able to show that using cooperative estimation algorithms yield improved guidance and control performance. Shaferman and Shima [8] combined adaptive control laws and multiple model filtering algorithms for collaborative interception. Liu et al. [9] also considered using multiple EKFs to improve guidance performance; in particular they discovered that the estimation performance improves as the relative line of sight (LOS) between two intercepting missiles gets larger and designed a control law that enforces separation between two missiles. Recently, Shaferman and Shima [10] also developed guidance laws for enforcing relative intercept angles. Another recent development was provided by Wang et al. [11], where the authors used a probabilistic framework to maximize hit-to-kill probability for two-missile cooperative interception.

These previous works showed that fusing the estimation process between multiple missiles almost certainly leads to improved control performance and hence reduced miss distance. Most of these works only study the scenario where missiles are about to close on the target; the launch conditions are largely ignored. However, as explained in the beginning of this section, estimation and control performance might vary significantly under different launch conditions, such as different initial heading and elevation angles of the missiles. Moreover, launch conditions could also require adjustments based on relative heading and velocity of the ballistic target. In particular, the sensitivity of miss distance based on different conditions would be higher for targets with high speeds, such as ballistic missiles.

1.2. Previous Work on Missile Allocation. The existing algorithms for multiple missile interception assume that a fixed number of missiles have been assigned for the target interception. In real scenarios, usually a larger set of missiles are available on the ground, and a certain subset of these missiles should be allocated and launched for interception. Deciding on which missiles to allocate depends on several factors, such as the altitude and the velocity of the target. Since increasing the number of allocated missiles improves the estimation performance and hence the probability of successful interception, it might be desirable to launch many missiles as possible. However, launching more missiles than necessary would result in inefficient use of resources; hence a trade-off exists between the kill probability and the number of launched missiles.

To the best of our knowledge, there are no previous works that directly address the allocation problem for multiple missiles that utilize collaborative estimation and control algorithms. However, previous researches on weapon and task allocation methods are certainly relevant to this problem, and they are reviewed in this subsection.

Weapon and task allocation algorithms generally address the problem of assigning a finite number of weapons to multiple targets. The search space for such problems are usually huge and the objective functions lack useful properties such as linearity or convexity; hence evolutionary algorithms and probabilistic search approaches are very popular for these problems. Lee et al. [12] used a genetic algorithm based on greedy eugenics to solve the assignment problem, whereas Teng et al. [13] utilized particle swarm optimization approach for the coordinated air combat missile-target assignment. Ant colony optimization [14] and simulated annealing [15] have also been applied to the target assignment problem. Besides genetic algorithms, game theoretic [16] and rule-based [17] approaches have also been applied to missile allocation problems.

These existing works either completely ignore the vehicle dynamics or use simplified missile and target models for solving the allocation problem. However, utilization of collaborative estimation and control algorithms is integral to the interception of ballistic target problem studied in this paper; hence the dynamics of estimation and control algorithms cannot be ignored by the allocation algorithm. Existing works usually remedy this problem by assigning a probability of success to each possible assignment; however overestimating or underestimating these probabilities might lead to serious performance problems.

1.3. Overview and Contribution of This Work. The main idea behind the developed algorithm is to cast the missile allocation problem as a mixed integer nonlinear program (MINP), where the discrete decision variables correspond to selection of which missiles to launch, and the continuous decision variables correspond to launch conditions such as elevation and heading angles for each missile. In order to solve this optimization problem, a novel probabilistic search method is developed, which extends the well-known continuous optimization algorithm Covariance Matrix Adaptation (CMA) [18] to solve problems that contain both continuous and discrete variables. The optimization problem also embeds the full dynamics of the missiles, where we have extended an existing collaborative EKF algorithm [7, 10] to work in three-dimensional settings.

With respect to the state of the art, our algorithm offers the following contributions:

(i) The previous work on missile control assumes fixed initial launch conditions for collaborative targets. Our algorithm automates the process of selecting missile heading and elevation angles, which impacts the estimation and control performance in later stages and hence reduces the average miss distance.

(ii) Besides optimizing the launch conditions, the developed algorithm also handles the selection of the subset of missiles for interception among the available larger set of missiles by examining the trade-off between the missile launch cost and achievable miss distance. This feature is captured in MINP formulation, and the results show that the developed algorithm yields lesser miss distances using lesser number of missiles, compared to alternative approaches.

(iii) The developed optimization algorithm works by propagating missile and target dynamics equipped with estimation and control laws. Hence unlike existing missile allocation algorithms, there are no simplifications or approximations in dynamical models, which translates into better prediction of overall system performance.

(iv) As a minor contribution, we would also like to note that most existing collaborative missile control papers consider two-dimensional (planar) missile and target dynamics. In this work we have extended the collaborative EKF to work in three dimensions. Although this is a straightforward extension, having the results in three dimensions leads to more realistic miss distances and hence enables a better assessment of the fitness of the developed algorithms on real battlefield operations.

The paper is structured as follows. In Section 2, the ballistic target interception problem is formulated and the missile modes for both the interceptors and the ballistic target are provided. Section 3 explains the collaborative estimation and control algorithms used in this work, and Section 4 gives the details of the novel optimization algorithm used for solving the allocation problem. Finally, Section 5 studies the performance of the algorithm by examining the results of Monte-Carlo simulations that involve various different target initial conditions.

2. Problem Formulation and Missile Models

In this section we provide the complete formulation of the optimization problem for the multiple missile ballistic target interception. Since the problem formulation involves the dynamics of the missiles, the model of the interceptor missiles and the ballistic target are also given in this section.

2.1. Multiple Missile Ballistic Target Interception Problem. Let $\mathbf{x}_T = [\mathbf{p}_T, \dot{\mathbf{p}}_T] \in \mathbb{R}^6$ denote the state vector of the ballistic target and let $\mathbf{x}_M^i = [\mathbf{p}_M^i, \dot{\mathbf{p}}_M^i] \in \mathbb{R}^6, i = 1, 2, \ldots, N_{\text{SAM}}$, denote the state vectors of the SAMs, where $\mathbf{p}$ denotes the position, $\dot{\mathbf{p}}$ denotes the velocity, and N_{SAM} denotes the number of missiles. It is assumed that SAMs initial positions are fixed at the launch site and the ballistic target is detected by a radar in its terminal phase. We assume that closed loop guidance and control algorithms are implemented on SAMs (see Section 3) and the decision maker only needs to decide on the initial launching conditions.

Let $\Gamma^0 = \{\gamma^i(0),\ i = 1, \ldots, N_{\text{SAM}}\}$ and $\Psi^0 = \{\psi^i(0),\ i = 1, \ldots, N_{\text{SAM}}\}$ denote the initial launching pitch and heading angles for each SAM. Let t_f be the terminal time for the scenario. Define the miss distance as

$$\text{miss}\left(\Gamma^0, \Psi^0\right) = \min_{i=1,\ldots,N_{\text{SAM}}, t \in [0, t_f]} \left\| \mathbf{p}_T(t) - \mathbf{p}_M^i(t) \right\|. \quad (1)$$

In other words, for a given set of launch conditions, miss distance is computed by propagating the dynamics of the SAMs and ballistic target over the horizon $[0, t_f]$ and finding the minimum distance achieved between the ballistic target and all launched SAMs.

The objective of the multiple missile ballistic target interception problem is to minimize the miss distance by optimizing over the launch conditions Γ^0, Ψ^0:

$$\begin{aligned} \min_{\Gamma^0, \Psi^0} \quad & \text{miss}\left(\Gamma^0, \Psi^0\right) \\ \text{subject to} \quad & \dot{\mathbf{x}}_T = f_T(\mathbf{x}_T) \\ & \dot{\mathbf{x}}_M^i = f_M\left(\mathbf{x}_T^i\right), \quad i = 1, \ldots, N_{\text{SAM}}, \end{aligned} \quad (2)$$

where f_T and f_M are the dynamical models of the ballistic target and interceptor missiles (see Section 2.2). For the sake of simplicity, we assume that all SAMs share the same model.

Although the optimization problem above reflects the main objective of the mission, it assumes that all of the SAMs will be used for interception. However, in many scenarios only a strict subset of the available SAMs might be sufficient for successful interception. Hence we should modify the objective function to reflect the cost of launching a missile in order to yield solutions that minimize the miss distance while avoiding inefficient use of resources.

Let $\mathbf{z} = \{z_i \in \{0, 1\},\ i = 1, \ldots, N_{\text{SAM}}\}$ denote the binary decision vector, where $z_i = 1$ means that SAM i is launched for intercepting the ballistic target and $z_i = 0$ means that SAM i is not launched and will stay in its initial position throughout the mission. The modified optimization problem is formulated as

$$\begin{aligned} \min_{\mathbf{z},\Gamma^0,\Psi^0} \quad & \text{miss}\left(\Gamma^0,\Psi^0\right) + \sum_{i=1}^{N_{\text{SAM}}} z_i r \\ \text{subject to} \quad & \dot{\mathbf{x}}_T = f_T\left(\mathbf{x}_T\right) \\ & \dot{\mathbf{x}}_M^i = f_M\left(\mathbf{x}_T^i\right), \quad \text{for } z_i = 1 \\ & \dot{\mathbf{x}}_M^i = 0, \quad \text{for } z_i = 0, \end{aligned} \tag{3}$$

where $r > 0$ is a user defined parameter that reflects the cost for launching a missile. Larger values of r favor launching lesser number of missiles. Since the optimization problem in (3) contains both continuous variables Γ^0, Ψ^0 and discrete variables $\mathbf{z}$ and since the dynamics f_M and f_T are nonlinear, it can be classified as a nonlinear mixed integer programming (MINP) problem. MINP are known to be very challenging to solve, and the major contribution of the paper is the development of a probabilistic search method (explained in Section 4) for solving the optimization problem in (3).

Note that we are only interested in deciding the allocation of the missiles and their launch conditions, and no further trajectory optimization is needed, since it is assumed that SAMs have closed loop estimation and guidance laws embedded in their dynamics. These estimation and control laws are further detailed in Section 3. However, before examining the control laws, we describe the open loop dynamics of the SAMs and ballistic target in the next subsection.

2.2. Missile Models

2.2.1. Interceptor SAM Model. The SAM interceptor used in this study is a 3-degree-of-freedom (DoF) point mass model of a missile, which is controlled via aerodynamic forces generated by the fins. Three-dimensional equations of motion of the interceptor can be stated as shown in [2]

$$\begin{aligned} \dot{x} &= V_m \cos\gamma \cos\psi, \\ \dot{y} &= V_m \cos\gamma \sin\psi, \\ \dot{z} &= V_m \sin\gamma, \\ \dot{\gamma} &= \left(a_{\text{pitch}} - \cos\gamma\right)\left(\frac{g}{V_m}\right), \\ \dot{\psi} &= \left(\frac{a_{\text{yaw}}}{\cos\gamma}\right)\left(\frac{g}{V_m}\right), \end{aligned} \tag{4}$$

where x is the downrange displacement of the SAM, y is the cross-range displacement of the SAM, z is the altitude of the SAM, g is the gravity acceleration, γ is the flight path angle, V_m is SAM velocity, and $a_{\text{pitch}}, a_{\text{yaw}}$ are the vertical and horizontal load factors.

2.2.2. Ballistic Target Model. In the 3D interception scenario, SCUD-B short range ballistic missile (SRBM) is used as the incoming ballistic threat. Simulation starts from a given

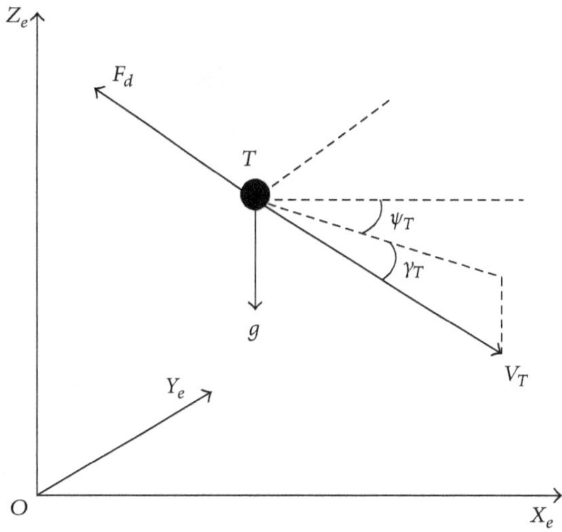

Figure 1: Forces acting on the ballistic target model.

apogee altitude, velocity, reentry angle, and heading. Aerodynamic data of the SCUD-B is obtained by using Missile DATCOM software [19].

The drag force and the gravity acceleration components that act on the ballistic target during reentry phase are shown in Figure 1. Drag force acts in the opposite direction of the velocity vector. Hence, if the effect of the drag force is greater than that of gravity, the ballistic target decelerates during its flight. The perpendicular component of the deceleration vector to the line of sight is shown as a target maneuver to the pursuing interceptor. Thus, the interceptor guidance law must take this deceleration maneuver of the ballistic threat into account [20].

The approximate mathematical model of the target in the reentry phase is given in [20]

$$\begin{aligned} \dot{V}_{Tx} &= \frac{-F_d}{m}\cos\left(\psi_T\right)\cos\left(\gamma_T\right), \\ \dot{V}_{Ty} &= \frac{F_d}{m}\sin\left(\psi_T\right)\cos\left(\gamma_T\right), \\ \dot{V}_{Tz} &= \frac{F_d}{m}\cos\left(\psi_T\right)\sin\left(\gamma_T\right) - g, \end{aligned} \tag{5}$$

where F_d is the drag force, V_{Tx}, V_{Ty}, and V_{Tz} are the velocity components, m is the target missile mass in the reentry phase, g is the gravity acceleration, γ_T is the reentry angle, and ψ_T is the heading angle.

3. Control and Estimation Algorithms

3.1. Radar Measurement for Initialization. For initialization of EoMs of the missiles and the estimation process, a ground based radar is used. The radar is fixed in a prespecified position. Relative positions of the interceptors and the target according to the radar in the 3D space are shown in Figure 2.

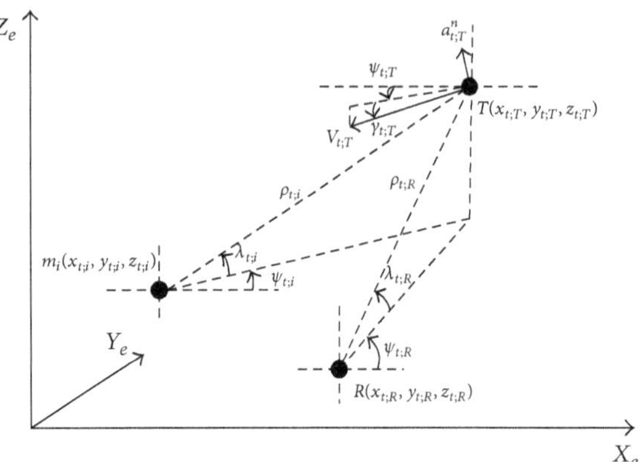

FIGURE 2: Radar geometry.

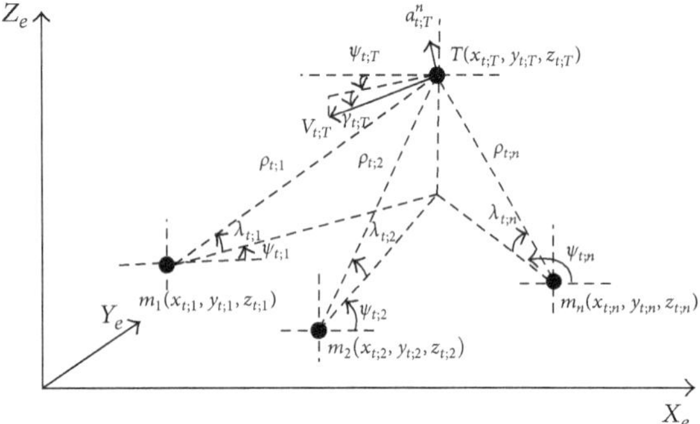

FIGURE 3: 3D engagement geometry.

We assume that the radar estimates the initial state vector $\widehat{x}_{0;R}$ without any error and send the initialization data to the interceptors without any delay

$$\widehat{\mathbf{x}}_{0;R} = \left[\rho_{0;R} \ \ \lambda_{0;R} \ \ \gamma_{0;T} \ \ a_{0;T}\right]. \tag{6}$$

The geometric relation between the ith interceptor and the target is given in (7). In this work, it is assumed that $\gamma_{0,T}$ and $a_{0,T}$ are estimated perfectly by the radar

$$\begin{aligned} \rho_{0;i} &= \sqrt{\left(x_{0;T} - x_{0,i}\right)^2 + \left(y_{0;T} - y_{0,i}\right)^2 + \left(z_{0;T} - z_{0,i}\right)^2}, \\ \lambda_{0;i} &= \arctan\left[\frac{z_{0,T} - z_{0,i}}{\sqrt{\left(x_{0;T} - x_{0,i}\right)^2 + \left(y_{0;T} - y_{0,i}\right)^2}}\right]. \end{aligned} \tag{7}$$

3.2. Relative Kinematics. 3D engagement geometry between the target and interceptors is shown in Figure 3. It is assumed that each interceptor missile can measure its own inertial state vector as shown in (8) and measurements are exact

$$\mathbf{x}^I_{t;i} = \left[x_{t;i} \ \ y_{t;i} \ \ z_{t;i} \ \ \gamma_{t;i} \ \ \psi_{t;i}\right]^T. \tag{8}$$

The state vector of the ith missile according to the target at time t is given in (9) where $i = 1, 2, \ldots, n$. In the information sharing mode, each missile could transmit and receive the estimated target data without any loss and delay

$$\mathbf{x}^R_{t;i} = \left[\rho_{t;i} \ \ \lambda_{t;i} \ \ \gamma_{t;T} \ \ a_{t;T}\right]^T. \tag{9}$$

Relative kinematics between the interceptors and the ballistic missile in S_{xz}, S_{xy}, and S_{yz} planes are defined by using

$$\begin{aligned} \dot{\rho}_{xz_i} &= V_{\rho_{xz}}, \\ \dot{\lambda}_{xz_i} &= \frac{V_{\lambda_{xz_i}}}{\rho_{xz_i}}, \\ \dot{\gamma}_{T_{xz}} &= -\frac{a_{T_{xz}}}{V_{T_{xz}}}, \\ \dot{a}_{T_{xz}} &= g\frac{a_{T_{xz}}}{V_{T_{xz}}}\sin\left(\gamma_{T_{xz}}\right), \\ \dot{\rho}_{xy_i} &= V_{\rho_{xy}}, \\ \dot{\lambda}_{xy_i} &= \frac{V_{\lambda_{xy_i}}}{\rho_{xy_i}}, \\ \dot{\gamma}_{T_{xy}} &= 0, \\ \dot{a}_{T_{xy}} &= 0, \\ \dot{\rho}_{yz_i} &= V_{\rho_{yz}}, \\ \dot{\lambda}_{yz_i} &= \frac{V_{\lambda_{yz_i}}}{\rho_{yz_i}}, \\ \dot{\gamma}_{T_{yz}} &= -\frac{a_{T_{yz}}}{V_{T_{yz}}}, \\ \dot{a}_{T_{yz}} &= g\frac{a_{T_{yz}}}{V_{T_{yz}}}\sin\left(\gamma_{T_{yz}}\right), \end{aligned} \tag{10}$$

where

$$\begin{aligned} V_{\rho_{xz}} &= -V_{m_{xz}}\cos\left(\gamma_{m_{xz}} - \lambda_{xz}\right) + V_{T_{xz}}\cos\left(\lambda_{xz} - \gamma_{T_{xz}}\right), \\ V_{\lambda_{xz}} &= -V_{m_{xz}}\sin\left(\gamma_{m_{xz}} - \lambda_{xz}\right) - V_{T_{xz}}\sin\left(\lambda_{xz} - \gamma_{T_{xz}}\right), \\ V_{\rho_{xy}} &= -V_{m_{xy}}\cos\left(\gamma_{m_{xy}} - \lambda_{xy}\right) \\ &\quad + V_{T_{xy}}\cos\left(\lambda_{xy} - \gamma_{T_{xy}}\right), \\ V_{\lambda_{xy}} &= -V_{m_{xy}}\sin\left(\gamma_{m_{xy}} - \lambda_{xy}\right) \\ &\quad - V_{T_{xy}}\sin\left(\lambda_{xy} - \gamma_{T_{xy}}\right), \\ V_{\rho_{yz}} &= -V_{m_{yz}}\cos\left(\gamma_{m_{yz}} - \lambda_{yz}\right) \\ &\quad + V_{T_{yz}}\cos\left(\lambda_{yz} - \gamma_{T_{yz}}\right), \end{aligned}$$

$$V_{\lambda_{yz}} = -V_{m_{yz}} \sin\left(\gamma_{m_{yz}} - \lambda_{yz}\right) - V_{T_{yz}} \sin\left(\lambda_{yz} - \gamma_{T_{yz}}\right). \tag{11}$$

The relative kinematics equations can be described in the discrete-time by using

$$\mathbf{x}_k = f_{k-1}\left(x_{k-1}\right), \tag{12}$$

where $\mathbf{x}_k$ is state vector of the ith missile at time t_k and f_{k-1} is obtained by integrating the relative kinematics EoMs in (10) [7].

3.3. Measurement Model. Each interceptor is equipped with the infrared seeker that measures the line of sight (LOS) angle $\lambda_{k;i}$. This measurement has a zero-mean Gaussian noise with standard deviation σ_{λ_i}. LOS angle measurements are performed by each interceptor missile separately; therefore $E(v_{k;i}, v_{k;j}) = 0, \forall i \neq j$. The ith missile LOS measurement is

$$z_{k;i} = h_{k;i}\left(x_k\right) + v_{k;i}, \tag{13}$$

where

$$v_{k;i} \sim \aleph\left(0, \sigma_{\lambda_i}^2\right). \tag{14}$$

The interceptor missiles can operate in two modes. The first mode is information nonsharing mode. In this mode, each interceptor measures only its own LOS angle and uses it in the estimation process of the relative states. The measurement vector in the information nonsharing mode is given as shown in (13).

The second mode is information sharing mode in which each interceptor not only measures its own LOS angle but also calculates the LOS angle of the other interceptors as shown in (15) [7]. Also, to improve the estimation quality, each interceptor calculates the range-to-go distance by using position and LOS angle data as shown in (16):

$$z_{\lambda_{i,j}} = \begin{bmatrix} z_{\lambda_{i,1}} \\ z_{\lambda_{i,2}} \\ \vdots \\ z_{\lambda_{i,n}} \end{bmatrix}, \tag{15}$$

$$z_{\rho_{i,j}} = \begin{bmatrix} z_{\rho_{i,1}} \\ z_{\rho_{i,2}} \\ \vdots \\ z_{\rho_{i,n}} \end{bmatrix}. \tag{16}$$

Here, i is the missile which performs measurement and n is the missile observed by the missile i.

The LOS angle of the missile j measured by the missile i is calculated by using

$$z_{\lambda_{i,j}} = \arctan \frac{\rho_i \sin\lambda_i - \left(z_j - z_i\right)}{\rho_i \cos\lambda_i + \left(x_i - x_j\right)}. \tag{17}$$

In addition, the range-to-go of the jth missile measured by the ith missile is obtained by using

$$z_{\rho_{i,j}} = \sqrt{\left(\rho_i \sin\lambda_i - \left(z_j - z_i\right)\right)^2 + \left(\rho_i \cos\lambda_i + \left(x_i - x_j\right)\right)^2}. \tag{18}$$

In the information sharing mode, the measurement model is given by

$$\mathbf{z}_{i,j} = \mathbf{h}\left(x_{i,j}\right) + \mathbf{v}_{i,j}, \tag{19}$$

where

$$\mathbf{z}_{i,j} = \left[z_{\rho(i,j)} \;\; z_{\lambda(i,j)}\right]^T, \qquad \mathbf{v}_{i,j} = \left[v_{\rho(i,j)} \;\; v_{\lambda(i,j)}\right]^T. \tag{20}$$

3.4. Extended Kalman Filter. Because of the nonlinear relative kinematics between the interceptors and the ballistic target, an extended Kalman filter (EKF) is used for estimating the unmeasured data and to filter the noisy LOS angle measurements [7]. The prediction error covariance matrix is given in

$$\mathbf{P}_{k|k-1} = \boldsymbol{\phi}_{k|k-1}\mathbf{P}_{k-1|k-1}\boldsymbol{\phi}_{k|k-1}^T + \mathbf{Q}_{k-1}, \tag{21}$$

where

$$\boldsymbol{\phi}_{k|k-1} = e^{\mathbf{F}_{k-1}T} \cong \mathbf{I} + \mathbf{F}_{k-1}T \tag{22}$$

is the transition matrix associated with the relative kinematics, T is sampling time, and $\mathbf{I}$ is the identity matrix with appropriate dimensions. $\mathbf{Q}_{k-1}$ is the covariance matrix of the equivalent discrete process noise and it is calculated as shown in

$$\mathbf{Q}_{k-1} = \int_0^T \boldsymbol{\phi}_{k|k-1}\mathbf{Q}\boldsymbol{\phi}_{k|k-1}^T dT. \tag{23}$$

$\mathbf{F}_{k-1}$ is the Jacobian matrix associated with the nonlinear relative kinematics

$$\mathbf{F}_{k-1} = \left.\frac{\partial f(x)}{\partial x}\right|_{x=\widehat{x}_{k-1}}. \tag{24}$$

The state estimation is performed by using

$$\widehat{\mathbf{x}}_{k|k} = \widehat{\mathbf{x}}_{k|k-1} + \mathbf{K}_k\left[\mathbf{z}_k - \mathbf{h}_k\widehat{\mathbf{x}}_{k|k-1}\right], \tag{25}$$

where $\widehat{\mathbf{x}}_{k|k-1}$ is the predicted state vector and $\mathbf{K}$ is the Kalman gain as shown in

$$\widehat{\mathbf{x}}_{k|k-1} = \boldsymbol{\phi}_{k|k-1}\widehat{\mathbf{x}}_{k-1|k-1}, \qquad \mathbf{K} = \mathbf{P}_{k|k-1}\mathbf{H}_k^T\left[\mathbf{H}_k\mathbf{P}_{k|k-1}\mathbf{H}_k^T + \mathbf{R}_k\right]^{-1}. \tag{26}$$

Here, $\mathbf{H}_k$ is the measurement Jacobian matrix and $\mathbf{R}_k$ is the measurement noise covariance matrix. The covariance matrix is updated as shown in

$$\mathbf{P}_{k|k} = \left[\mathbf{I} - \mathbf{K}_k\mathbf{H}_k\right]\mathbf{P}_{k|k-1}\left[\mathbf{I} - \mathbf{K}_k\mathbf{H}_k\right]^T + \mathbf{K}_k\mathbf{R}_k\mathbf{K}_k^T. \tag{27}$$

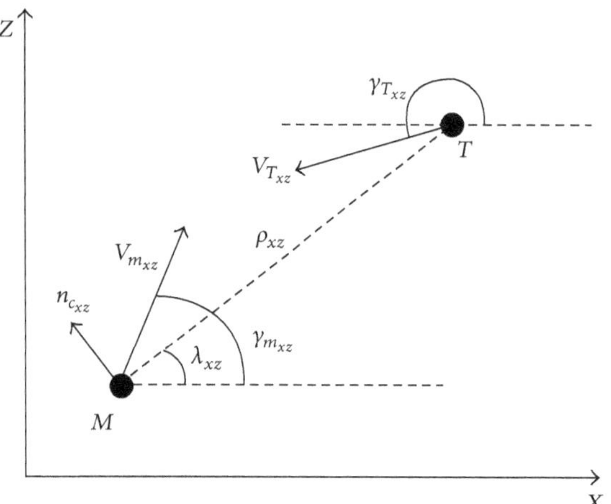

Figure 4: Relative kinematics on S_{xz} plane.

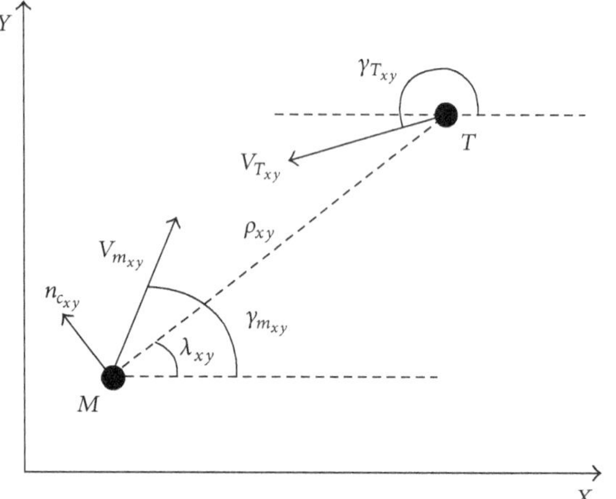

Figure 5: Relative kinematics on S_{xy} plane.

3.5. 3D True Proportional Navigation Algorithm. The proportional algorithm is one of the most common and effective guidance techniques because of its simple structure and implementation. The true proportional navigation (TPN) system generates the acceleration command perpendicular to the LOS. As shown in (29), the acceleration command is a function of closing velocity V_c and LOS rate $\dot{\lambda}$

$$n_c = NV_c\dot{\lambda}, \tag{28}$$

where n_c is the acceleration command perpendicular to the LOS, V_c is closing velocity, and N is navigation ratio which is generally between 3 and 5 [20]. In this 3D interception study, TPN algorithm is applied for S_{xz}, S_{xy}, and S_{yz} planes separately [21]. Geometry of relative kinematics for each different plane is displayed in Figures 4, 5, and 6.

Acceleration commands in S_{xz}, S_{xy}, and S_{yz} are obtained as shown in

$$\begin{aligned} n_{c_{xz}} &= NV_{c_{xz}}\dot{\lambda}_{xz}, \\ n_{c_{xy}} &= NV_{c_{xy}}\dot{\lambda}_{xy}, \\ n_{c_{yz}} &= NV_{c_{yz}}\dot{\lambda}_{yz}. \end{aligned} \tag{29}$$

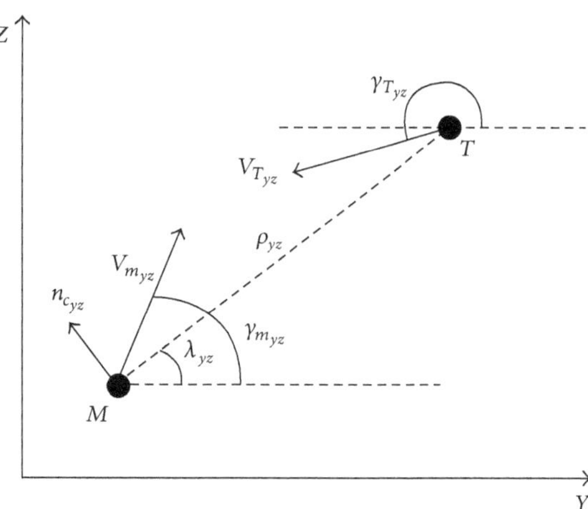

Figure 6: Relative kinematics on S_{yz} plane.

The acceleration components of the interceptor in the x-, y-, and z-axis $(a_{m_x}, a_{m_y}, a_{m_z})$ can be obtained from (30) by using the trigonometric relations:

$$\begin{aligned} a_{m_x} &= -n_{c_{xy}} \sin(\lambda_{xy}) - n_{c_{xz}} \sin(\lambda_{xz}), \\ a_{m_y} &= -n_{c_{xy}} \cos(\lambda_{xy}) - n_{c_{yz}} \sin(\lambda_{yz}), \\ a_{m_z} &= n_{c_{xz}} \cos(\lambda_{xz}) + n_{c_{yz}} \cos(\lambda_{yz}). \end{aligned} \tag{30}$$

Before applying the control commands to the interceptor, vertical and horizontal components a_{pitch} and a_{yaw} should be calculated. Here, a_{pitch} is in the pitch plane and perpendicular to the velocity vector of the interceptor and a_{yaw} is perpendicular to both velocity vector and vertical acceleration vector. For TPN, these acceleration components are calculated using

$$\begin{aligned} a_{\text{pitch}} &= a_{mz} \cos(\gamma_m) - a_{mx} \sin(\gamma_m) - g\cos(\gamma_m), \\ a_{\text{yaw}} &= a_{my} \cos(\psi_m) - a_{mx} \sin(\psi_m). \end{aligned} \tag{31}$$

4. Optimization Algorithm

4.1. CMA Algorithm. Consider the general form unconstrained optimization problem:

$$\text{minimize} \quad f(\mathbf{x}), \quad f : \mathbb{R}^n \longrightarrow \mathbb{R}. \tag{32}$$

It is well known that when f possess a certain structure (such as being continuous, linear, or convex), there are variety of local search algorithms that can be applied to solve this optimization problem efficiently. However, when f does not possess these desirable properties, local search methods either fail to find an answer or get stuck in local minima. Global search methods [22] remedy this problem by generalizing the search over the entire state space. Although global methods can also exploit the structure of f, many global methods treat f as a black box function, and hence the solution is found entirely by examining the input-output pairs $(\mathbf{x}, f(\mathbf{x}))$.

Input: Objective Function $f : \mathbb{R}^n \rightarrow \mathbb{R}$, Number of Samples per Iteration λ, Number of Iterations n_{iter}, Weights $w_1, \ldots, w_\mu$
Output: Approximate optimal solution $\mathbf{x}^*$
`// Initialization`
(1) $\mathbf{m}_0, \sigma_0, \mathbf{C}_0 \leftarrow \mathbf{I}, \mathbf{p}_\sigma, \mathbf{p}_c, k \leftarrow 0, \mu_w \leftarrow 1/\sum_{j=1}^{\mu} w_j^2, c_1 \leftarrow 2/n^2, c_\mu \leftarrow \mu_w/n^2, c_\sigma^{-1} \leftarrow n/3, c_c^{-1} \leftarrow n/4, \alpha \leftarrow 1.5\ \mu \leftarrow \lambda/2$
(2) **while** $k < n_{\text{iter}}$ **do**
(3) **for** i in $\{1, \ldots, \lambda\}$ **do**
`// Sample Candidate Solutions`
(4) $\mathbf{x}_i \sim \mathcal{N}(\mathbf{m}, \sigma_k^2 \mathbf{C}_k),\ f_i \leftarrow f(\mathbf{x}_i)$
`// Sort the Candidate Solutions Based on Their Cost`
(5) $\mathbf{x}_{1,\ldots,\lambda} \leftarrow \mathbf{x}_{t(1)}, \ldots, \mathbf{x}_{t(\lambda)}$, such that $f_{t(1)} \leq \cdots \leq f_{t(\lambda)}$
`// Move the mean to low cost solutions`
(6) $\mathbf{m}_{k+1} \leftarrow \mathbf{m}_k + \sum_{i=1}^{\mu} w_i(\mathbf{x}_i - \mathbf{m}_k)$
`// Update Evolution Path Variables`
(7) $\mathbf{p}_\sigma \leftarrow (1 - c_\sigma)\mathbf{p}_\sigma + \sqrt{1 - (1 - c_\sigma)^2}\sqrt{\mu_w} C_k^{-1/2}((\mathbf{m}_{k+1} - \mathbf{m}_k)/\sigma_k)$
(8) $\sigma_{k+1} \leftarrow \sigma_k \times \exp\left(c_\sigma(\|\mathbf{p}_\sigma\|/\mathbb{E}\|\mathcal{N}(\mathbf{0}, \mathbf{I})\| - 1)\right)$
`// Update The Covariance Matrix`
(9) **if** $\|\mathbf{p}_\sigma\| < \alpha\sqrt{(n)}$ **then**
(10) $d_k \leftarrow 1$
(11) **else**
(12) $d_k \leftarrow 0$
(13) $c_s \leftarrow (1 - d_k^2)c_1 c_c(2 - c_c)$
$\mathbf{p}_c \leftarrow (1 - c_c)\mathbf{p}_c + d_k\sqrt{(1 - c_c)^2}\sqrt{\mu_w} C_k^{-1/2}((\mathbf{m}_{k+1} - \mathbf{m}_k)/\sigma_k)$
(14) $\mathbf{C}_{k+1} \leftarrow (1 - c_1 - c_\mu + c_s)\mathbf{C}_k + c_1 \mathbf{p}_c^\top \mathbf{p}_c + c_\mu \sum_{i=1}^{\mu} w_i((\mathbf{x}_i - \mathbf{m}_k)/\sigma_k)((\mathbf{x}_i - \mathbf{m}_k)/\sigma_k)^\top$
(15) $k \leftarrow k + 1$
`// After the algorithm stops, output the best sample`
(16) $\mathbf{x}^* \leftarrow \mathbf{x}_{t(1)}$

ALGORITHM 1: Covariance Matrix Adaptation (CMA).

Covariance Matrix Adaptation (CMA) [18] is a popular global search method that usually ranks among the best solvers in global search benchmarks [23]. The basic idea behind CMA is to place a multivariate normal distribution over the search space $\mathbb{R}^n$ and sample candidate solutions $(\mathbf{x}_i, f(\mathbf{x}_i))$ from this distribution. The mean vector and covariance matrix of the distribution are incrementally updated at each step based on the values of the sampled solutions. The objective is to eventually steer the mean vector to the optimal solution $\mathbf{x}^*$ and shrink the covariance matrix to identity matrix; hence in the limit the distribution will yield the optimal solution when it is sampled.

For completeness, we provide the pseudocode for the CMA algorithm in Algorithm 1, taken from [24]. The algorithm starts by initializing its internal parameters (Line (1)). At the kth iteration, the algorithm samples λ number of samples from a multivariate Gaussian distribution with mean $\mathbf{m}_k$ and covariance $\mathbf{C}_k$ (Line (4)). Next, the samples $\mathbf{x}_i$ are sorted according to their costs f_i. The weighted average of top μ number of solutions is computed to find the mean vector $\mathbf{m}_{k+1}$ for the next iteration (Line (6)), which moves the mean of the distribution towards samples with lower costs. Next, algorithm updates the covariance matrix with the help of the evolution path variables which are $\mathbf{p}_\sigma$ (Line (7)) and $\mathbf{p}_c$ (Line (13)), which ensures that the adaptation steps are conjugate directions. The interested reader is referred to [24] for the full derivation of the algorithm and the intuition for updating the path parameters.

4.2. CMA-MV Algorithm. Unfortunately, CMA algorithm is only applicable to continuous optimization problems; hence we cannot use it to solve the missile launch condition setting problem given in (3), since the allocation of missiles is determined by the integer variables.

To overcome this issue, we develop a novel algorithm named Covariance Matrix Adaptation with Mixed Variables (CMA-MV), which extends the classical CMA algorithm to work on nonlinear mixed integer optimization problems. The generic nonlinear mixed integer programming problem is of the form:

$$\text{minimize} \quad f(\mathbf{x}, \mathbf{z}), \quad f : \mathbb{R}^n \times \mathbb{Z}^d \longrightarrow \mathbb{R}, \tag{33}$$

where $\mathbb{Z}$ is the set of integers. The special case we are interested in is the problem where the discrete variable $\mathbf{z}$ is a binary vector; that is, $\mathbf{z} \in \{0, 1\}^d$. This is also the case for the missile allocation problem in (3), where $z_i = 1$ refers to missile i being launched.

The main idea behind the CMA-MV algorithm is to define and update two probability distributions for sampling continuous and discrete variables. For the continuous variables $\mathbf{x}$, we use a multivariate normal distribution and we use the exact same procedure followed in the CMA algorithm (Algorithm 1) to update the mean and covariance of the distribution. For the discrete variables, we use a multivariate Bernoulli distribution and update the mean and covariance of this distribution based on the costs of sampled variables.

Input: Mean $\mathbf{m}' \in [0,1]^d$, Covariance $\mathbf{C}'$
Output: Sample $\mathbf{z} \in \{0,1\}^d$
// Compute the corresponding multivariate Normal distrubtion
(1) **for** i in $\{1,\ldots,d\}$ **do**
(2) $\quad \gamma_i \leftarrow \Phi^{-1}(m_i')$
(3) $\quad$ **for** j in $\{1,\ldots,d\}$ **do**
(4) $\quad\quad$ **if** $i \neq j$ **then**
(5) $\quad\quad\quad \Lambda_{ij} \leftarrow \text{Solve } \mathbf{C}_{ij}' - \Psi(\gamma_i, \gamma_j, \Lambda_{ij}) = 0$
(6) $\quad\quad$ **else**
(7) $\quad\quad\quad \Lambda_{ij} = 1$
// Sample from the corresponding multivariate Normal distrubtion and transform the results
(8) $u \sim \mathcal{N}(\gamma, \Lambda)$
(9) **for** l in $\{1,\ldots,d\}$ **do**
(10) $\quad$ **if** $u_l > 0$ **then**
(11) $\quad\quad z_l \leftarrow 1$
(12) $\quad$ **else**
(13) $\quad\quad z_l \leftarrow 0$

ALGORITHM 2: Sample from a multivariate Bernoulli distribution.

However, sampling from the multivariate Bernoulli distribution is not as straightforward as sampling from a multivariate normal distribution. We use the method described in [25] for this purpose. The pseudocode for the sampling process is given in Algorithm 2. The algorithm takes the given mean vector $\mathbf{m}'$ and the covariance matrix for $\mathbf{C}'$ and computes a corresponding multivariate distribution with mean $\boldsymbol{\gamma}$ and covariance $\boldsymbol{\Lambda}$ by solving the equations given on Lines (2) and (4). In these equations Φ is the cumulative distribution of a univariate normal variable with zero mean and unit variance, $\Psi(x, y, z) = \Phi_2(x, y, z) - \Phi(x)\Phi(y)$, where $\Phi_2(x, y, z)$ is the cumulative distribution of a bivariate normal variable with mean $[x, y]$ and correlation z. After solving these equations using numerical techniques, we sample the normal variable in Line (8). Then we loop through the components of the sample and set $z_l = 1$ if the components are positive and set $z_l = 0$ otherwise. It can be shown that the multivariate sample generated via this fashion comes from a distribution with first and second moments, $\mathbf{m}'$ and $\mathbf{C}'$, respectively.

The pseudocode for the complete CMA-MV algorithm is given in Algorithm 3. In the kth iteration, algorithm fixes the value of the discrete variables and hence recovers the function $\tilde{f}(\mathbf{x}) = f(\mathbf{x}, \mathbf{z}^k)$. Note that $\tilde{f}$ is a function of a continuous variable; hence we can apply the CMA algorithm (Algorithm 1) to obtain a solution (Line (3)). Next, we fix the value of the continuous variable to $\mathbf{x}^{k+1}$ to recover the function $\hat{f}(\mathbf{z}) = f(\mathbf{x}^{k+1}, \mathbf{z})$. Then we sample λ' solution candidates from the Bernoulli distribution with mean $\mathbf{m}_k'$ and $\mathbf{C}_k'$ using the sampling algorithm given in Algorithm 2 (Line (5)). Then we sort the solution from the lowest cost to highest cost (Line (6)). Next, we use a weighted average of the low cost solutions to compute the updated mean $\mathbf{m}^{k+1}$ of the Bernoulli distribution (Line (7)). Similarly, we use the weighted sample covariance estimate of the low cost solution candidates to compute the updated covariance matrix $\mathbf{C}_{k+1}'$. After each update, the distribution puts more mass on low cost solution candidates and hence with each iteration probability of sampling the optimal solution increases.

5. Simulation Results

In this section we fuse our optimization algorithm (Algorithm 3) with the control and estimation methods given in Section 3 to create an integrated solution to multiple missile allocation and control for ballistic target interception. We first give detailed results for two specific missions in order to give a better understanding of how the algorithm works, and then we demonstrate the effectiveness of the algorithm by comparing its performance to Heuristic and noncollaborative methods in Monte-Carlo simulations.

In all experiments, we use the following parameters for SAM defense system:

(i) The number of SAMs N_{SAM} is 5, and they are arranged in two parallel lines with the back line containing 3 SAMs and the front line containing 2 SAMs. The arrangement can be seen in the upper left corners of Figures 7 and 9. Radar is placed in front of the front line.

(ii) SAM velocity is set to Mach = 3.5.

(iii) Maximum number of iterations for CMA-MV is set to 50. The number of samples is set to 100 for both continuous and discrete variables. The rest of the parameters are tuned manually.

5.1. Results for a High Altitude-Low Velocity Target. First we examine a mission where the ballistic target has relatively low kinetic energy. The ballistic target's initial conditions are set to 80000 meters of altitude and speed equivalent to Mach number 5. This is a less challenging scenario, since the ballistic threat has relatively longer time till it hits the ground, giving the enough time for the filters of the SAM defense system to converge. Resulting trajectory of the target and

Input: Objective Function $f : \mathbb{R}^n \times \{0,1\}^d \rightarrow \mathbb{R}$,
Number of Continuous Samples per Iteration λ,
Number of Discrete Samples per Iteration λ,
Number of Iterations n_{iter}, Weights for continuous samples $w_1, \ldots, w_\mu$, Weights for discrete samples $w'_1, \ldots, w'_{\mu'}$, Initial guess $\mathbf{x}^0, \mathbf{z}^0$
Output: Approximate optimal solution $\mathbf{x}^*, \mathbf{z}^*$
// Initialization
(1) $\mathbf{m}_0, \sigma_0, \mathbf{C}_0 \leftarrow \mathbf{I}, \mathbf{p}_\sigma, \mathbf{p}_c, k \leftarrow 0, \mu_w \leftarrow 1/\sum_{j=1}^{\mu} w_j^2$
(2) **while** $k < n_{\text{iter}}$ **do**
// Fix the discrete variables to $\mathbf{z}^k$ and use CMA to solve for $\mathbf{x}$
(3) $\mathbf{x}^{k+1} \leftarrow$ Call CMA (Algorithm 1) with $\tilde{f}(\mathbf{x}) = f(\mathbf{x}, \mathbf{z}^k)$
// Fix the continous variables and sample from the multivariate Bernoulli distrubution
(4) **for** i in $\{1, \ldots, \lambda'\}$ **do**
// Sample
(5) $\mathbf{z}_i \sim \mathscr{B}(\mathbf{m}'_k, \mathbf{C}'_k)$ using Algorithm 2 $\hat{f}_i \leftarrow f(\mathbf{x}^{k+1}, \mathbf{z}_i)$
// Sort the candidate Solutions Based on Their Cost
(6) $\mathbf{z}_{1,\ldots,\lambda'} \leftarrow \mathbf{z}_{t(1)}, \ldots, \mathbf{z}_{t(\lambda')}$, such that $\hat{f}_{t(1)} \leq \cdots \leq \hat{f}_{t(\lambda')}$
// Move the mean to low cost solutions
(7) $\mathbf{m}'_{k+1} \leftarrow \sum_{i=1}^{\mu'} w'_i \mathbf{z}_i$
// Update The Covariance Matrix
(8) $\mathbf{C}'_{k+1} \leftarrow \frac{1}{\lambda' - 1} \sum_{i=1}^{\mu'} w'_i (\mathbf{z}_i - \mathbf{m}'_{k+1})(\mathbf{z}_i - \mathbf{m}'_{k+1})^\top$
(9) $\mathbf{z}^{k+1} \leftarrow \mathbf{z}_t^k(1)$
(10) $k \leftarrow k + 1$
// After the algorithm stops, output the best sample
(11) $\mathbf{x}^* \leftarrow \mathbf{x}_{t(1)}^k, \mathbf{z}^* \leftarrow \mathbf{z}_{t(1)}^k$

Algorithm 3: Covariance Matrix Adaptation with Mixed Variables (CMA-MV).

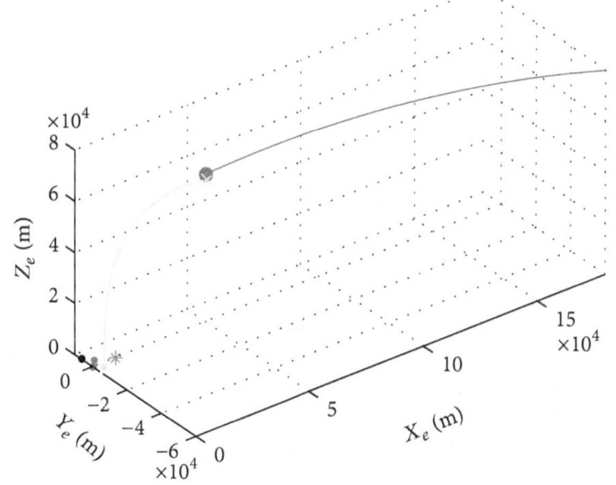

Figure 7: Interception of a high altitude-low velocity ballistic target. Red line depicts the trajectory of the ballistic target. Algorithm chooses only 1 missile for interception.

the launched missile is shown in Figure 7. The optimization algorithm also recognizes that filters have enough time to converge in this case and launches only a single missile. The missile intercepts the ballistic threat with a miss distance less than 1 meter.

Figures 8(a), 8(b), and 8(c) show the estimation performance of the filter of the missile, for range-to-go, line of sight, and target acceleration estimations. It can be seen that filters converged rapidly in the terminal phase of the mission. These plots justify the decision of the algorithm to launch only a single missile; in this case algorithm recognized that a single filter would yield sufficient performance and did not choose to allocate more missiles, in order to keep the cost as close to minimum as possible. Also note that no collaborative filtering is performed in this mission, since only a single missile is launched.

5.2. Results for a Low Altitude-High Velocity Target. To complement the results of the previous subsection, now we look at a mission that corresponds to high kinetic energy target. For this simulation the ballistic target's initial conditions are set to 60000 meters of altitude and speed equivalent to Mach number 7. This scenario is much more challenging, since the target's established time of impact is much shorter. Resulting trajectory of the target and the launched missile is shown in Figure 9. In this case it is seen that the algorithm launches 3 collaborative missiles to intercept the target. The interception is achieved with a miss distance of approximately 1 meter.

Figures 10(a), 10(b), and 10(c) show the estimation performance of the filter of the missiles, for range-to-go, line of sight, and target acceleration estimations averaged over the 3 launched missiles. For comparison, performance of

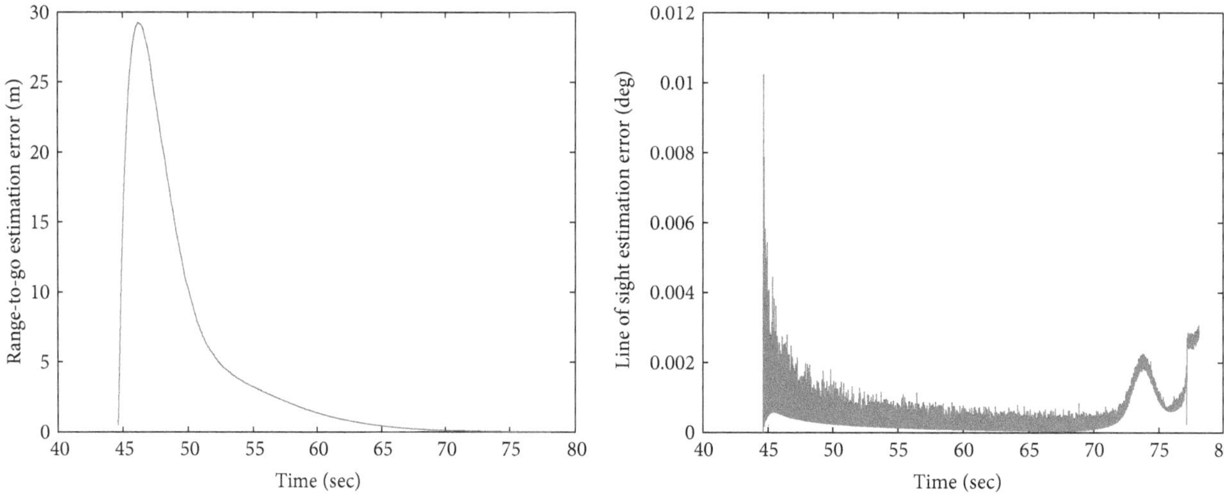

(a) Norm of the range-to-go estimation error versus time in terminal phase

(b) Norm of the line of sight estimation error versus time in terminal phase

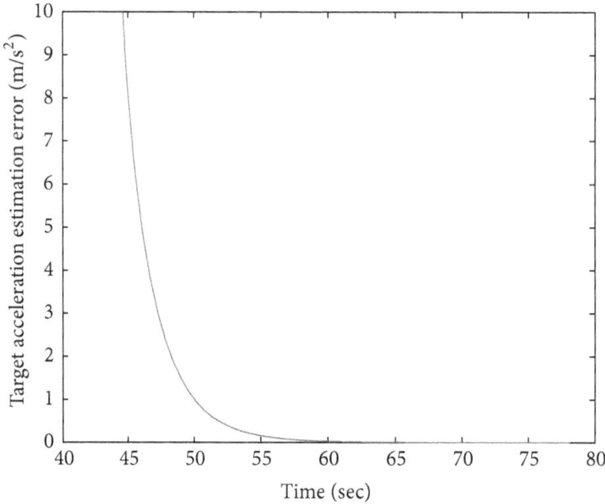

(c) Norm of the target acceleration estimation error versus time in terminal phase

FIGURE 8: Filter performance versus time in terminal phase for intercepting high altitude-low velocity target.

the individual filters is also plotted in these figures. These individual filter performances correspond to the case where missiles do not cooperate; hence no information is shared between them. Examining these plots gives us a good insight on algorithm's decision to launch 3 missiles. It is seen that individual filters did not have enough time to converge for this high kinetic energy target; hence launching a single or even two missiles would result in high miss distances. Algorithm made the necessary trade-off analysis and found out that launching 3 collaborative missiles would generate enough information flow for estimators to converge.

5.3. Monte-Carlo Results for Multiple Scenarios. The previous simulation results demonstrated that the algorithm yields sound decisions on selected scenarios. However, in order to truly assess the performance of the algorithm, a wide range of initial conditions that corresponds to different ballistic threat should be analyzed. Also we need to compare the performance of the algorithm to alternative methods. For this purpose we conducted a Monte-Carlo test over 100 randomly sampled initial conditions for the ballistic target. The initial altitude of the target was sampled in the interval [40000, 80000] meters and the speed was sampled in the Mach number interval [5, 8]. The following alternative methodologies are compared:

(i) *Heuristic Collaborative Interception.* In this simple algorithm no launch condition or missile allocation optimization is conducted. This method always launches the same number of missiles that are closest to the ballistic target at the beginning of the simulation. Launch conditions of the missiles are always set to 0 heading and 90 degrees of pitch angle. Missiles use collaborative filtering for interception.

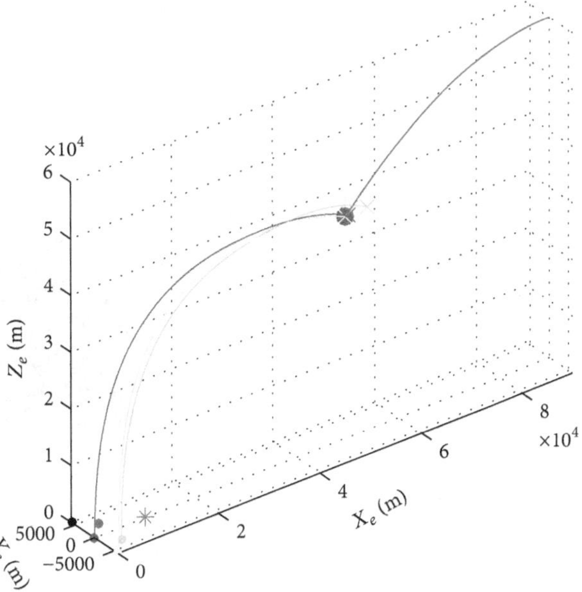

FIGURE 9: Interception of a low altitude-high velocity ballistic target. Red line depicts the trajectory of the ballistic target. Algorithm launches 3 missiles to intercept the target.

TABLE 1: Average number of missiles and average miss distance obtained by different methods, averaged over 100 random initial conditions of the ballistic target.

	Average number of missiles launched	Average miss distance (m)
Heuristic method 1 missile	1	5020.78 ± 600.21
Heuristic method 2 missiles collaborative	2	490.22 ± 41.08
Heuristic method 3 missiles collaborative	3	60.55 ± 11.2
Optimized noncollaborative	1.05 ± 0.23	320.55 ± 50.77
Optimized collaborative	1.56 ± 0.34	1.43 ± 0.22

(ii) *Optimized Noncollaborative Interception.* In this methodology optimization algorithm CMA-MV is used for optimizing the launching conditions and the allocation of the missiles. However missiles do not run collaborative filtering algorithms on-board.

(iii) *Optimized Collaborative Interception.* This is the approach developed in this paper. The CMA-MV algorithm (Algorithm 3) is used for optimization of launch conditions and missile allocation and the missiles run collaborative filtering algorithms.

Table 1 depicts the results of the Monte-Carlo analysis. We see that Heuristic method's performance gets better as the number of missiles used by the method increases. This is expected, since increased number of missiles translates to improved estimation performance. However, even using 3 missiles for all conditions does not reduce the average miss distance substantially. This is due to fact that Heuristic method does not optimize the launch conditions; hence the missile autopilots do not have enough time to restore the missiles into the desired trajectories in the terminal phase. On the other hand, optimized noncollaborative method yields substantially lower miss distances than Heuristic methods that use 1 or 2 missiles, while launching only 1.05 missiles on average. This is because the optimized noncollaborative method optimizes the launch conditions for the missiles, which leads to improved interception performance. However, this method is outperformed by Heuristic method that uses 3 missiles, because optimized noncollaborative method does not utilize collaborative filters; hence the algorithm can not take advantage of improved estimation performance gained by launching multiple missiles against high kinetic energy ballistic targets.

Finally, we see that the approach developed in this paper, the optimized collaborative method, outperforms the compared approaches in terms of both resource management efficiency and miss distance. This is because, unlike the compared approaches, the developed method optimizes the launch conditions and missile allocation simultaneously, and hence it is able to assess the right trade-off between the number of missiles launched and attainable miss distance.

6. Conclusions and Future Work

In this work we have developed a novel probabilistic search algorithm for allocation and launch condition optimization

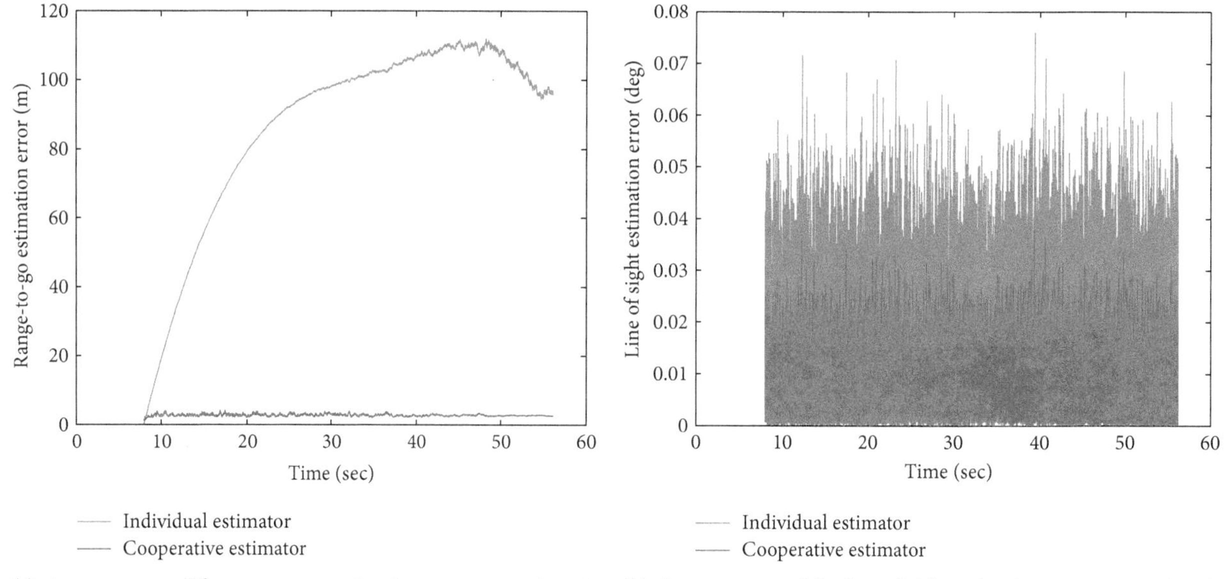

(a) Average norm of the range-to-go estimation error versus time in terminal phase for individual and collaborative estimators

(b) Average norm of the line of sight estimation error versus time in terminal phase for individual and collaborative estimators

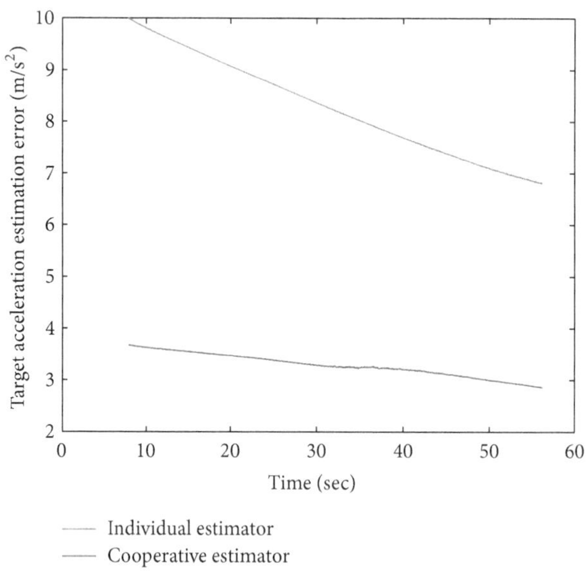

(c) Average norm of the target acceleration estimation error versus time in terminal phase for individual and collaborative estimators

FIGURE 10: Filter performance versus time in terminal phase for intercepting low altitude-high velocity target.

of multiple missiles intercepting a ballistic target. Through simulation studies, we have verified that the algorithm makes sound decisions and yields more efficient resource management and lower miss distance compared to alternative approaches.

Future works involve generalization of the problem into intercepting multiple ballistic targets, developing alternative cooperative estimation/control algorithms, and investigation of theoretical properties of the CMA-MV algorithm.

Competing Interests

The authors declare that they have no competing interests.

References

[1] B.-Z. Naveh and A. Lorber, "Theater ballistic missile defense," *Progress in Astronautics and Aeronautics*, vol. 192, pp. 1–397, 2001.

[2] G. M. Siouris, *Missile Guidance and Control Systems*, Springer, New York, NY, USA, 2004.

[3] R. H. Chen and J. L. Speyery, "Terminal and boost phase intercept of ballistic missile defense," in *Proceedings of the AIAA Guidance, Navigation and Control Conference and Exhibit*, Honolulu, Hawaii, USA, August 2008.

[4] T.-K. Wang and L.-C. Fu, "A guidance strategy for multi-player pursuit and evasion game in maneuvering target interception,"

in *Proceedings of the 9th Asian Control Conference (ASCC '13)*, June 2013.

[5] I.-S. Jeon, J.-I. Lee, and M.-J. Tahk, "Homing guidance law for cooperative attack of multiple missiles," *Journal of Guidance, Control, and Dynamics*, vol. 33, no. 1, pp. 275–280, 2010.

[6] E. Daughtery and Z. Qu, "Optimal design of cooperative guidance law for simultaneous strike," in *Proceedings of the 53rd IEEE Annual Conference on Decision and Control (CDC '14)*, pp. 988–993, Los Angeles, Calif, USA, December 2014.

[7] V. Shaferman and Y. Oshman, "Cooperative interception in a multi-missile engagement," in *Proceedings of the AIAA Guidance, Navigation, and Control Conference*, Chicago, Ill, USA, August 2009.

[8] V. Shaferman and T. Shima, "Cooperative multiple-model adaptive guidance for an aircraft defending missile," *Journal of Guidance, Control, and Dynamics*, vol. 33, no. 6, pp. 1801–1813, 2010.

[9] Y. Liu, N. Qi, and J. Shan, "Cooperative interception with double-line-of-sight measuring," in *Proceedings of the AIAA Guidance, Navigation and Control Conference*, Boston, Mass, USA, August 2013.

[10] V. Shaferman and T. Shima, "Cooperative optimal guidance laws for imposing a relative intercept angle," *Journal of Guidance, Control, and Dynamics*, vol. 38, no. 8, pp. 1395–1408, 2015.

[11] L. Wang, H. Fenghua, and Y. Yu, "Guidance law design for two flight vehicles cooperative interception," in *Proceedings of the AIAA Guidance, Navigation and Control Conference*, Kissimmee, Fla, USA, January 2015.

[12] Z.-J. Lee, S.-F. Su, and C.-Y. Lee, "Efficiently solving general weapon-target assignment problem by genetic algorithms with greedy eugenics," *IEEE Transactions on Systems, Man, and Cybernetics, Part B: Cybernetics*, vol. 33, no. 1, pp. 113–121, 2003.

[13] P. Teng, H. Lv, J. Huang, and L. Sun, "Improved particle swarm optimization algorithm and its application in coordinated air combat missile-target assignment," in *Proceedings of the 7th World Congress on Intelligent Control and Automation (WCICA '08)*, pp. 2833–2837, IEEE, Chongqing, China, June 2008.

[14] J. Wang, X.-G. Gao, Y.-W. Zhu, and H. Wang, "A solving algorithm for target assignment optimization model based on ACO," in *Proceedings of the 6th International Conference on Natural Computation (ICNC '10)*, vol. 7, pp. 3753–3757, IEEE, Yantai, China, August 2010.

[15] J. Wang and Y.-W. Zhu, "A solving algorithm for target assignment optimization model based on SA," in *Proceedings of the International Conference on Artificial Intelligence and Computational Intelligence (AICI '10)*, pp. 489–493, Sanya, China, October 2010.

[16] M. Wei, G. Chen, K. Pham, E. Blasch, and Y. Wu, "Game theoretic target assignment approach in ballistic missile defense," in *Proceedings of the SPIE Defense and Security Symposium*, International Society for Optics and Photonics, 2008.

[17] B. Xin, J. Chen, Z. Peng, L. Dou, and J. Zhang, "An efficient rule-based constructive heuristic to solve dynamic weapon-target assignment problem," *IEEE Transactions on Systems, Man, and Cybernetics Part A:Systems and Humans*, vol. 41, no. 3, pp. 598–606, 2011.

[18] N. Hansen and A. Ostermeier, "Adapting arbitrary normal mutation distributions in evolution strategies: the covariance matrix adaptation," in *Proceedings of the IEEE International Conference on Evolutionary Computation (ICEC '96)*, pp. 312–317, May 1996.

[19] Z. B. Tatas, *Modeling and autopilot design for a SCUD Type ballistic missile [M.S. thesis]*, Department of Electrical Engineering, Hacettepe University, Ankara, Turkey, 2006.

[20] P. Zarchan, *Tactical and Strategic Missile Guidance*, vol. 199 of *Progress in Aeronautics and Astronautics*, AIAA, 4th edition, 2002.

[21] I. Moran and D. T. Altilar, "Three plane approach for 3D true proportional navigation," in *Proceedings of the AIAA Guidance, Navigation and Control Conference*, San Francisco, Calif, USA, August 2005.

[22] E. K. P. Chong and H. Stanislaw, *An Introduction to Optimization*, vol. 76, John Wiley & Sons, New York, NY, USA, 2013.

[23] N. Hansen, A. Auger, R. Ros, S. Finck, and P. Pošík, "Comparing results of 31 algorithms from the black-box optimization benchmarking BBOB-2009," in *Proceedings of the 12th Annual Conference Companion on Genetic and Evolutionary Computation (GECCO '10)*, ACM, July 2010.

[24] N. Hansen, S. D. Müller, and P. Koumoutsakos, "Reducing the time complexity of the derandomized evolution strategy with covariance matrix adaptation (CMA-ES)," *Evolutionary Computation*, vol. 11, no. 1, pp. 1–18, 2003.

[25] J. H. Macke, P. Berens, A. S. Ecker, A. S. Tolias, and M. Bethge, "Generating spike trains with specified correlation coefficients," *Neural Computation*, vol. 21, no. 2, pp. 397–423, 2009.

12

Autonomous Orbit Determination for a Hybrid Constellation

Muzi Li,[1] **Bo Xu**,[2] **and Jun Sun**[3]

[1]*School of Astronomy and Space Science, Nanjing University, Nanjing 210023, China*
[2]*School of Aeronautics and Astronautics, Sun Yat-Sen University, Guangzhou 510006, China*
[3]*Shanghai Aerospace Control Technology Institute, Shanghai 201199, China*

Correspondence should be addressed to Bo Xu; xubo@nju.edu.cn

Academic Editor: Jose Carlos Páscoa

A new orbit determination scheme targeting communication and remote sensing satellites in a hybrid constellation is investigated in this paper. We first design one such hybrid constellation with a two-layer configuration (LEO/MEO) by optimizing coverage and revisit cycle. The main idea of the scheme is to use a combination of imagery, altimeter data, and inter-satellite range data as measurements and determine orbits of the satellites in the hybrid constellation with the help of the extended Kalman filter (EKF). The performance of the new scheme is analyzed with Monte Carlo simulations. We first focus on an individual remote sensing satellite and compared the performance of orbit determination using only imagery with its counterpart using both imagery and altimeter measurements. Results show that the performance improves when imagery is used with altimeter data pointing to geometer calibration sites but declines when used with ocean altimeter data. We then expand the investigation to the whole constellation. When inter-satellite range data is added, orbits of all the satellites in the hybrid constellation can be autonomously determined. We find that the combination of inter-satellite range data with remote sensing observations lead to a further improvement in orbit determination precision for LEO satellites. Our results also show that the performance of the scheme would be affected when remote sensing observations on certain satellites are absent.

1. Introduction

A hybrid constellation refers to a group of satellites at different orbital regimes working in concert. A widely known example is the BEIDOU navigation satellite constellation of China. The hybrid constellation concept commonly appears in studies on navigation constellation design and optimization [1, 2]. It has also been introduced to the fields of satellite communication and remote sensing [3, 4], and Fahnestock and Erwin [5] presented a kind of hybrid constellation to meet space situation awareness requirements. However, few research tries to design a hybrid constellation as a multifunction constellation.

At present, orbit determination for satellites depends heavily on the earth stations and navigation constellations (for example, GPS constellation). In order to reduce maintenance cost of these systems and enhance survivability of satellites in cases of emergencies, autonomous orbit determination methods which depend on instruments on board were proposed. Some of these works focus on autonomous orbit determination based on either optical imagery or altimeter data or inter-satellite range data. White et al. [6] used line-of-sight (LOS) measurements of stars and landmarks to estimate the attitude and orbit of satellites. Straub and Christian [7] used observations of coastlines on the Earth's surface as inputs to autonomously determine the orbits of earth-observing satellites with different orbit inclinations and altitudes. Li, Xu and Zhang [8] proposed a scheme using images of ground objects and analyzed the influence of image resolution, pointing accuracy and lighting constraints on the orbit determination performance. Following that study, Li and Xu [9] presented an orbit and attitude determination (OAD) scheme using images of regular-shaped ground landmarks to overcome the disadvantages using ground point features. As another high-precision measurement, the altimeter can provide highly accurate altitude information which helps to effectively improve the orbit determination accuracy. Born et al. [10] determined the orbit of the N-ROSS satellite using the altimetric crossing arc residuals between the TOPEX and N-ROSS orbits and demonstrated that submeter radial accuracy can be attained. Lemoine et al. [11] showed that altimeter crossover data can significantly modify the gravity field and improve the radial orbit accuracy of POD to 4-5 cm for the GEOSAT Follow-On spacecraft when used in combination with SLR data. For satellites in constellations, intersatellite links can be established and pseudo-range observations of these links can be used for orbit determination. Markley and Naval [12] investigated orbit determination performance using landmarks and intersatellite data. Psiaki [13] proposed an autonomous orbit determination system based on the relative position measurement of a pair of satellites and analyzed the observability and orbit estimation accuracy of the system. Li et al. [14] verified the possibility of reducing the errors resulting from constellation rotation by using cameras to obtain the direction between satellites. Kai et al. [15] evaluated the performance of a navigation scheme which uses relative bearing measurements from navigation star sensors combined with relative range measurements from intersatellite links. Besides, Kai et al. [16] introduced a scheme using the time difference of arrival (TDOA) measurements to X-ray pulsars and inter-satellite range measurements to determine the absolute position of satellites. Wang and Cui [17] also achieved autonomous navigation using the X-ray pulsars and inter-satellite range measurements for Mars obiters.

In this paper, unlike previous studies on hybrid constellations which focus on satisfying a specific requirement of communication, navigation, or remote sensing, a hybrid constellation containing two layers (MEO/LEO) was proposed to meet both the requirements of satellite communication and remote sensing. The LEO layer satellites with optical cameras and altimeters onboard are designed for Earth observation, and the MEO layer is designed in combination with the LEO layer to be a communication constellation. For the hybrid constellation, a new orbit determination scheme is proposed. Without other observation data external to the constellation, only optical imagery and altimeter data can be used as high-precision observations for autonomous orbit determination of LEO satellites. Two usage patterns are considered for the altimetry data. One is the ocean altimeter data generated with nadir-pointing altimeters. The other is the range data generated with altimeters pointing to the geometer calibration sites which can be captured and recognized by the camera systems. When inter-satellite range data is considered, orbits of MEO layer satellites can also be determined, which in turn has an effect on the LEO layer satellites. As a result, autonomous orbit determination of the constellation containing communication satellites and remote sensing satellites can be achieved.

Under such a constellation, the performance of autonomous orbit determination using optical imagery, altimeter data, and inter-satellite range data is evaluated. For altimeter data-based orbit determination, the influence of different usage patterns on orbit accuracy is compared. For orbit determination using all three observation data, the performance is also assessed in the circumstance when certain remote sensing observations are absent.

To this end, the remainder of this paper is organized as follows: Section 2 introduces the optimal hybrid constellation. In Section 3, a detailed description of the orbit determination algorithm is given, including the dynamic model, the measurement model, and the filter model. For different observation data, orbit determination simulations and performance analysis are shown in Section 4. Finally, some brief conclusions and discussions are provided in Section 5.

2. Hybrid Constellation Design

In this section, a hybrid constellation consisting of MEO/LEO two layer satellites is proposed. The LEO layer is designed to implement an Earth optical observation mission. Besides, the LEO layer cooperating with the MEO layer can provide continuous regional communication coverage. Considering orbit characteristics of the hybrid constellation and related constraints, an efficient design procedure is presented below.

2.1. LEO Layer. To ensure the accuracy of obtained data, satellites performing earth observation missions are mostly placed on LEO. In the paper, the LEO layer is designed as a remote sensing constellation satisfying coverage and revisit cycle requirements. Given that the satellites placed on sun-synchronous orbit pass over a given sub-satellite point at a fixed local solar time, the sun-synchronous orbit is appropriate for earth observation satellites and satisfies the nondimensional form [18]:

$$\dot{\Omega} = -\frac{3}{2}J_2\frac{n\cos i}{p^2} = n_s, \tag{1}$$

where Ω is the right ascension of ascending node, J_2 is the second zonal harmonic of the gravitational field, n is the mean motion of the satellite, p is the semi-latus rectum of the orbit, and n_s is the mean angular velocity of the Earth orbiting the Sun. On the basis of the sun-synchronous orbit, a further assumption is made that the orbits meet the conditions of a repeat circular orbit:

$$N_p T_\Omega = N_d T_E, \tag{2}$$

$$T_E = \frac{2\pi}{n_e - \dot{\Omega}}, \tag{3}$$

$$T_\Omega = \frac{2\pi}{n}\left(1 + \frac{3}{2}J_2\frac{1 - 4\cos^2 i}{p^2}\right), \tag{4}$$

where N_p represents the number of revolutions in one repetition, N_d is the number of days to repeat (revisit cycle), n_e is the inertial rotational velocity of the Earth, T_Ω is the nodal period of the orbit, and T_E is the rotational period of the Earth. Substituting (2), (3), and (4) into (1), a nonlinear mathematical equation with a single variable "the semi-major axis a" can be written as

$$A_1a^7 + A_2a^2 + A_3a^{0.5} + A_4 = 0, \tag{5}$$

where the expressions of the coefficients are as follows:

$$\begin{cases} A_1 = \dfrac{8(n-n_s)n_s^2}{3J_2}, \\ A_2 = n_e - n_s, \\ A_3 = \dfrac{-N_d}{N_p}, \\ A_4 = \dfrac{3J_2(n_e-n_s)}{2}. \end{cases} \quad (6)$$

For (5), there are some constraint conditions: (1) In order to ensure good revisit performance and less computation burden, the revisit cycle is set no more than 10 days; (2) For a remote sensing satellite bus orbiting in LEO, the orbit altitude normally ranges between 500 km and 1000 km. Therefore, the orbit revolution per day is limited to more than 14 and less than 15; (3) Satellites with a total number of N_s are uniformly distributed on the same orbit. To make sure that intersatellite links can be established between adjacent satellites, the number of satellites is not less than the minimum number which keeps adjacent satellites visible to each other, meanwhile the number of satellites is required to be no larger than 2 times of this minimum number in order to limit the constellation size; (4) In the nadir viewing case, the single-plane constellation can provide complete coverage of the Earth (except polar regions). The field of view (FOV) of every satellite is set as 2.06° which equals the FOV of satellites in the high-resolution satellite constellation "DMC-3G" [19]. The corresponding mathematical expressions (the derivation is provided in Appendix A) are shown as

$$\begin{cases} N_d \le 10, \\ 14 \le \dfrac{N_p}{N_d} \le 15, \\ \dfrac{\pi}{\arccos(r_e/a)} \le N_s \le \dfrac{2\pi}{\arccos(r_e/a)}, \\ \dfrac{2(a-r_e)\tan(\mathrm{FOV}/2)}{|\sin i|} > 2\pi r_e \dfrac{N_d}{\mathrm{LCM}(N_d, N_s)N_p}, \end{cases} \quad (7)$$

where r_e is the equatorial radius. Under these constraints and repeat sun-synchronous orbit equation, the number of feasible solutions is finite. By going through all the possible combinations of N_d and N_p, all the feasible solutions can be generated. Then the solution with the minimum sum of N_d and N_s is identified as the optimal solution (shown in Table 1). Among the two solutions, the first solution with the minimal orbit altitude is the optimal choice.

In order to guarantee good illumination, the descending node local time is set around 10:30 am or 1:30 pm for most LEO Earth observation satellites. Here the single plane determined by (5) is set with a descending node local time of 10:30 am. For further decreasing the revisit cycle, an orbit plane with a descending node local time of 1:30 pm is added under the Flower Constellation concept. The Flower

TABLE 1: The optimal results of parameters for a single plane of the LEO layer.

Altitude (km)	Inclination (°)	N_p	N_d	N_s
811.17	98.66	128	9	11
848.91	98.82	127	9	11

Constellation has the property that all satellites share identical repeat ground tracks [20], which helps to reduce revisit cycle. For satellites in a Flower Constellation, the difference of the right ascension of ascending node and the difference of mean anomaly ΔM satisfies

$$\begin{cases} \Delta\Omega = -2\pi \dfrac{F_n}{F_d}, \\ \Delta M = 2\pi \dfrac{F_n}{F_d}\left(\dfrac{n+\dot{M}_0}{n_e+\dot{\Omega}}\right), \end{cases} \quad (8)$$

in which F_n and F_d are the phasing parameters and $\dot{M}_0$ is the rate of change in the mean anomaly due to perturbations. $\Delta\Omega$ is 45° which can be calculated by subtracting between the descending node local time of two orbit planes, and then the ratio F_n/F_d and ΔM can be obtained. A total of N_s satellites are uniformly distributed in the first plane, so the same number of N_s satellites are placed in the second plane to repeat the corresponding satellite ground track. From (8), it can be inferred that there exist N_d positions in one plane which share the same ground track [21]; the mean anomaly separation between adjacent positions is $2\pi/N_d$. The minimum and the maximum revisit cycles for these N_d positions are shown in Table 2. In order to minimize the maximum revisit cycle, $\Delta M = 213.89°$ is chosen. Finally, the maximum revisit cycle of LEO layer satellites is reduced to 117 hours.

2.2. MEO Layer. The MEO layer is constructed to work in conjunction with the LEO layer for continuous regional communication. As communication satellites, GEO satellites do well in covering low latitudes; however, their long round-trip time make it difficult for GEO satellites to provide quality-guaranteed service for real-time need. What is more, the position resource on GEO satellites is scant. The MEO and LEO satellites greatly overcome the disadvantages of GEO satellites. When compared with MEO satellites, LEO satellites have a better quality of timeliness, while more LEO satellites would be needed to achieve continuous global or regional coverage. Therefore, a hybrid constellation including the MEO layer and LEO layer would be suitable to be used for regional communication.

For the design of the MEO layer, the Walker Delta constellation concept [22] is adopted. This constellation can provide an excellent coverage and is defined by four parameters ($i/T/W/F$). The first parameter i is the inclination of the satellite orbit. T is the total number of satellites. W is the number of orbital planes, and F ($0 \le F \le W-1$) is the phase factor which denotes the relative phases of the satellites in any two adjacent planes. In the paper, the MEO layer is

TABLE 2: The minimum and the maximum revisit cycles.

	ΔM (degree)	Minimum revisit cycle (hours)	Maximum revisit cycle (hours)
1	82.98	3	213
2	115.71	27	189
3	148.43	51	165
4	181.16	75	141
5	213.89	99	117
6	246.61	93	123
7	279.34	69	147
8	312.07	45	171
9	344.80	21	195

required to cooperate with the LEO layer to realize continuous coverage of the latitude zone including the China mainland (3.86°N~53.55°N). Parameters T, W, and F can be empirically chosen as $T = 8$, $W = 2$, and $F = 1$. The orbit altitude and inclination are important factors affecting the coverage rate and minimum elevation angle in this design. The higher the altitude is, the better the visibility of the communication satellite is. Here the orbit altitude can be set according to the 8-hour circular periodic orbit. As for the inclination, Figure 1 shows the relation between the minimum elevation angle and the inclination under the constraint of continuous coverage. Given the requirement of a minimum elevation angle higher than 20 degrees, choice of the inclination is limited between 36 and 42. Figure 2 shows the variation of coverage rate with inclination. When inclination is set around 40, coverage rate reaches a peak. Here the inclination is set at 40, which guarantees a larger minimum elevation angle and a continuous coverage at the same time. The overall design configuration of the hybrid constellation is summarized in Table 3.

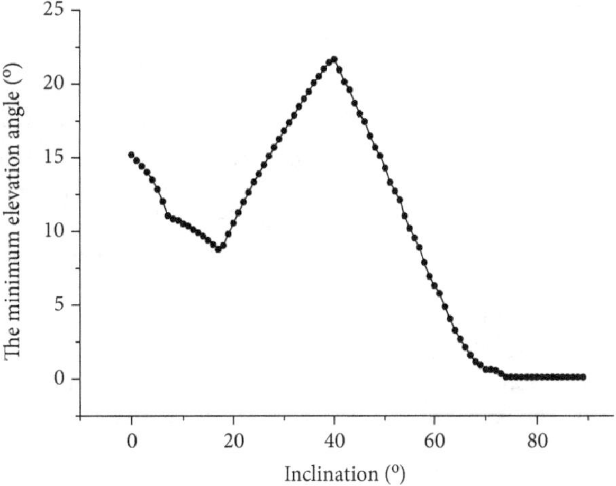

FIGURE 1: The relation between the minimum elevation angle and the inclination.

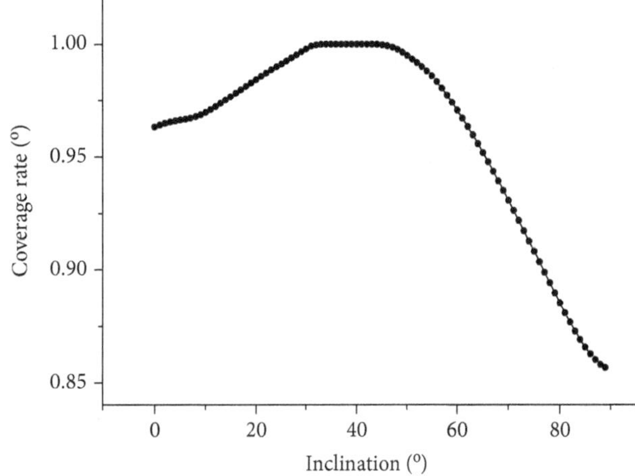

FIGURE 2: Variation of the coverage rate with the inclination.

3. Orbit Determination Scheme

This section presents the orbit determination scheme for the constellation designed in Section 2, including a description of the dynamic model, the measurement model, and the filter model to estimate orbit states.

3.1. Dynamic Model. For the constellation, the state vector is represented as $\mathbf{x} = [\mathbf{x}_1^T \mathbf{x}_2^T \cdots \mathbf{x}_l^T]$ where $\mathbf{x}_i = [\mathbf{r}_i\, \dot{\mathbf{r}}_i]^T$, $\mathbf{r}_i = [\mathbf{r}_{x,i}\, \mathbf{r}_{y,i} \mathbf{r}_{z,i}]$, and $\dot{\mathbf{r}}_i = [\mathbf{v}_{x,i}\, \mathbf{v}_{y,i}\, \mathbf{v}_{z,i}]$ are position and velocity vectors of the satellites in J2000 Earth-centered inertial coordinate system. The subscripts i and l indicate the i-th satellite and the total number of satellites, respectively. The equation of motion for a satellite is described as

$$\dot{\mathbf{x}} = f(\mathbf{x}, t), \tag{9}$$

where

$$f(\mathbf{x}, t) = \begin{bmatrix} \dot{\mathbf{r}} \\ \dfrac{-\mu}{\|\mathbf{r}\|^3}\mathbf{r} + \mathbf{F}_\varepsilon \end{bmatrix}, \tag{10}$$

TABLE 3: Initial orbit element of the hybrid constellation.

	LEO layer	MEO layer
Satellite number	22	8
Plane number	2	2
Semi-major axis (km)	7189.31	20268.43
Eccentricity	0	0
Inclination (°)	98.66	40
Right ascending node (°)	257.18/302.05	0/180
Argument of perigee plus mean anomaly for the first satellite of every plane (°)	0/213.90	0/45

μ is the earth gravitational constant, and F_ε represents perturbation accelerations. The sources of perturbations include the nonspherical gravitation of the Earth (the WGS84 gravity model), the atmospheric drag (1976 standard atmosphere model), third-body perturbations (DE405 numerical planetary ephemeris), solar radiation pressure (spherical model), Earth radiation pressure, the tide effects, and the effect of general relativity.

3.2. Measurement Model. In this paper, optical imagery, altimeter data, and inter-satellite range data are used to provide the orientation, altitude, and relative distance information, respectively, for the constellation satellites.

For remote sensing satellites with an optical camera, images of known ground features can be captured. By the process of image recognition and matching, the focal plane coordinates of a ground feature are obtained, which imply the orientation information of the cameras to ground features. The focal plane coordinates (u, v) satisfy

$$\begin{bmatrix} \frac{u}{f} \\ \frac{v}{f} \\ 1 \end{bmatrix} = (\mathbf{T}_A \mathbf{T}_I)^{-1} [\mathbf{r} - \mathbf{r}_g]. \quad (11)$$

$\mathbf{T}_I$ and $\mathbf{T}_A$ are the camera installation matrix and the rotation matrix from the satellite body coordinate frame to the J2000 initial coordinate frame, respectively. It is assumed that a good estimation of $\mathbf{T}_A$ can be available from a star tracker. f is the focal length of the optical camera, and $\mathbf{r}_g$ is the ground feature position. Simplifying (11) yields

$$\begin{bmatrix} u \\ v \end{bmatrix} = \mathbf{g}(\mathbf{x}) + \boldsymbol{\eta}, \quad (12)$$

where $\boldsymbol{\eta}$ denotes the measurement noise satisfying the zero-mean Gaussian distribution

$$\mathrm{E}\left(\boldsymbol{\eta}\boldsymbol{\eta}^{\mathrm{T}}\right) = \sigma_i^2 \mathbf{I}_{2\times 2}, \quad (13)$$

in which σ_i is the standard deviation of the focal plane coordinate error and $\mathbf{I}_{2\times 2}$ is a two-order identity matrix.

The laser altimeter measurement belongs to the two-way range measurement. The laser altimeter sends a laser beam and measures the round-trip time. The range can be described as

$$\rho_a = \frac{ct_a}{2}, \quad (14)$$

in which c is the speed of light, t_a is the round-trip time, and ρ_a is the altimeter measurement. By adjusting the orientation of the altimeter antenna boresight, the altimeter can be directed toward the nadir to obtain height information above the Earth's surface or toward interested regions to measure the range between each other.

When the altimeter is directed toward the nadir, the range measurement is described as

$$\rho_a = |\mathbf{r} - \mathbf{r}_u| + \zeta = |\mathbf{r}| - r_e\left(1 - f_e \sin^2\theta\right) + \zeta, \quad (15)$$

in which $\mathbf{r}_u$ is the position of the sub-satellite point, f_e is the flattening of the Earth, and θ is the geocentric latitude. The range to interested regions measured by the altimeter is expressed as

$$\rho_a = |\mathbf{r} - \mathbf{r}_p| + \zeta, \quad (16)$$

where $\mathbf{r}_p$ is the position of the interested point and ζ is the measurement noise which satisfies

$$E\left(\zeta^2\right) = \sigma_a^2, \quad (17)$$

where σ_a is the standard deviation of the altimeter measurement error.

For inter-satellite range measurement, the observed inter-satellite range ρ_s can be represented by

$$\rho_{s,ij} = |\mathbf{r}_j - \mathbf{r}_i| + \chi, \quad (18)$$

in which the subscripts j and i represent the j-th and i-th satellites, respectively, and χ presents the random measurement noise which satisfies

$$E\left(\chi^2\right) = \sigma_s^2. \quad (19)$$

In (19), σ_s is the standard deviation of the inter-satellite range error.

3.3. Filter Model. Taking into account that the satellite orbit dynamics and measurement model are nonlinear, an EKF is used to estimate the state of the system. Equation (9) is used as the state equation, and (12), (15), and (18) represent the observation equations for different observations. The EKF works by a two-step cycle: a time update step and a measurement update step [23].

The covariance $\mathbf{P}$ propagation from time $k-1$ to the next observation epoch k is processed with the "time update" equation:

$$\bar{\mathbf{P}}_k = \boldsymbol{\Phi}(t, t_k)\mathbf{P}_{k-1}\boldsymbol{\Phi}^T(t, t_k), \quad (20)$$

where $\boldsymbol{\Phi}(t, t_k)$ is the state transformation matrix. The state $\mathbf{x}$ and covariance $\mathbf{P}$ are updated with the observation using the "measurement update" equation:

$$\begin{aligned} \mathbf{K}_k &= \bar{\mathbf{P}}_k \mathbf{H}_k^T \left(\mathbf{H}_k \bar{\mathbf{P}}_k \mathbf{H}_k^T + \mathbf{R}_k\right)^{-1}, \\ \hat{\mathbf{x}}_k &= \mathbf{K}_k \mathbf{y}_k, \\ \mathbf{x}_k &= \mathbf{x}_k + \hat{\mathbf{x}}_k, \\ \mathbf{P}_k &= (\mathbf{I} - \mathbf{K}_k \mathbf{H}_k)\bar{\mathbf{P}}_k, \end{aligned} \quad (21)$$

where $\mathbf{H}_k$ is the measurement sensitivity matrix which is the partial derivative of measurements with respect to the state vector, $\mathbf{K}_k$ is the gain matrix, $\hat{\mathbf{x}}_k$ is the state deviation, and $\mathbf{y}_k$ is the observation residual. The measurement sensitivity matrixes can be derived according to the observation equations. For different measurement models, the measurement sensitivity matrixes can be expressed as follows:

For image measurement (the detailed formula is in Appendix B),

$$\mathbf{H}_{\mathbf{k}}^{\mathbf{i}} = \frac{\partial(u, v)}{\partial \mathbf{r}}. \tag{22}$$

For the altimeter measurement to interested regions,

$$\mathbf{H}_{\mathbf{k}}^{\mathbf{a}} = \left[\frac{(\mathbf{r} - \mathbf{r}_p)^T}{|\mathbf{r} - \mathbf{r}_p|} \quad 0_{1\times3} \right]. \tag{23}$$

For altimeter measurement toward the nadir,

$$\mathbf{H}_{\mathbf{k}}^{\mathbf{a}} = \left[\frac{r_x}{|r|} - \frac{2r_e f_e r_x r_z^2}{|r|^4} \quad \frac{r_y}{|r|} - \frac{2r_e f_e r_y r_z^2}{|r|} \quad \frac{r_z}{|r|} - \frac{2r_e f_e r_z \left(r_x^2 + r_y^2\right)}{|r|} \quad 0_{1\times3} \right]. \tag{24}$$

For inter-satellite range measurement (the homologous part of one satellite in the link),

$$\mathbf{H}_{\mathbf{k}}^{\mathbf{s}} = \left[\frac{(\mathbf{r} - \mathbf{r}_i)^T}{|\mathbf{r} - \mathbf{r}_i|} \quad 0_{1\times3} \right]. \tag{25}$$

Once a new observation is obtained, the state is updated. The updated state is propagated forward to the next observation time, and the filter continues until all measurements are processed.

4. Simulation Results

In the section, orbit determination performance of simulations using different strategies for the hybrid constellation are evaluated and compared.

4.1. Initial Condition. The initialization was set at the epoch (1 July 2009, 10:30:00), and the initial orbit elements are shown in Table 3. Nominal orbit data of the hybrid constellation was generated considering all perturbations referred to in Section 3.1. The dynamic model used in orbit determination includes (1) the two-body gravitation, (2) the non-spherical gravitation (the WGS84 20×20 gravity), (3) the atmospheric perturbation (1976 standard atmosphere model), (4) the gravitation of the Moon and Sun (DE405 numerical planetary ephemeris), (5) the solar radiation pressure (spherical model), and (6) the perturbation of rigid tides.

For MEO layer satellites, only inter-satellite range data can be used to determine orbit, while for LEO layer satellites, the inter-satellite range data, optical imagery, and altimeter data can all be used to determine orbit. The visibility limitation was taken into consideration while generating inter-satellite range data. For optical imagery, when the sub-satellite region is at night or covered by clouds, no effective images can be

FIGURE 3: The distribution of the geometric calibration sites on the Earth's surface.

achieved by satellites. The probability that the imaged region is covered by clouds is set as 50%. Measurement errors from different sources were simulated by Gaussian zero-mean random error for three measurement data. The uncertainty of inter-satellite range was set as 1 m (1σ). For optical imagery, the measurement error of focal plane coordinates mainly results from three sources: the image resolution was set as 1 m (1σ), the pointing accuracy was chosen as 0.001°, and the coastline and rivers were used as ground features with a 10 m (1σ) position error [8]. For the altimeter data, although the altimeter precision can reach the level of centimeters, the height measuring accuracy used in orbit determination is also influenced by several other factors. For example, the difference between the actual Earth ground and the geodetic reference ellipsoid is the main factor leading to the height measurement error [23]. Besides, due to beam divergence, the altimeter range measurement is the average of the ranges from the altimeter to the footprint of each laser beam on the Earth surface. The effect of this averaging could be decreased when only ocean altimeter data is used, as ocean topography is flatter than that of land. Considering these facts, a 10 m (1σ) error was added to ocean altimeter data with the help of the precise Earth shape and tide model. Unlike the traditional usage pattern of the altimeter, a laser altimeter can be used to measure the range between satellites and some interested regions when used in combination with optical cameras. This is achieved by making the laser altimeter point to the ground features that can be captured and recognized by optical cameras. The geometer calibration sites applied in mission SPOT6 and provided by ESA [24] were used as the ground targets in this paper. The site distribution is shown in Figure 3. The small number of ground sites would make it hard to capture them. Therefore, it is assumed that the satellites can swing (the swing angle is no more than 30°) to point to these sites, when approaching them. In such a situation, the altimeter measurement error was set as 1 m (1σ).

In these scenarios, all measurements were obtained with an interval of 1 min. The inter-satellite range data and altimeter data are generated by the nominal orbit data and Reference Earth Model (WGS84). Due to the use of synthetic data instead of real satellite imagery, the image processing stage is not considered in this article. Thus, related latency is not involved in the simulations. The observed coordinates of ground features on the focal plane are generated through geometric relationships between the positions of ground features and the actual orbit, under the nadir imaging model.

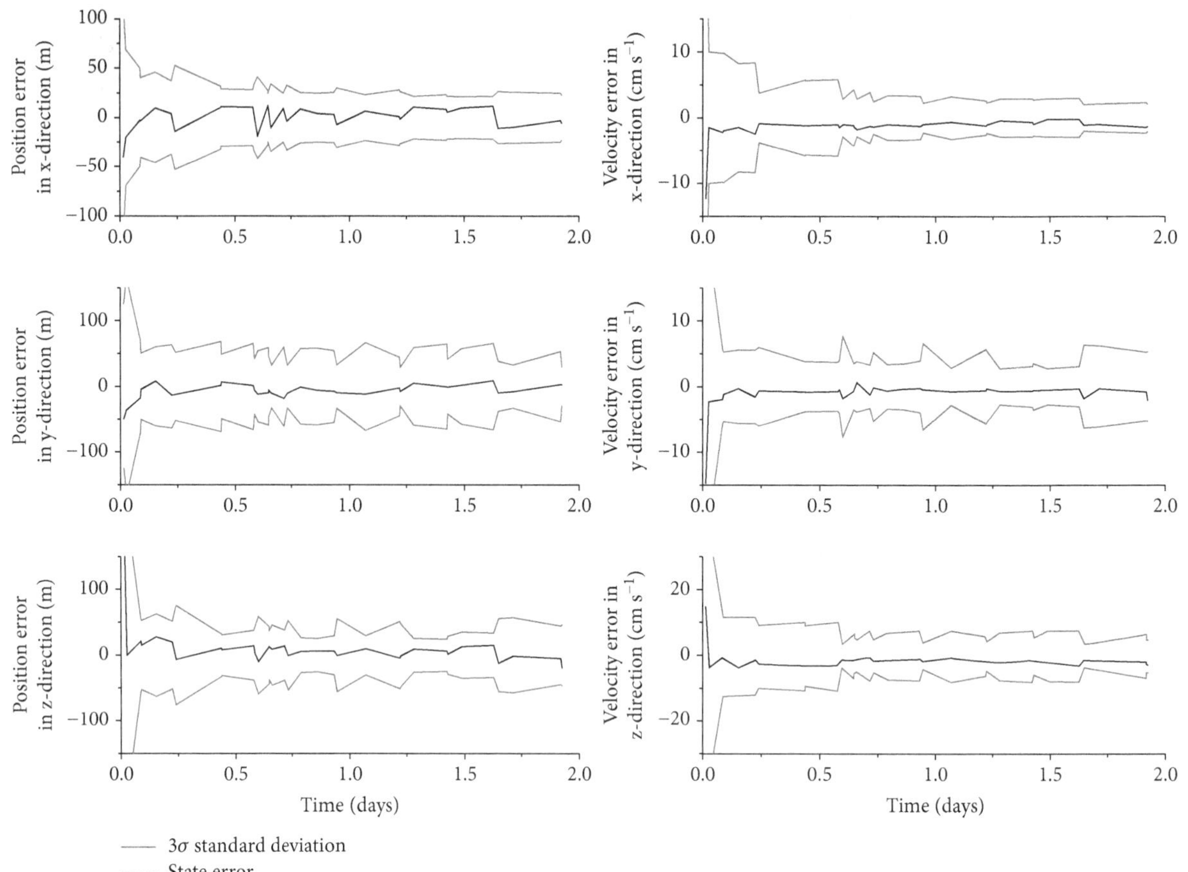

Figure 4: Time history of position and velocity errors of a LEO layer satellite in the J2000 coordinate frame for the case using only optical imagery.

For the satellite orbit, the initial position error was 100 m and the initial velocity error was set as 0.1 cm/s.

4.2. Results and Analysis. In this part, results of case simulations using different orbit determination strategies are analyzed in detail.

The comparison was first conducted between the case using only optical imagery and the case using either altimeter data. As the orbit determination results are similar for all LEO layer satellites under these two scenarios, only results of the first satellite in the LEO layer are shown. Figure 4 shows the time history of orbit error only using optical imagery, and the orbit determination results using optical imagery and altimeter data as inputs are presented in Figures 5 and 6. In these figures, the black line in each plot denotes the actual orbit error, and the red line indicates the 3σ standard deviation of the estimated error. It can been seen that for all cases, the actual errors are well-bounded by the 3-sigma standard deviations and drop significantly to a steady state in less than 0.2 days. Because the measurement and dynamic model are nonlinear, a Monte Carlo test of 100 runs was performed to further compare orbit determination accuracy, and the average RMS errors are given in Table 4. In the case using ocean altimeter data and optical imagery, the results get worse compared with the case using only optical imagery. The average RMS errors increase by a factor of 2.30 for position error and 1.90 for velocity error. In general, the EKF should be able to filter out measurements characterized by higher error; thus, the orbit determination accuracy would not decrease when more measurements are added to an EKF. However, the result draws the opposite conclusion in this simulation. This is mainly due to the small amount of valid images. Although the optical imagery is set to be captured at a high frequency with an interval of 1 min, only a small fraction of the images that contain a ground feature can be used as valid observation. Factors including small FOV, the discontinued distribution of the ground feature, and existence of light constraints would lead to the absence of a ground feature in some images and result in a lower measurement frequency of valid imagery compared to that of the ocean altimeter data. What is more, the ocean altimeter data only provides radial measurement and is not precise enough. Therefore, the orbit precision decreases when the ocean altimeter data is added. On the contrary, in the case using altimeter data pointing to the geometer calibration sites and optical imagery, a better performance than the case using only optical imagery is achieved, with a 41.35% enhancement in position accuracy and 45.14% enhancement in velocity accuracy.

In the case using optical imagery, altimeter data, and inter-satellite range data, the orbits of both LEO layer

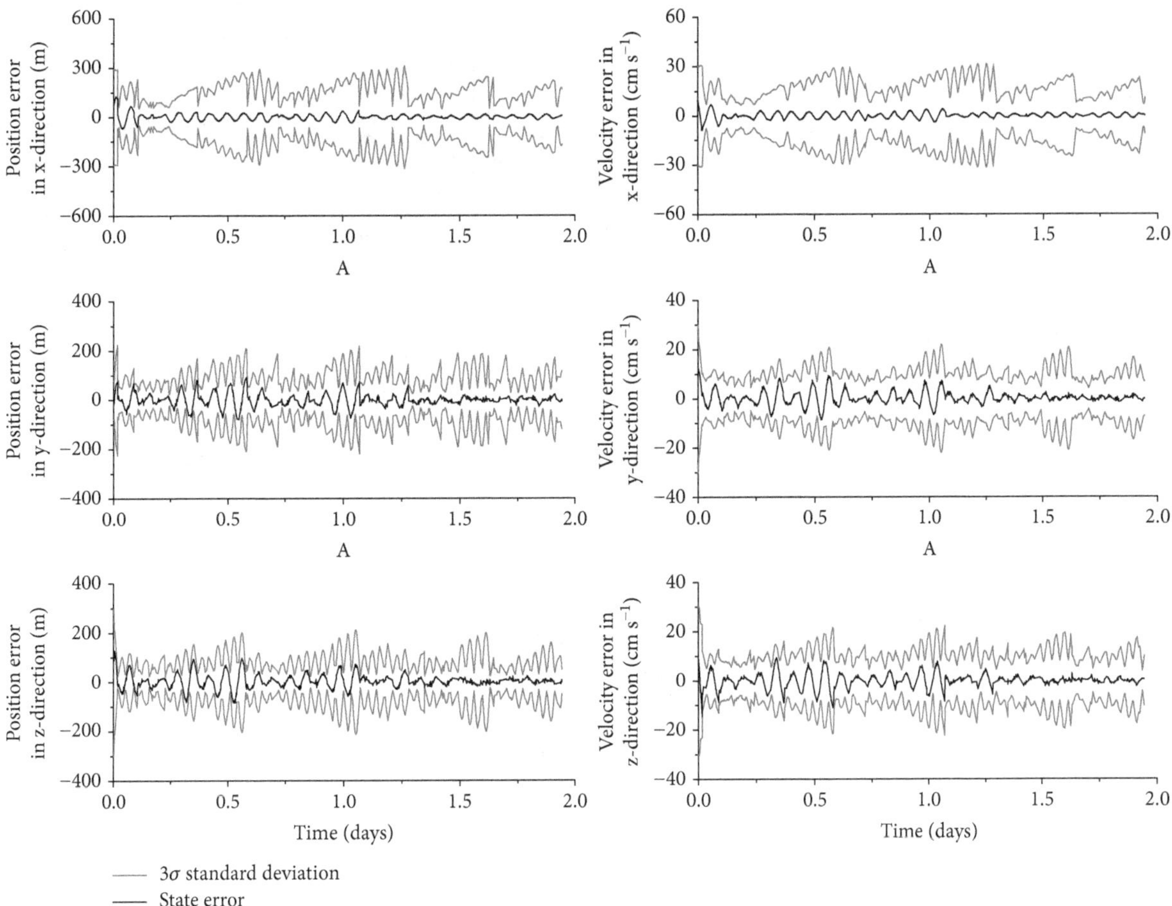

FIGURE 5: Time history of position and velocity errors of a LEO layer satellite in the J2000 coordinate frame for the case using optical imagery and ocean altimeter data.

satellites and MEO layer satellites can be determined simultaneously with the help of inter-satellite range data. The time history of orbit errors for a LEO satellite and a MEO satellite is show in Figures 7 and 8. It is obvious that the position error and velocity error further decrease compared with the cases discussed before. This conclusion is also confirmed by results from the Monte Carlo runs in Table 4. The average RMS errors of position and velocity drop to 47.24% and 42.70%, respectively, compared to those in the case using optical imagery and altimeter data. The improvement in orbit accuracy can be explained as follows: to the multiplane, multilayer structure of the hybrid constellation, a good geometry configuration for relative range measurement is established. Under this configuration, the high-precision inter-satellite range data can effectively improve orbit determination accuracy in all directions.

In the above-mentioned case, all three data including optical imagery, altimeter data, and inter-satellite range data are used for orbit determination. However, in actual missions, the remote sensing instruments may have malfunction or get disturbed, and there are circumstances that LEO layer satellites are not loaded with both optical cameras and altimeters due to their mission requirements. These situations would make it impossible for certain LEO layer satellites to generate effective remote sensing observations. Thus, another two scenarios where there is a lack of remote sensing observation for certain LEO layer satellites are discussed. One (scenario 1) assumes that satellites on the first plane of the LEO layer fail to obtain altimeter measurement while other LEO satellites can get all three measurement data. The other (scenario 2) assumes that the satellites on the first plane of the LEO layer fail to obtain both imagery and altimeter data while other satellites can get all three measurement data. The same Monte Carlo runs were applied to these two scenarios, and simulated results are compared with the ideal case mentioned above in which all satellites in the LEO layer can obtain three measurement data. As is shown in Table 5, for the same scenario, there is no significant difference on the orbit accuracy between LEO satellites with and without a loss of remote sensing observation. Taking scenario 1 as an example, the change of position error (from 9.03 to 9.05) and velocity error (from 0.94 to 0.94) is almost negligible. However, between different scenarios, there exists significant difference on the average RMS of position and velocity errors. As long as there is absence of remote sensing observations, the orbit determination precisions for all satellites in the hybrid constellation decline significantly compared to the ideal scenario.

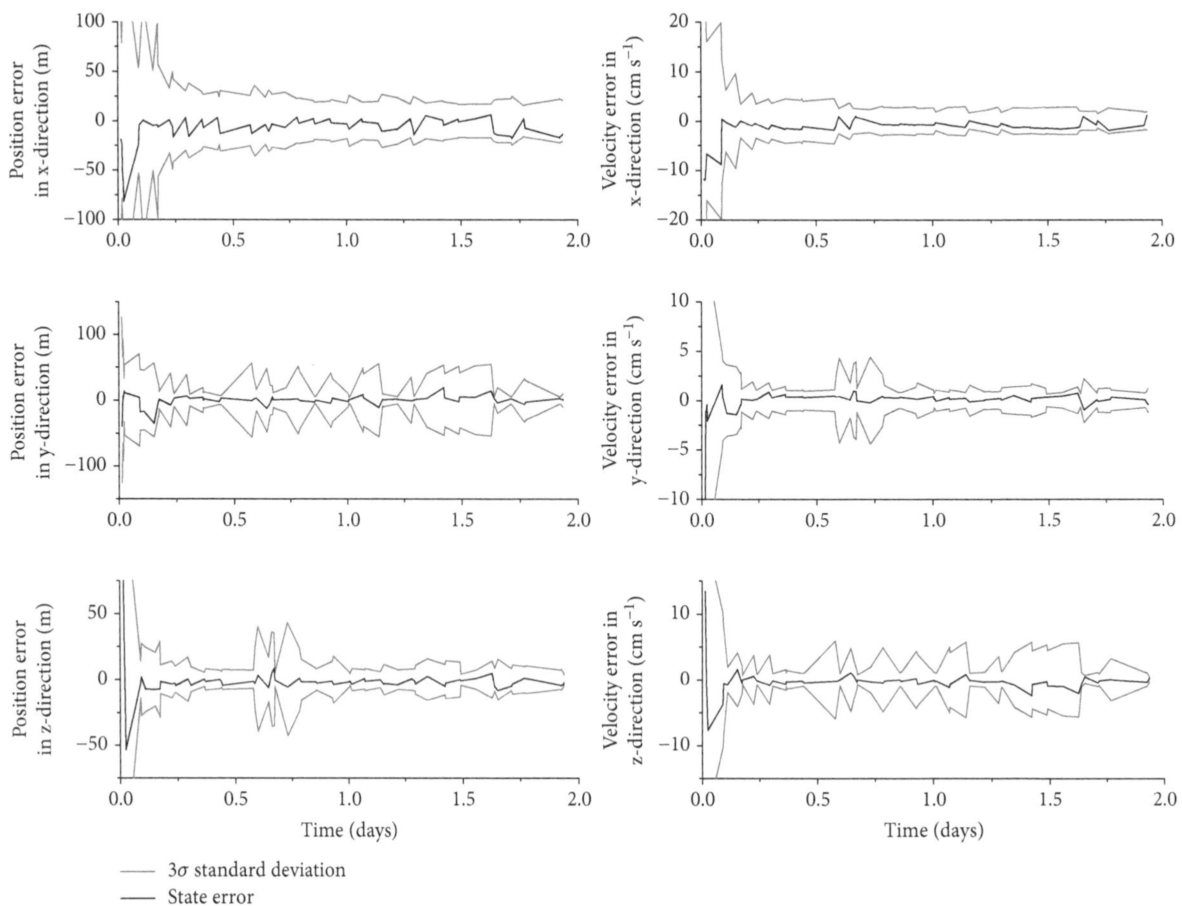

FIGURE 6: Time history of position and velocity errors of a LEO layer satellite in the J2000 coordinate frame for the case using optical imagery and range data to the geometric calibration sites.

TABLE 4: Average RMS errors of 100 Monte Carlo runs for different orbit determination scenarios.

	RMS error			
	LEO layer satellite		MEO layer satellite	
	Position (m)	Velocity (cm/s)	Position (m)	Velocity (cm/s)
Image	14.22	1.75	—	—
Image + altimeter (ocean)	32.77	3.32	—	—
Image + altimeter (the geometer calibration site)	8.34	0.96	—	—
Image + altimeter (the geometer calibration site) + inter-satellite range	3.94	0.41	9.06	0.38

5. Conclusions

In this paper, we proposed an autonomous orbit determination scheme for hybrid constellations by using a combination of three measurement data including optical imagery, altimeter data, and inter-satellite range data. By applying the scheme to a constructed hybrid constellation, the performance of the scheme is then investigated. The hybrid constellation consists of satellites in two LEO/MEO layers. The LEO layer is designed for Earth observation, which is capable of providing complete coverage of the Earth (except polar regions) and has a maximum revisit cycle of 117 hours. The MEO layer is designed in conjunction with the LEO laye for communication and is capable of providing continuou coverage of latitude zone including China mainland.

With the help of measurement from onboard optica cameras and altimeters, orbits of individual LEO satellite can be autonomously determined. Comparison results shov that a better performance is achieved when optical imager is used in combination with altimeter data pointing to th geometer calibration sites than with ocean altimeter data For MEO satellites, orbits can be determined autonomousl by considering inter-satellite range data in combination wit remote sensing observations, which in turn lead to a furthe

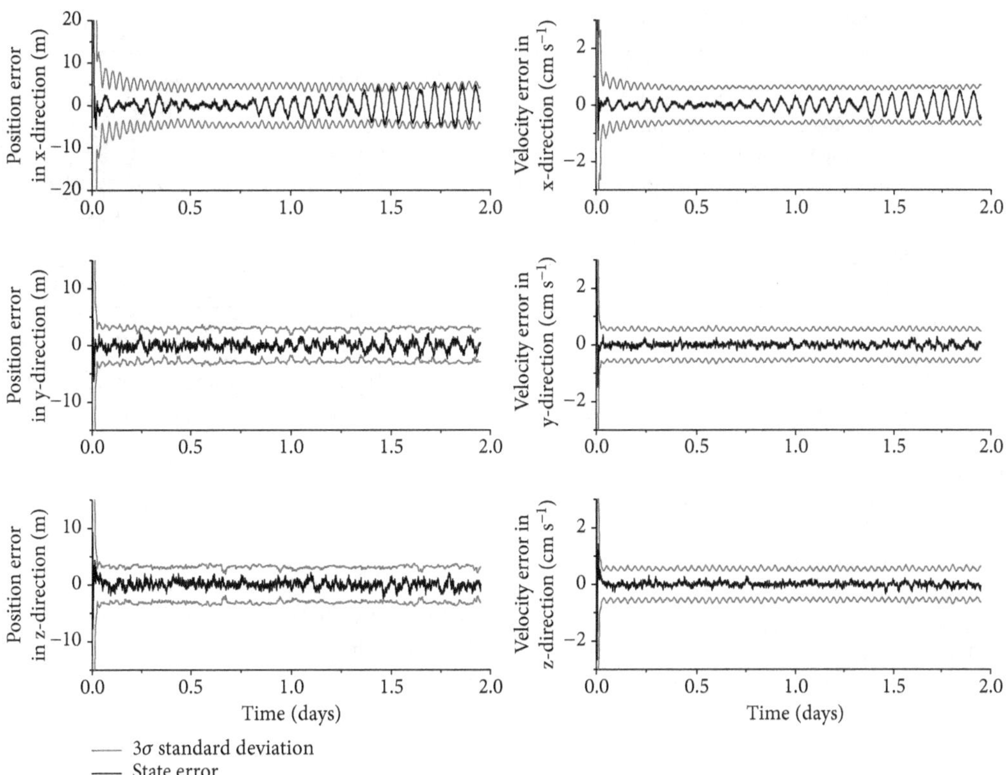

FIGURE 7: Time history of position and velocity errors of a LEO layer satellite in the J2000 coordinate frame for the case using all three observation data.

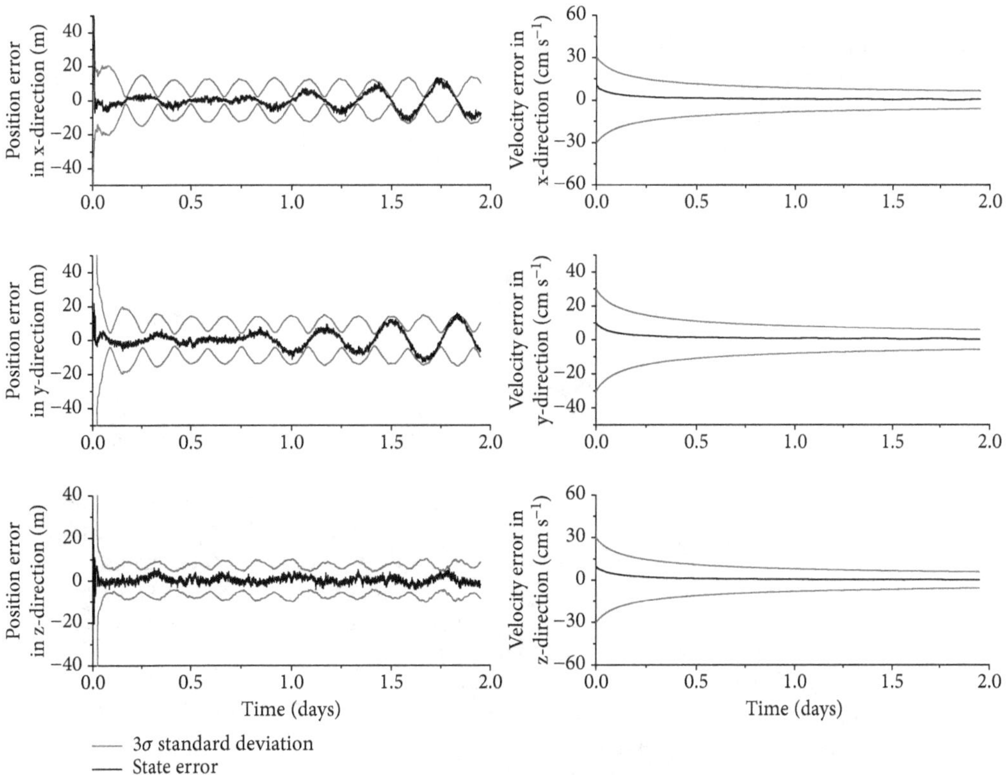

FIGURE 8: Time history of position and velocity errors of a MEO layer satellite in the J2000 coordinate frame for the case using all three observation data.

Table 5: Average RMS errors of 100 Monte Carlo runs for the scenarios losing certain remote sensing observation.

	RMS error					
	LEO layer				MEO layer	
	Satellites without a loss of remote sensing observations		Satellites with a loss of remote sensing observations			
	Position (m)	Velocity (cm/s)	Position (m)	Velocity (cm/s)	Position (m)	Velocity (cm/s)
Ideal scenario	3.94	0.41	—	—	9.06	0.38
Scenario 1	9.03	0.94	9.05	0.94	25.29	0.66
Scenario 2	11.14	1.15	11.12	1.15	35.42	0.83

improvement of orbit accuracy for LEO satellites. To approach real-world simulations, we investigated the orbit determination performance in situations where certain remote sensing observations are absent. Two scenarios in which only altimeter data or both imagery and altimeter data are absent on half of the LEO satellites are considered. Results show that compared to the ideal scenario, orbit determination performance of the hybrid constellation declines in both scenarios. However, for each scenario, orbit precision between LEO satellites with and without lack of measurements shows no difference. These results provide a reference for autonomous operation of constellations containing remote sensing satellites and communication satellites in future missions.

In the paper, the design of the constellation is based mainly on experience and deterministic algorithm. When more constraints are considered or more parameters need to be optimized, a heuristic algorithm could be used instead. Additionally, in order to further improve radial orbit accuracy, the altimeter crossover data could be used. How these changes would influence the orbit determination performance is worthy of in-depth analysis in future work.

Appendix

A. The Derivation of (7)

The expression of equation (7) is

$$\begin{cases} N_d \leq 10, \\ 14 \leq \dfrac{N_p}{N_d} \leq 15, \\ \dfrac{\pi}{\arccos (r_e/a)} \leq N_s \leq \dfrac{2\pi}{\arccos (r_e/a)}, \\ \dfrac{2(a-r_e)\tan(\text{FOV}/2)}{|\sin i|} > 2\pi r_e \dfrac{N_d}{\text{LCM}(N_d, N_s)N_p}. \end{cases} \tag{A.1}$$

The first two mathematical formulations in (7) are comprehensible, which present the constraints: the revisit cycle is less than 10 days and the orbit revolution per day is more than 14 and less than 15.

For the third constraint, in order to make sure that it is visible between adjacent satellites, the distance of the Earth center to the connection line of two adjacent satellites should be larger than the Earth radius. The corresponding mathematical expression is

$$\frac{\varphi}{2} \leq \arccos \frac{r_e}{a}, \tag{A.2}$$

in which φ is the geocentric angle between two adjacent satellites. As there are N_s satellites in an orbit plane, the geocentric angle can be given as

$$\varphi = \frac{2\pi}{N_s}. \tag{A.3}$$

Substituting (A.3) into (A.2) and considering that the number of satellites required is no larger than 2 times this minimum number, the third inequality relation in (7) can be obtained:

$$\frac{\pi}{\arccos (r_e/a)} \leq N_s \leq \frac{2\pi}{\arccos (r_e/a)}. \tag{A.4}$$

The fourth constraint is that the single-plane constellation can provide complete coverage of the Earth (except polar regions). For analyzing the coverage of satellites, the swath width W_s of a satellite needs to be known, which is

$$W_s = 2(a - r_e) \tan \frac{\text{FOV}}{2}. \tag{A.5}$$

With (A.5), the width for every satellite sweep on the equator is

$$W_e = \frac{W_s}{|\sin i|}. \tag{A.6}$$

Since the equator is the longest latitude circle, the complete coverage of the Earth (except polar regions) is equivalent with the complete coverage of the equator region. Thus, in a revisit period, the constraint condition that satellites on an orbit plane can provide complete coverage of the Earth can be expressed as

$$\frac{\text{LCM}(N_d, N_s)N_p}{N_d} W_e > 2\pi r_e, \tag{A.7}$$

where $\text{LCM}(N_d, N_s)$ represents the least common multiple of N_d and N_s. Substituting (A.5) and (A.6) into (A.7) yields the fourth inequality relation in (7).

B. The Detailed Formula of (22)

The detailed formula of equation (22) is ($(a_{11}, \ldots, a_{33})$ are elements of $(\mathbf{T}_A \mathbf{T}_I)^{-1}$)

$$\mathbf{H}_\mathbf{k}^\mathbf{i} = \begin{bmatrix} -f\frac{(a_{31}a_{12}-a_{11}a_{32})(r_y-r_y^g)+(a_{31}a_{13}-a_{11}a_{33})(r_z-r_z^g)}{\left(a_{31}(r_x-r_x^g)+a_{32}(r_y-r_y^g)+a_{33}(r_z-r_z^g)\right)^2} & -f\frac{(a_{32}a_{12}-a_{12}a_{31})(r_x-r_x^g)+(a_{32}a_{13}-a_{11}a_{33})(r_z-r_z^g)}{\left(a_{31}(r_x-r_x^g)+a_{32}(r_y-r_y^g)+a_{33}(r_z-r_z^g)\right)^2} & -f\frac{(a_{33}a_{11}-a_{13}a_{31})(r_x-r_x^g)+(a_{32}a_{13}-a_{12}a_{33})(r_y-r_y^g)}{\left(a_{31}(r_x-r_x^g)+a_{32}(r_y-r_y^g)+a_{33}(r_z-r_z^g)\right)^2} & (\mathbf{0})_{1\times 3} \\ -f\frac{(a_{31}a_{22}-a_{21}a_{32})(r_y-r_y^g)+(a_{31}a_{23}-a_{21}a_{33})(r_z-r_z^g)}{\left(a_{31}(r_x-r_x^g)+a_{32}(r_y-r_y^g)+a_{33}(r_z-r_z^g)\right)^2} & -f\frac{(a_{32}a_{22}-a_{22}a_{31})(r_x-r_x^g)+(a_{32}a_{23}-a_{21}a_{33})(r_z-r_z^g)}{\left(a_{31}(r_x-r_x^g)+a_{32}(r_y-r_y^g)+a_{33}(r_z-r_z^g)\right)^2} & -f\frac{(a_{33}a_{21}-a_{23}a_{31})(r_x-r_x^g)+(a_{32}a_{23}-a_{22}a_{33})(r_y-r_y^g)}{\left(a_{31}(r_x-r_x^g)+a_{32}(r_y-r_y^g)+a_{33}(r_z-r_z^g)\right)^2} & (\mathbf{0})_{1\times 3} \end{bmatrix}. \tag{A.8}$$

Conflicts of Interest

The authors declare that there is no conflict of interest regarding the publication of this paper.

Acknowledgments

This work was carried out with financial support from the National Basic Research Program 973 of China (2015CB857100), the Base of National Defence Scientific Research Fund (Nos. 2016110C019 and JCKY2016203C067), and the National Natural Science Foundation (No. 11603011).

References

[1] A. L. Jennings and H. Diniz, "Global navigation satellite system design exploration using a multi-objective genetic algorithm," in *AIAA SPACE 2015 Conference and Exposition*, p. 4622, Pasadena, CA, USA, 2015.

[2] H. Xu, X. Zhang, and X. Zhan, "Analyzing RAIM performance for an MEO/GEO hybrid GNSS constellation," in *IET International Communication Conference on Wireless Mobile and Computing (CCWMC 2011)*, pp. 465–467, Shanghai, China, November 2011.

[3] S. Chan, A. T. Samuels, N. B. Shah, J. E. Underwood, and O. L. De Weck, "Optimization of hybrid satellite constellations using multiple layers and mixed circular-elliptical orbits," in *22nd AIAA International Communications Satellite Systems Conference & Exhibit 2004*, pp. 9–12, Monterey, CA, USA, May 2004.

[4] M. Luglio and W. Pietroni, "Optimization of double-link transmission in case of hybrid orbit satellite constellations," *Journal of Spacecraft and Rockets*, vol. 39, no. 5, pp. 761–770, 2002.

[5] E. Fahnestock and R. S. Erwin, "Optimization of hybrid satellite and constellation design for GEO-belt space situational awareness using genetic algorithms," in *Proceedings of the 2005, American Control Conference, 2005*, pp. 2110–2115, Portland, OR, USA, June 2005.

[6] R. L. White, M. B. Adams, E. G. Geisler, and F. D. Grant, "Attitude and oribit estimation using stars and landmarks," *IEEE Transactions on Aerospace and Electronic Systems*, vol. AES-11, no. 2, pp. 195–203, 1975.

[7] M. Straub and J. Christian, "Autonomous optical navigation for Earth-observing satellites using coastline matching," in *AIAA Guidance, Navigation, and Control Conference*, Kissimmee, FL, USA, 2015.

[8] M. Li, B. Xu, and L. Zhang, "Orbit determination for remote-sensing satellites using only optical imagery," *International Journal of Remote Sensing*, vol. 38, no. 5, pp. 1350–1364, 2017.

[9] M. Li and B. Xu, "Autonomous orbit and attitude determination for Earth satellites using images of regular-shaped ground objects," *Aerospace Science and Technology*, vol. 80, pp. 192–202, 2018.

[10] G. H. Born, B. D. Tapley, and M. L. Santee, "Orbit determination using dual crossing arc altimetry," *Acta Astronautica*, vol. 13, no. 4, pp. 157–163, 1986.

[11] F. G. Lemoine, D. D. Rowlands, S. B. Luthcke et al., "Precise orbit determination for Geosat follow-on using satellite laser ranging data and intermission altimeter crossovers," in *2001 Flight Mechanics Symposium*, Greenbelt, MD, USA, 2001.

[12] F. Markley and U. Naval, "Autonomous navigation using landmark and intersatellite data," in *AIAA/AAS Astrodynamics Conference*, Seattle, WA, USA, 1984.

[13] M. L. Psiaki, "Autonomous orbit determination for two spacecraft from relative position measurements," *Journal of Guidance, Control, and Dynamics*, vol. 22, no. 2, pp. 305–312, 1999.

[14] R. Li, H. Qiu, and K. Xiong, "Autonomous navigation for constellation based on inter-satellite ranging and directions," in *IECON 2017 - 43rd Annual Conference of the IEEE Industrial Electronics Society*, pp. 2985–2990, Beijing, China, November 2017.

[15] X. Kai, W. Chunling, and L. Liangdong, "Autonomous navigation for a group of satellites with star sensors and inter-satellite links," *Acta Astronautica*, vol. 86, pp. 10–23, 2013.

[16] X. Kai, W. Chunling, and L. Liangdong, "The use of X-ray pulsars for aiding navigation of satellites in constellations," *Acta Astronautica*, vol. 64, no. 4, pp. 427–436, 2009.

[17] S. Wang and P. Cui, "Autonomous orbit determination using pulsars and inter-satellite ranging for Mars orbiters," in *2018 IEEE Aerospace Conference*, Big Sky, MT, USA, March 2018.

[18] M. Capderou, *Handbook of Satellite Orbits: From Kepler to GPS*, Springer Science & Business, 2014.

[19] https://directory.eoportal.org/web/eoportal/satellite-missions/d/dmc-3.

[20] D. Mortari, M. P. Wilkins, and C. Bruccoleri, "The Flower Constellations," *Journal of Astronautical Sciences*, vol. 52, pp. 107–127, 2004.

[21] M. P. Wilkins and D. Mortari, "Flower Constellation set theory part ii: secondary paths and equivalency," *IEEE Transactions on Aerospace and Electronic Systems*, vol. 44, no. 3, pp. 964–976, 2008.

[22] J. G. Walker, *Continuous Whole-Earth Coverage by Circular-Orbit Satellite Patterns*, Royal Aircraft Establishment Farnborough (United Kingdom), 1977.

[23] B. Schutz, B. Tapley, and G. H. Born, *Statistical Orbit Determination*, Academic Press, 2004.

[24] http://calvalportal.ceos.org/calibration-test-sites.

13

The Coupled Orbit-Attitude Dynamics and Control of Electric Sail in Displaced Solar Orbits

Mingying Huo,[1] He Liao,[2] Yanfang Liu,[1] and Naiming Qi[1]

[1] *Department of Aerospace Engineering, Harbin Institute of Technology, Harbin 150001, China*
[2] *Shanghai Satellite Engineering Research Institute, Shanghai 200240, China*

Correspondence should be addressed to Mingying Huo; huomingying123@gmail.com

Academic Editor: Christian Circi

Displaced solar orbits for spacecraft propelled by electric sails are investigated. Since the propulsive thrust is induced by the sail attitude, the orbital and attitude dynamics of electric-sail-based spacecraft are coupled and required to be investigated together. However, the coupled dynamics and control of electric sails have not been discussed in most published literatures. In this paper, the equilibrium point of the coupled dynamical system in displaced orbit is obtained, and its stability is analyzed through a linearization. The results of stability analysis show that only some of the orbits are marginally stable. For unstable displaced orbits, linear quadratic regulator is employed to control the coupled attitude-orbit system. Numerical simulations show that the proposed strategy can control the coupled system and a small torque can stabilize both the attitude and orbit. In order to generate the control force and torque, the voltage distribution problem is studied in an optimal framework. The numerical results show that the control force and torque of electric sail can be realized by adjusting the voltage distribution of charged tethers.

1. Introduction

Displaced solar orbit is a kind of non-Keplerian orbits, which is lifted above the ecliptic plane by applying a continuous thrust to counterbalance the gravity. From displaced orbit, one would have a continuous view to the polar region of the sun, or a long time scale uninterrupted helioseismological coverage [1]. As a result of that the maintenance of displaced orbit requires continuous propulsive thrust; this mission is impossible for most of conventional (either chemical or electrical) propulsion systems. The solar sail is firstly proposed to maintain the displaced orbits as it utilizes solar radiation pressure to generate continuous and propellant-less thrust. As early as 1929, Oberth mentioned that the solar radiation pressure can be used to generate a displacement between the orbital plane and the ecliptic plane. More recently, large families of displaced solar sail orbits are proposed by McInnes and Simmons [2–4]. The dynamics, stability, and control of displaced solar sail orbits were investigated by considering a solar sail in a rotating frame. Gong et al. did a lot of work on the coupled attitude-orbit dynamics and control of a solar sail in displaced orbits [5, 6] and proposed the solar sail formation flying around displaced orbits [7–10].

However, the thrust acceleration level of solar sails cannot meet the requirements of maintaining high displacement orbits, because its reflection film is not light enough [11]. In addition, as the thrust acceleration of solar sail cannot be adjusted at will between zero and some maximum, the displaced orbit maintained by solar sail is not flexible enough. In light of these problems, the electric solar wind sail (electric sail for short) is used, as an alternative to the use of solar sail, to maintain displaced orbits in this paper. The electric sail, which is first proposed by Janhunen [12] in 2004, is an innovative concept for spacecraft propulsion. Similar to solar sails, electric sails can produce continuous thrust without the need of propellant. Unlike solar sails, electric sails are propelled by the solar wind dynamic pressure, instead of the solar photon momentum.

As shown in Figure 1, the electric sail consists of many tethers, which are held at a high positive potential by a solar-powered electron gun. The electrostatic field generated by the charged tethers can reflect the solar wind protons to

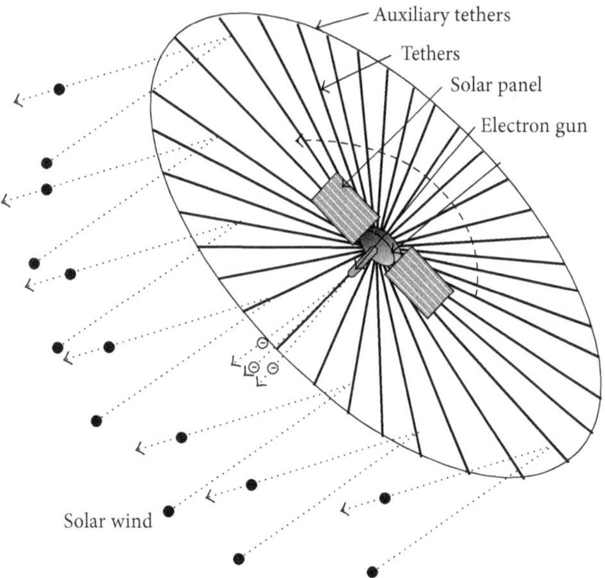

FIGURE 1: Conceptual sketch of an electric sail.

generate a continuous thrust without any propellant. The deployment and maintenance of electric sail are implemented by spinning the sailcraft about the symmetry axis. It is noticed that an electrostatic field potential structure with a spatial scale larger than 100 m can be created around a thin tether with thickness of a few tens of micrometers. Therefore, the characteristic acceleration of electric sails can be higher than that of solar sails. Recent results show that electric sails can generate 1 N thrust with only 100 kg propulsion system mass [13]. In addition, the thrust of electric sail can be adjusted at will between zero and some maximum by controlling the power of the electron gun [1]. Consequently, the displaced orbit maintained by electric sail is more flexible than that maintained by solar sail. The displaced non-Keplerian orbits for electric sails have been studied by Mengali and Quarta [14]. In their paper, the electric sail capabilities of generating a class of displaced non-Keplerian orbits are analyzed, and a comparison with a solar sail is made. Qi et al. [15] investigated displaced electric sail orbits and the transition trajectory optimization. However, in the above literatures, the displaced orbits of electric sail are researched based on a classical thrust model, and the coupled orbit-attitude dynamics and control of electric sail are not considered. In the classical thrust model, the effects of electric sail attitude on the propulsive thrust modulus and direction were neglected. The thrust modulus was assumed to be invariable with the change of pitch angle, and the thrust cone angle was assumed to be approximately equal to one-half of the pitch angle. Obviously, the above models are not accurate enough to describe the thrust vector of an electric sail for mission analysis. In this paper, the coupled orbit-attitude dynamics and control of electric sail will be considered together based on an advanced thrust model.

Since the propulsive thrust is induced by the sail attitude, the orbital and attitude dynamics of electric-sail-based spacecraft are coupled. However, the coupled orbit-attitude control of electric sails has not been discussed in most published

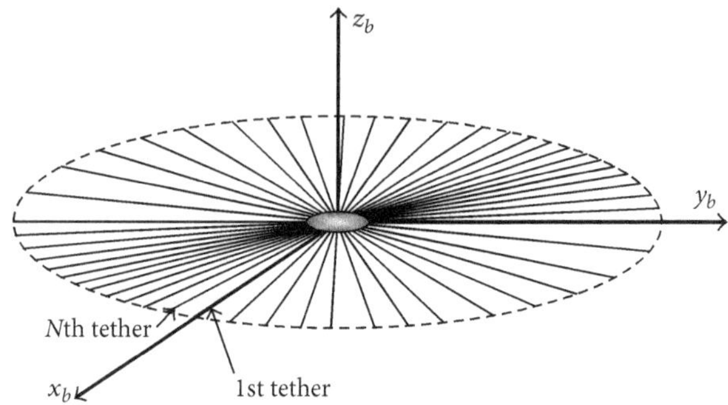

FIGURE 2: Body reference frame.

literatures. In our previous research [16], the propulsive thrust of electric sail is described as a function of the orbital radius and the sail angle. Consequently, the coupled attitude-orbit dynamics of an electric-sail-based spacecraft are obtained. For the heliocentric transfer mission, the flight control is investigated, wherein the orbital control is studied in an optimal framework via a hybrid optimization method and the attitude controller is designed based on feedback linearization control. In fact, the above attitude and orbital control are carried out separately. Differently, in this paper, the attitude and orbital control of electric sail in displaced solar orbits will be investigated together, instead of being implanted separately.

This paper is organized as follows. First of all, the coupled orbit-attitude dynamics of electric sail are discussed in the heliocentric-ecliptic inertial frame and the body frame. Secondly, the equilibrium point of the coupled dynamical system in displaced orbit is obtained, and its stability is analyzed through a linearization. Finally, the linear quadratic regulator is employed to control the coupled orbit-attitude system for unstable displaced orbits.

2. Coupled Orbit-Attitude Dynamics of Electric Sail

In our previous research [16], the coupled orbit-attitude dynamics of electric sail are obtained. For displaced orbits design and coupled control of electric sail, the coupled dynamics are discussed briefly in this paper.

2.1. Reference Frame. Before the description of coupled dynamics, three reference frames are introduced, which are the body frame $o_b x_b y_b z_b$, the orbital frame $o_o x_o y_o z_o$, and the heliocentric-ecliptic inertial frame $o_i x_i y_i z_i$. Considering an electric sail consists of N tethers as seen in Figure 2, these tethers can be numbered counterclockwise. The origin of the body frame $o_b x_b y_b z_b$ is at the center of mass of the sail, and the x_b-axis is in the direction of a given reference tether. The z_b-axis is along the normal of the sail, and the y_b-axis forms a right-handed system.

The orbital frame $o_o x_o y_o z_o$ and the inertial frame $o_i x_i y_i z_i$ are shown in Figure 3. The origin of the orbital frame $o_o x_o y_o z_o$ is at the center of mass of the sail and the z_o-axis is along the sun-spacecraft direction. The y_o-axis is perpendicular to the

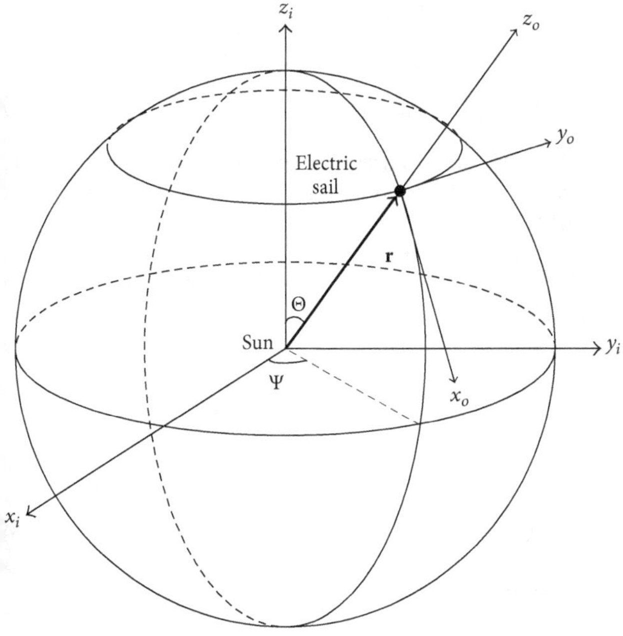

FIGURE 3: Heliocentric-ecliptic inertial frame and orbital reference frame.

normal of the ecliptic plane and the z_o-axis, and the x_o-axis forms a right-handed triad. The origin of the inertial frame $o_i x_i y_i z_i$ is at the center of mass of the sun, and the x_i-axis is in the direction of sun-equinox. The z_i-axis is along the normal of the ecliptic plane, and the y_i-axis forms a right-handed system. The attitude of the sail in the orbital frame can be described by three angles, ϕ, θ, and ψ, and the rotation sequence from the orbital frame to the body frame is $x(\phi) \rightarrow y(\theta) \rightarrow z(\psi)$.

2.2. Coupled Orbit-Attitude Dynamical Equations. In order to get greater propulsive acceleration and reduce the launch cost, the electric sail must employ large-scale light-weight charged tethers. With these tethers, the electric sail has some special properties of extreme flexibility, low damping, and low frequency. The flexibility of tethers will influence the attitude operation. When the flexible tethers are in motion, the vibration of the structure influences the dynamic characteristics of the entire structure and the coupled control, especially when the structure undergoes large-angle maneuvering. Attitude manoeuvres combined with gyroscopic effects might induce motions that cannot be captured by the rigid body model. However, the coupling mechanism of orbit-attitude-deformation of flexible electric sails is quite complicated. For simplicity, the flexibility of charged tethers is ignored, because the angle of maneuver is small in this paper. As discussed in our previous research [16], the coupled orbit-attitude dynamical scalar equations of electric-sail-based spacecraft, under the assumption of rigid body model, can be written as

$$
\begin{aligned}
\dot{r} &= v_r \\
\dot{\Theta} &= \omega_\Theta \\
\dot{\Psi} &= \omega_\Psi \\
\dot{v}_r &= r\omega_\Psi^2 \sin^2\Theta + r\omega_\Theta^2 - \frac{\mu_\odot}{r^2} \\
&\quad + \frac{\kappa a_\oplus r_\oplus}{2r}\left(\cos^2\phi\cos^2\theta + 1\right) \\
\dot{\omega}_\Theta &= \omega_\Psi^2 \sin\Theta\cos\Theta - \frac{2v_r\omega_\Theta}{r} \\
&\quad + \frac{\kappa a_\oplus r_\oplus}{2r^2}\left(\cos\phi\sin\theta\cos\theta\right) \\
\dot{\omega}_\Psi &= -\frac{2v_r\omega_\Psi}{r} - 2\omega_\Theta\omega_\Psi\cot\Theta \\
&\quad - \frac{\kappa a_\oplus r_\oplus}{2r^2}\left(\sin\phi\cos\phi\cos^2\theta\right) \\
\dot{\phi} &= \frac{\omega_{bx}\cos\psi}{\cos\theta} - \frac{\omega_{by}\sin\psi}{\cos\theta} + \omega_\Psi\sin\Theta \\
&\quad - \omega_\Theta\sin\phi\tan\theta + \omega_\Psi\cos\Theta\cos\phi\tan\theta \\
\dot{\theta} &= \omega_{bx}\sin\psi + \omega_{by}\cos\psi - \omega_\Theta\cos\phi \\
&\quad - \omega_\Psi\cos\Theta\sin\phi \\
\dot{\psi} &= -\omega_{bx}\tan\theta\cos\psi + \omega_{by}\tan\theta\sin\psi + \omega_{bz} \\
&\quad + \frac{\omega_\Theta\sin\phi}{\cos\theta} - \frac{\omega_\Psi\cos\Theta\cos\phi}{\cos\theta} \\
\dot{\omega}_{bx} &= \frac{\omega_{by}\omega_{bz}\left(I_y - I_z\right)}{I_x} + \frac{T_x}{I_x} \\
\dot{\omega}_{by} &= \frac{\omega_{bx}\omega_{bz}\left(I_z - I_x\right)}{I_y} + \frac{T_y}{I_y} \\
\dot{\omega}_{bz} &= \frac{\omega_{bx}\omega_{by}\left(I_x - I_y\right)}{I_z} + \frac{T_z}{I_z},
\end{aligned}
\tag{1}
$$

where r is the distance between sun and sailcraft, Ψ is the ecliptic longitude and Θ is the ecliptic latitude as shown in Figure 3, $\mu_\odot$ is sun's gravitational parameter, thrust control coefficient $\kappa \in [0, 1]$, as the thrust of the electric sail can be adjusted by the electron gun, $a_\oplus$ is the characteristic acceleration of the electric sail, when the sun-spacecraft distance is $r_\oplus = 1$ au (the characteristic acceleration of the electric sail can be adjusted by the electron gun within certain range, $[\omega_{bx}\ \ \omega_{by}\ \ \omega_{bz}]$ is the vector of angular velocity described in the body frame $o_b x_b y_b z_b$, $[T_x\ \ T_y\ \ T_z]$ is the vector of control torque, which can be generated by adjusting the voltage distribution of charged tethers [16], and the inertia matrix of sailcraft $\mathbf{I} = \mathrm{diag}(I_x, I_y, I_z)$, wherein $I_z = 2I_x = 2I_y = 14.666 \times 10^8$ kgm^2 in this paper.

3. Displaced Electric Sail Orbits

3.1. Displaced Orbits in General. Solutions of displaced electric sail orbits can be found by seeking equilibrium solutions to the two-body problem. Considering the definition of displaced orbits in [2], the orbital radius r, the ecliptic latitude Θ, and the ecliptic longitude angle velocity ω_Ψ should be constant and can be chosen to be some particular fixed value. Therefore, the displaced orbits in general should have the following characteristics:

$$r = r_d$$
$$\Theta = \Theta_d$$
$$\omega_\Psi = \omega_d$$
$$\dot{r}_d = \ddot{r}_d = \dot{\Theta}_d = \ddot{\Theta}_d = \dot{\omega}_d = 0, \quad (2)$$

where r_d, Θ_d, and ω_d are the selected orbital radius, ecliptic latitude, and ecliptic longitude velocity of displaced orbit, respectively.

Substituting (2) in the orbital dynamical scalar equations (1), the required characteristic acceleration $a_{\oplus d}$ and required attitude angles can be obtained as

$$a_{\oplus d} = \frac{3\mu_\odot - 3r_d^3\omega_d^2\sin^2\Theta_d - \chi}{2r_d r_\oplus}$$
$$\phi_d = 0 \quad (3)$$
$$\sin\theta_d = \sqrt{\frac{\chi + \mu_\odot^2 + r_d^6\omega_d^4\sin^2\Theta_d\left(3\sin^2\Theta_d - 2\right) - r_d^3\omega_d^2\sin^2\Theta_d\left(\chi + 2\mu_\odot\right)}{2r_d^6\omega_d^4\sin^2\Theta_d - 4r_d^3\omega_d^2\sin^2\Theta_d + 2\mu_\odot^2}}\left(\frac{\mu_\odot + \chi - r_d^3\omega_d^2\sin^2\Theta_d}{2r_d^3\omega_d^2\sin\Theta_d\cos\Theta_d}\right),$$

where $\chi = \sqrt{r_d^6\omega_d^4\sin^2\Theta_d(9\sin^2\Theta_d - 8) - 2\mu_\odot r_d^3\omega_d^2\sin^2\Theta_d + \mu_\odot^2}$.

As shown in (3), the attitude angle ϕ_d should be equal to zero for keeping the required displaced orbit. It means that the normal of the electric sail should be in the same plane as the position vector $\mathbf{r}$ and the orbital angular velocity vector $\boldsymbol{\omega}$. This requirement is consistent with the requirement for keeping displaced orbit using a solar sail [5]. In addition, through a series of analyses, it is known that the chosen parameters of displaced orbit $(r_d, \Theta_d, \omega_d)$ should meet the following criteria:

$$r_d^6\omega_d^4\sin^2\Theta_d\left(9\sin^2\Theta_d - 8\right) - 2\mu_\odot r_d^3\omega_d^2\sin^2\Theta_d + \mu_\odot^2 \geq 0. \quad (4)$$

If the selected parameters cannot meet the requirements in (4), electric sails would not be used in keeping the selected displaced orbit, due to the limitation of the maximum thrust cone angle of the electric sail. As discussed in our previous research [16], the thrust cone angle of the electric sail reaches the maximum value 19.47°, when the light incident angle β is 54.75°.

3.2. Displaced Geostationary Orbits. If the orbital period of displaced orbit is chosen to be consistent with the orbital period of Earth, this kind of displaced orbits is named displaced geostationary orbits. Displaced geostationary orbits have the following characteristics:

$$\omega_d = \sqrt{\frac{\mu_\odot}{r_\oplus^3}}. \quad (5)$$

Substituting (5) in (3), the requirements of displaced geostationary orbits are obtained as follows:

$$a_\oplus = \frac{3\mu_\odot r_\oplus^3 - 3\mu_\odot r_d^3\sin^2\Theta_d - \chi r_\oplus^3}{2r_d r_\oplus^4}$$
$$\phi_d = 0 \quad (6)$$
$$\sin\theta = \sqrt{\frac{\left(\chi + \mu_\odot^2\right)r_\oplus^6 + \mu_\odot^2 r_d^6\sin^2\Theta_d\left(3\sin^2\Theta_d - 2\right) - \mu_\odot r_d^3 r_\oplus^3\sin^2\Theta_d\left(\chi + 2\mu_\odot\right)}{2\mu_\odot^2 r_d^6\sin^2\Theta_d - 4\mu_\odot r_d^3 r_\oplus^3\sin^2\Theta_d + 2\mu_\odot^2 r_\oplus^6}}\left(\frac{\left(\mu_\odot + \chi\right)r_\oplus^3 - \mu_\odot r_d^3\sin^2\Theta_d}{2\mu_\odot r_d^3\sin\Theta_d\cos\Theta_d}\right),$$

where $\chi = \sqrt{\mu_\odot^2 r_d^6\sin^2\Theta_d(9\sin^2\Theta_d - 8)/r_\oplus^6 - 2\mu_\odot^2 r_d^3\sin^2\Theta_d/r_\oplus^3 + \mu_\odot^2}$.

According to the previous discussion, the chosen parameters of displaced geostationary orbit (r_d, Θ_d) should meet the following criteria:

$$\frac{\mu_\odot^2 r_d^6\sin^2\Theta_d\left(9\sin^2\Theta_d - 8\right)}{r_\oplus^6} - \frac{2\mu_\odot^2 r_d^3\sin^2\Theta_d}{r_\oplus^3} + \mu_\odot^2 \geq 0. \quad (7)$$

Based on (7), we can obtain the feasible region for the displaced geostationary orbit, which is shown in Figure 4.

Based on (6), the required characteristic acceleration $a_{\oplus d}$ and attitude angle θ_d, for keeping displaced geostationary orbits with different parameters (r_d, Θ_d) in feasible region, are calculated and shown in Figures 5 and 6, respectively. As shown in Figure 5, the required characteristic acceleration $a_{\oplus d}$ increases with the decrease of the sun-sailcraft distance and increases with the increase of $|\Theta_d - 90°|$.

4. Stability Analysis

4.1. Analysis of the Stability. In order to facilitate the analysis of the stability of displaced orbits,let $\mathbf{X}_1 = [r, \Theta, \Psi]^{\mathrm{T}}$, $\mathbf{X}_2 = [v_r, \omega_\Theta, \omega_\Psi]^{\mathrm{T}}$, $\mathbf{X}_3 = [\phi, \theta, \psi]^{\mathrm{T}}$, and $\mathbf{X}_4 = [\omega_{bx}, \omega_{by}, \omega_{bz}]^{\mathrm{T}}$. Then, the coupled orbit-attitude dynamical scalar equations of electric-sail-based spacecraft can be written as

$$\begin{aligned} \dot{\mathbf{X}}_1 &= \mathbf{X}_2 \\ \dot{\mathbf{X}}_2 &= \Gamma(\mathbf{X}_1, \mathbf{X}_2, \mathbf{X}_3) \\ \dot{\mathbf{X}}_3 &= \Lambda(\mathbf{X}_1, \mathbf{X}_2, \mathbf{X}_3, \mathbf{X}_4) \\ \dot{\mathbf{X}}_4 &= \Pi(\mathbf{X}_4), \end{aligned} \tag{8}$$

where

$$\Gamma = \begin{Bmatrix} r\omega_\Psi^2 \sin^2\Theta + r\omega_\Theta^2 - \dfrac{\mu_\odot}{r^2} + \dfrac{a_\oplus r_\oplus}{2r}\left(\cos^2\phi\cos^2\theta + 1\right) \\ \omega_\Psi^2 \sin\Theta\cos\Theta - \dfrac{2v_r\omega_\Theta}{r} + \dfrac{a_\oplus r_\oplus}{2r^2}\left(\cos\phi\sin\theta\cos\theta\right) \\ -\dfrac{2v_r\omega_\Psi}{r} - 2\omega_\Theta\omega_\Psi\cot\Theta - \dfrac{a_\oplus r_\oplus}{2r^2}\left(\sin\phi\cos\phi\cos^2\theta\right) \end{Bmatrix}$$

$$\Lambda = \begin{Bmatrix} \dfrac{\omega_{bx}\cos\psi}{\cos\theta} - \dfrac{\omega_{by}\sin\psi}{\cos\theta} + \omega_\Psi\sin\Theta - \omega_\Theta\sin\phi\tan\theta + \omega_\Psi\cos\Theta\cos\phi\tan\theta \\ \omega_{bx}\sin\psi + \omega_{by}\cos\psi - \omega_\Theta\cos\phi - \omega_\Psi\cos\Theta\sin\phi \\ -\omega_{bx}\tan\theta\cos\psi + \omega_{by}\tan\theta\sin\psi + \omega_{bz} + \dfrac{\omega_\Theta\sin\phi}{\cos\theta} - \dfrac{\omega_\Psi\cos\Theta\cos\phi}{\cos\theta} \end{Bmatrix} \tag{9}$$

$$\Pi = \begin{Bmatrix} \dfrac{\omega_{by}\omega_{bz}\left(I_y - I_z\right)}{I_x} \\ \dfrac{\omega_{bx}\omega_{bz}\left(I_z - I_x\right)}{I_y} \\ \dfrac{\omega_{bx}\omega_{by}\left(I_x - I_y\right)}{I_z} \end{Bmatrix}.$$

According to the previous discussion, there are equilibrium points of coupled orbit-attitude system in feasible region. By the linearization of the coupled system at the equilibrium point $\mathbf{X}_{\mathrm{ref}} = [\mathbf{X}_{\mathrm{ref1}}^{\mathrm{T}}, \mathbf{X}_{\mathrm{ref2}}^{\mathrm{T}}, \mathbf{X}_{\mathrm{ref3}}^{\mathrm{T}}, \mathbf{X}_{\mathrm{ref4}}^{\mathrm{T}}]^{\mathrm{T}}$, the variational equations are obtained as

$$\begin{aligned} \delta\dot{\mathbf{X}}_1 &= \delta\mathbf{X}_2 \\ \delta\dot{\mathbf{X}}_2 &= \left.\frac{\partial\Gamma}{\partial\mathbf{X}_1}\right|_{\substack{\mathbf{X}_1=\mathbf{X}_{\mathrm{ref1}}\\ \mathbf{X}_2=\mathbf{X}_{\mathrm{ref2}}\\ \mathbf{X}_3=\mathbf{X}_{\mathrm{ref3}}}} \delta\mathbf{X}_1 + \left.\frac{\partial\Gamma}{\partial\mathbf{X}_2}\right|_{\substack{\mathbf{X}_1=\mathbf{X}_{\mathrm{ref1}}\\ \mathbf{X}_2=\mathbf{X}_{\mathrm{ref2}}\\ \mathbf{X}_3=\mathbf{X}_{\mathrm{ref3}}}} \delta\mathbf{X}_2 + \left.\frac{\partial\Gamma}{\partial\mathbf{X}_3}\right|_{\substack{\mathbf{X}_1=\mathbf{X}_{\mathrm{ref1}}\\ \mathbf{X}_2=\mathbf{X}_{\mathrm{ref2}}\\ \mathbf{X}_3=\mathbf{X}_{\mathrm{ref3}}}} \delta\mathbf{X}_3 \\ &= \mathbf{A}_1\delta\mathbf{X}_1 + \mathbf{A}_2\delta\mathbf{X}_2 + \mathbf{A}_3\delta\mathbf{X}_3 \\ \delta\dot{\mathbf{X}}_3 &= \left.\frac{\partial\Lambda}{\partial\mathbf{X}_1}\right|_{\substack{\mathbf{X}_1=\mathbf{X}_{\mathrm{ref1}}\\ \mathbf{X}_2=\mathbf{X}_{\mathrm{ref2}}\\ \mathbf{X}_3=\mathbf{X}_{\mathrm{ref3}}\\ \mathbf{X}_4=\mathbf{X}_{\mathrm{ref4}}}} \delta\mathbf{X}_1 + \left.\frac{\partial\Lambda}{\partial\mathbf{X}_2}\right|_{\substack{\mathbf{X}_1=\mathbf{X}_{\mathrm{ref1}}\\ \mathbf{X}_2=\mathbf{X}_{\mathrm{ref2}}\\ \mathbf{X}_3=\mathbf{X}_{\mathrm{ref3}}\\ \mathbf{X}_4=\mathbf{X}_{\mathrm{ref4}}}} \delta\mathbf{X}_2 + \left.\frac{\partial\Lambda}{\partial\mathbf{X}_3}\right|_{\substack{\mathbf{X}_1=\mathbf{X}_{\mathrm{ref1}}\\ \mathbf{X}_2=\mathbf{X}_{\mathrm{ref2}}\\ \mathbf{X}_3=\mathbf{X}_{\mathrm{ref3}}\\ \mathbf{X}_4=\mathbf{X}_{\mathrm{ref4}}}} \delta\mathbf{X}_3 + \left.\frac{\partial\Lambda}{\partial\omega}\right|_{\substack{\mathbf{X}_1=\mathbf{X}_{\mathrm{ref1}}\\ \mathbf{X}_2=\mathbf{X}_{\mathrm{ref2}}\\ \mathbf{X}_3=\mathbf{X}_{\mathrm{ref3}}\\ \mathbf{X}_4=\mathbf{X}_{\mathrm{ref4}}}} \delta\mathbf{X}_4 \\ &= \mathbf{A}_4\delta\mathbf{X}_1 + \mathbf{A}_5\delta\mathbf{X}_2 + \mathbf{A}_6\delta\mathbf{X}_3 + \mathbf{A}_7\delta\mathbf{X}_4 \\ \delta\dot{\mathbf{X}}_4 &= \left.\frac{\partial\Pi}{\partial\mathbf{X}_4}\right|_{\mathbf{X}_4=\mathbf{X}_{\mathrm{ref4}}} \delta\mathbf{X}_4 = \mathbf{A}_8\delta\mathbf{X}_4, \end{aligned} \tag{10}$$

where $\mathbf{A}_i$ $(i = 1, 2, \ldots, 8)$ are 3×3 matrixes.

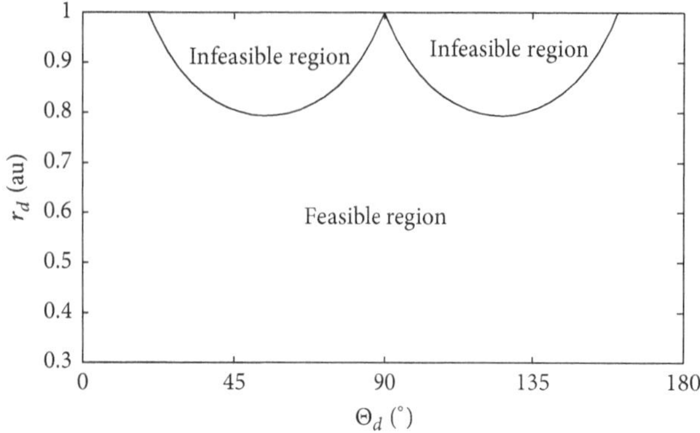

FIGURE 4: Feasible region for the displaced geostationary orbit.

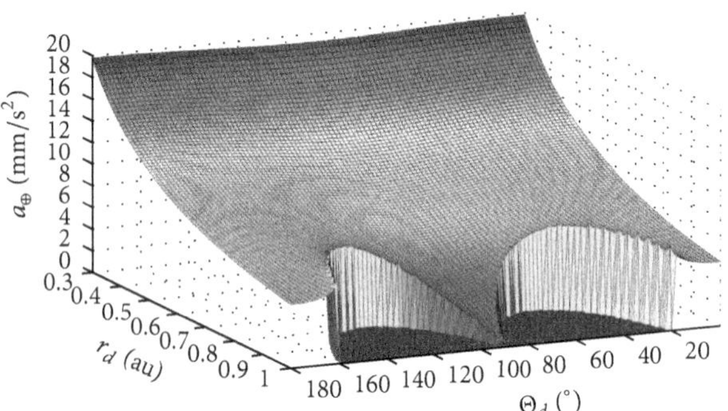

FIGURE 5: Required characteristic acceleration for the displaced geostationary orbit.

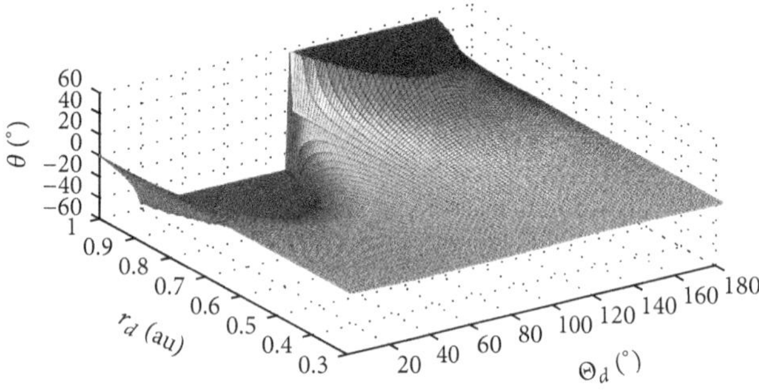

FIGURE 6: Required attitude angle θ for the displaced geostationary orbit.

Let $\boldsymbol{\delta} = \left[\mathbf{X}_1^{\mathrm{T}} - \mathbf{X}_{\mathrm{ref}1}^{\mathrm{T}}\ \mathbf{X}_2^{\mathrm{T}} - \mathbf{X}_{\mathrm{ref}2}^{\mathrm{T}}\ \mathbf{X}_3^{\mathrm{T}} - \mathbf{X}_{\mathrm{ref}3}^{\mathrm{T}}\ \mathbf{X}_4^{\mathrm{T}} - \mathbf{X}_{\mathrm{ref}4}^{\mathrm{T}}\right]^{\mathrm{T}} = \left[\delta\mathbf{X}_1^{\mathrm{T}}\ \delta\mathbf{X}_2^{\mathrm{T}}\ \delta\mathbf{X}_3^{\mathrm{T}}\ \delta\mathbf{X}_4^{\mathrm{T}}\right]^{\mathrm{T}}$; the variational equations can be written as

$$\dot{\boldsymbol{\delta}} = \begin{bmatrix} 0 & \mathbf{E}_{3\times3} & 0 & 0 \\ \mathbf{A}_1 & \mathbf{A}_2 & \mathbf{A}_3 & 0 \\ \mathbf{A}_4 & \mathbf{A}_5 & \mathbf{A}_6 & \mathbf{A}_7 \\ 0 & 0 & 0 & \mathbf{A}_8 \end{bmatrix} \boldsymbol{\delta} = \mathbf{H}_c \boldsymbol{\delta}, \tag{11}$$

where $\mathbf{E}_{3\times3}$ is a 3×3 identity matrix.

The stability of the equilibrium point can be checked by calculating the eigenvalues of the variational coefficient matrix of coupled system $\mathbf{H}_c$. To quantify the stability of the coupled system, a parameter labeled as characteristic index is defined as the maximal real part of all eigenvalues, given by

$$\Xi_c = \max_{1\le i\le 12} \operatorname{Re}(\lambda_i), \tag{12}$$

where λ_i $(i = 1, 2, \ldots, 12)$ are eigenvalues of matrix $\mathbf{H}_c$.

According to Lyapunov's stability theory, if $\Xi_c < 0$, the coupled system is asymptotically stable. If $\Xi_c = 0$, the coupled system is marginally stable. If $\Xi_c > 0$, the coupled system is unstable. Referring to (11), the orbital variational coefficient matrix $\mathbf{H}_o$ and the attitude variational coefficient matrix $\mathbf{H}_a$ can be written as

$$\mathbf{H}_o = \begin{bmatrix} 0 & \mathbf{E} \\ \mathbf{A}_1 & \mathbf{A}_2 \end{bmatrix}$$
$$\mathbf{H}_a = \begin{bmatrix} \mathbf{A}_6 & \mathbf{A}_7 \\ 0 & \mathbf{A}_8 \end{bmatrix}. \tag{13}$$

We can also define the characteristic index of orbit and attitude as

$$\Xi_o = \max_{1\le i\le 6} \operatorname{Re}(\lambda_{oi})$$
$$\Xi_a = \max_{1\le i\le 6} \operatorname{Re}(\lambda_{ai}), \tag{14}$$

where λ_{oi} $(i = 1, 2, \ldots, 6)$ are eigenvalues of the orbital variational coefficient matrix $\mathbf{H}_o$ and λ_{ai} $(i = 1, 2, \ldots, 6)$ are eigenvalues of the attitude variational coefficient matrix $\mathbf{H}_a$.

4.2. Numerical Analysis of the Stability. In this section, the linearized coupled dynamical equations are employed to investigate the stability of the coupled equilibrium numerically. An arbitrary displaced orbit is selected to investigate the relation between the stability of the coupled system and the parameter ω_d. The selected displaced orbit for simulation is determined by $r_d = 0.9$ au and $\Theta_d = 86^\circ$. The range of selected orbital angular velocity for simulation is $\omega_d \in [0.5\omega_\oplus, 5\omega_\oplus]$. Other displaced orbits can be analyzed in the same way. As shown in Figure 7, it is found that the orbital system of selected displaced orbit is unstable when the selected orbital angular velocity ω_d is less than a critical value (the critical value is $0.84\omega_\oplus$ for $r_d = 0.9$ au and $\Theta_d = 86^\circ$). It is marginally stable when the selected orbital angular velocity ω_d is more than the critical value. Figure 8 indicates that the attitude system is always marginally stable and ω_d does not influence the stability.

Figure 9 indicates that the coupled system of selected displaced orbit is unstable when the selected orbital angular velocity ω_d is less than a critical value (the critical value is $1.32\omega_\oplus$ in this simulation). The coupled system is marginally stable when the selected orbital angular velocity ω_d is more than the critical value. Therefore, it is easy to be known that the displaced geostationary orbit ($\omega_d = \omega_\oplus$) with $r_d = 0.9$ au and $\Theta_d = 86^\circ$ is unstable for a fixed attitude.

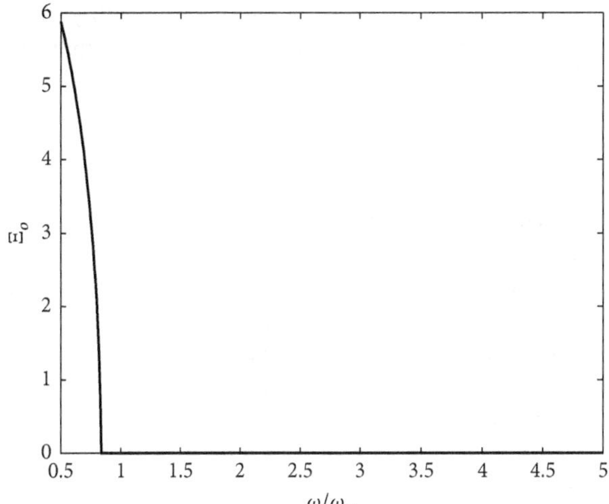

FIGURE 7: Orbital characteristic index Ξ_o as a function of the angular velocity ω_d.

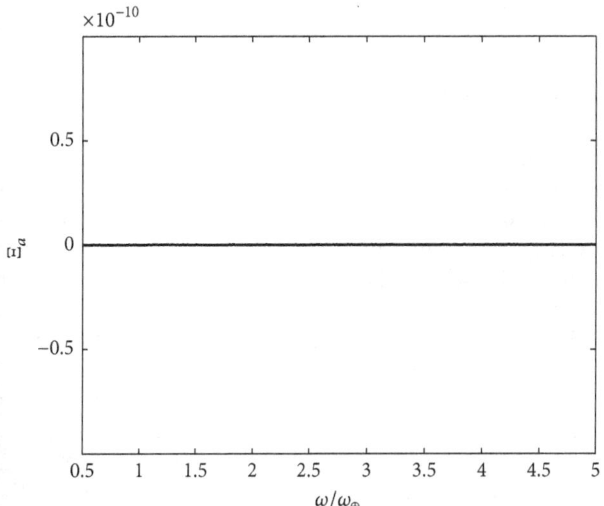

FIGURE 8: Attitude characteristic index Ξ_a as a function of the angular velocity ω_d.

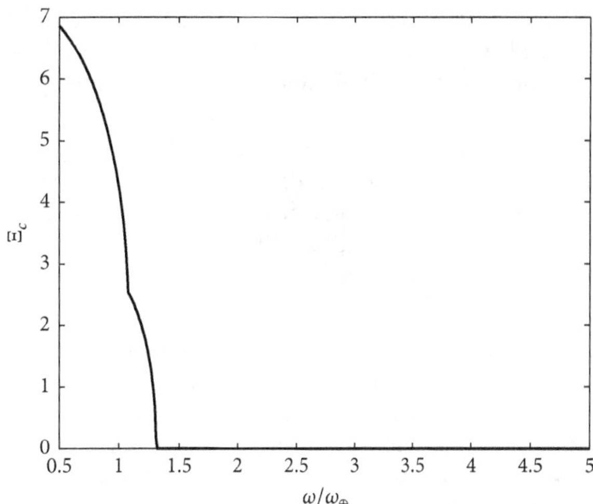

FIGURE 9: Coupled-system characteristic indexes Ξ_c as a function of the angular velocity ω_d.

5. Coupled Orbit-Attitude Control of Electric Sail

5.1. Control Algorithm. The stability of the displaced orbit has been discussed above. The numerical results show that only some of displaced orbits are marginally stable. Therefore, active control is necessary to stabilize the unstable displaced orbits. In this paper, only control torque is employed to stabilize both the attitude and orbit of sailcraft. Then, the coupled system can be linearized and the corresponding linear system is employed for control design. The coupled equations linearized at the equilibrium point with control torque can be written as

$$\dot{\boldsymbol{\delta}} = \mathbf{H}_c\boldsymbol{\delta} + \mathbf{B}\mathbf{u}, \tag{15}$$

where $\boldsymbol{\delta} = [\delta r, \delta\Theta, \delta\Psi, \delta v_r, \delta\omega_\Theta, \delta\omega_\Psi, \delta\phi, \delta\theta, \delta\psi, \delta\omega_x, \delta\omega_y, \delta\omega_z]^{\mathrm{T}}$ is state perturbation, $\mathbf{H}_c$ is the variational coefficient matrix of coupled system, $\mathbf{B} = [0_{3\times3}, 0_{3\times3}, 0_{3\times3}, 1/\mathbf{I}]^{\mathrm{T}}$ (**I** is the inertia matrix of sailcraft), and $\mathbf{u} = \mathbf{T} = [T_x, T_y, T_z]^{\mathrm{T}}$ is the vector of control torque.

In order to determine the controllability of this system, the rank of the controllability matrix of the linearized system has been calculated. The results show that the system is controllable, because the controllability matrix has full rank. In this paper, linear quadratic regulator (LQR) is employed to design the control torque to track the desired attitude and orbit. The performance function is selected as

$$J = \int_0^\infty \left(\boldsymbol{\delta}^{\mathrm{T}}\mathbf{Q}\boldsymbol{\delta} + \mathbf{u}^{\mathrm{T}}\mathbf{R}\mathbf{u}\right) \mathrm{d}t, \tag{16}$$

where the matrix $\mathbf{Q} \in \mathbb{R}^{12\times12}$ is symmetric positive definite and represents the weight of the state errors. The matrix $\mathbf{R} \in \mathbb{R}^{3\times3}$ is also a symmetric positive definite matrix and represents the weight of the control input.

The linear state feedback can be obtained by solving the following algebraic Riccati matrix:

$$\mathbf{P}\mathbf{H}_c + \mathbf{H}_c\mathbf{P} - \mathbf{P}\mathbf{B}\mathbf{R}^{-1}\mathbf{B}^{\mathrm{T}}\mathbf{P} + \mathbf{Q} = 0. \tag{17}$$

Then, the control torque can be written in the state-feedback form as

$$\mathbf{u} = -\mathbf{R}^{-1}\mathbf{B}^{\mathrm{T}}\mathbf{P}\boldsymbol{\delta}. \tag{18}$$

5.2. Numerical Simulation. In order to investigate the stability of the displaced orbit with linear quadratic regulator control, a numerical example is adopted in this section, using the full nonlinear dynamical equations. The selected displaced orbit for simulation is a displaced geostationary orbit ($\omega_d = \omega_\oplus$) and is determined by $r_d = 0.9\,\mathrm{au}$ and $\Theta_d = 86^\circ$. For keeping the selected displaced geostationary orbit, the required characteristic acceleration $a_{\oplus d}$ should be equal to $1.942\,\mathrm{mm/s^2}$, and attitude angle θ_d should be equal to -21.749°. As discussed above, the chosen displaced orbit is unstable without control. Assume that the orbit and the attitude have a small perturbation to the required orbit and attitude at the initial time. The initial values of the perturbation are given by $\delta \mathbf{X}_1 = [4 \times 10^4\,\text{m}, 0, 0]^{\mathrm{T}}$, $\delta \mathbf{X}_2 = [0,0,0]^{\mathrm{T}}$, $\delta \mathbf{X}_3 = [0, 0.2^\circ, 0]^{\mathrm{T}}$, and $\delta \mathbf{X}_4 = [0,0,0]^{\mathrm{T}}$. Figures 10 and 11 give the responses of attitude and orbit, respectively. Numerical simulation shows that the LQR control law can control the coupled system. As shown in Figure 12, a small control torque can stabilize both the attitude and orbit.

5.3. Preliminary Study of Voltage Distribution. The control torque of electric sail can be realized by adjusting the voltage distribution of charged tethers of electric sail [16, 17]. In this paper, the generation of control torques is preliminarily discussed in an optimal framework. Referring to Eq. (10) and Eq. (13) in [16], the vector of propulsive force $\mathbf{F}$ and the vector of control torque $\mathbf{T}$ can be written as functions of the vertical thrust magnitude of charged tethers $\sigma_{\oplus 1}, \ldots, \sigma_{\oplus N}$.

$$\begin{aligned} \mathbf{F}(\phi,\theta,\psi,r,l,\sigma_{\oplus 1},\ldots,\sigma_{\oplus N}) &= \sum_{k=1}^{N} \mathbf{F}_k(\phi,\theta,\psi,r,l,\sigma_{\oplus k}) \\ \mathbf{T}(\phi,\theta,\psi,r,l,\sigma_{\oplus 1},\ldots,\sigma_{\oplus N}) & \\ = \sum_{k=1}^{N} \mathbf{T}_k(\phi,\theta,\psi,r,l,\sigma_{\oplus k}), & \end{aligned} \tag{19}$$

where $\sigma_{\oplus k}$, $k \in [1, N]$, is the kth tether's magnitude of the force per unit length, when the sun-spacecraft distance is $r_\oplus = 1$ au and the solar wind is perpendicular to tether. As discussed in [17], $\sigma_{\oplus k}$ can be controlled by altering the voltage of kth charged tether V_k and is given by

$$\sigma_{\oplus k} = 0.18 \max(0, V_k - V_0) \sqrt{\varepsilon_0 p_{\text{dyn}}}, \tag{20}$$

where V_0 is the electric potential corresponding to the kinetic energy of the solar wind ions, ε_0 is the vacuum permittivity, and p_{dyn} is the solar wind dynamic pressure.

In order to generate the required control force $\mathbf{F}_c$ and control torque $\mathbf{T}_c$, the following equality constraints should be satisfied:

$$\begin{aligned} \sum_{k=1}^{N} \mathbf{F}_k(\phi,\theta,\psi,r,l,\sigma_{\oplus k}) &= \mathbf{F}_c \\ \sum_{k=1}^{N} \mathbf{T}_k(\phi,\theta,\psi,r,l,\sigma_{\oplus k}) &= \mathbf{T}_c. \end{aligned} \tag{21}$$

As shown in (21), the quantity of equality constraints is 6 in voltage distribution problem. Comparatively, the number of design variables $\sigma_{\oplus 1}, \ldots, \sigma_{\oplus N}$ is $N \in [20, 100]$ in general, which is decided by the quantity of charged tethers of electric sail. Therefore, the definite solution cannot be obtained by solving (21). In this paper, the voltage distribution problem is converted into a nonlinear programming problem (NLP). Equation (21) is processed into equality constraints in optimal framework. In order to reduce the required power, the optimization performance index is selected as follows:

$$J = \sum_{k=1}^{N} \sigma_{\oplus k}. \tag{22}$$

The above NLP can be solved within the MATLAB environment using *fmincon*, which is a constrained optimization routine available with the Optimization Toolbox. The selected optimization algorithm is sequential quadratic programming (SQP). Consider an electric sail, which is composed of $N = 100$ 20 km-long tether. At the beginning of the mission, the required control force can be calculated according to (1), with the required characteristic acceleration $a_{\oplus d} = 1.942\,\text{mm/s}^2$ and attitude angle $\theta_d = -21.949^\circ$. As seen in Figure 12, the required control torque is $T = [9.3941 \times 10^{-5}, -6.1 \times 10^{-3}, 1.8116 \times 10^{-6}]$ Nm at the beginning of the mission.

This NLP was solved in 2.453 seconds using a personal computer with 8 GB RAM running at 2.20 GHz processor speed. The obtained thrust magnitudes per unit length $\sigma_{\oplus k}$, $k \in [1, N]$, are shown in Figure 13. Considering (20), the required voltage distribution V_k, $k \in [1, N]$, can be obtained and is shown in Figure 14, where $V_{\text{aver}} = \sum_{k=1}^{N} V_k$ is the average value of V_k, $k \in [1, N]$. Through the preliminary study of voltage distribution, it can be found that the control force and torque of electric sail can be realized by adjusting the voltage distribution of charged tethers. This feature is quite different between traditional torque-generated methods, in which the actuator is at the bus center in general.

Even though a solar sail and an electric sail are both capable of producing a propulsive thrust without the need of any propellant, these two propulsion systems are substantially different in terms of performance, shape, and dimensions [18]. Compared to the dynamics and control of solar sail discussed in [6], the difference is mainly reflected in the following aspects. In terms of propulsion mechanism, electric sails are propelled by the solar wind dynamic pressure using charged tethers, instead of the solar photon momentum. In the thrust model of solar sail, the thrust cone angle of ideal solar sail is equal to the pitch angle. Differently, the thrust cone angle of electric sail reaches a maximum value of 19.47°, when pitch angle is equal to 54.75°. This feature causes the fact that there are infeasible regions for the displaced geostationary orbit using electric sail, as shown in Figure 4. About the difference on control, the control force and control torque of electric sail can be realized by adjusting the voltage distribution of charged tethers, as discussed in this section. This feature is quite different from solar sail.

6. Conclusions

In this paper, displaced solar orbits for spacecraft propelled by electric sails are investigated as an alternative to solar sails. Since the propulsive thrust is induced by the sail attitude, the orbital and attitude dynamics of electric-sail-based spacecraft are coupled and investigated together. The equilibrium

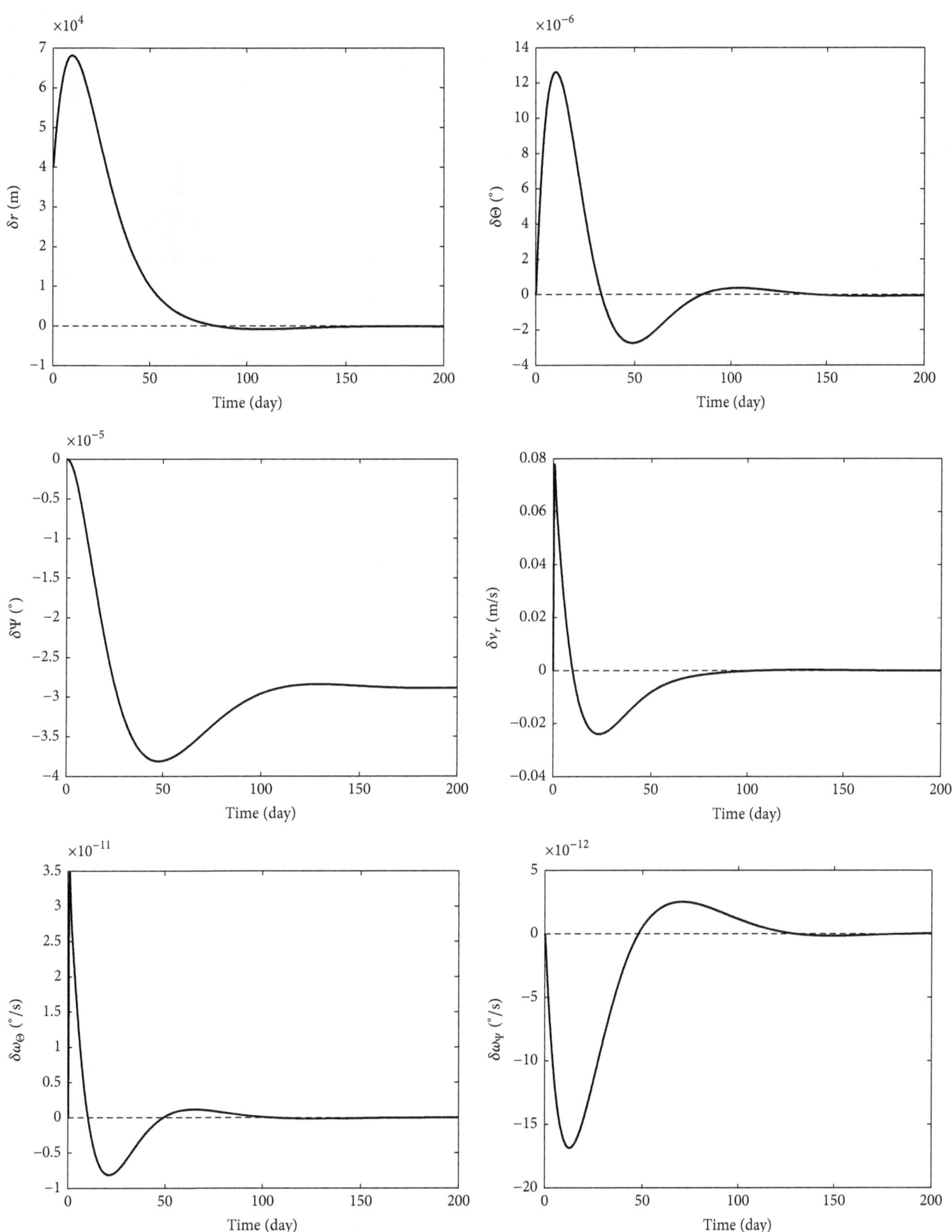

FIGURE 10: The responses of the orbit errors.

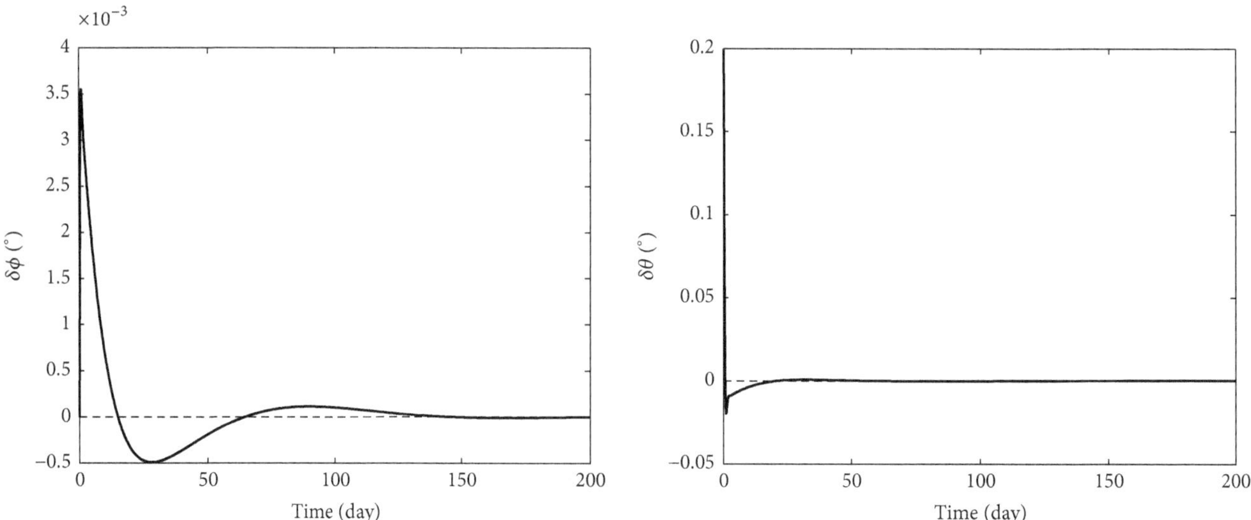

FIGURE 11: The responses of the attitude errors.

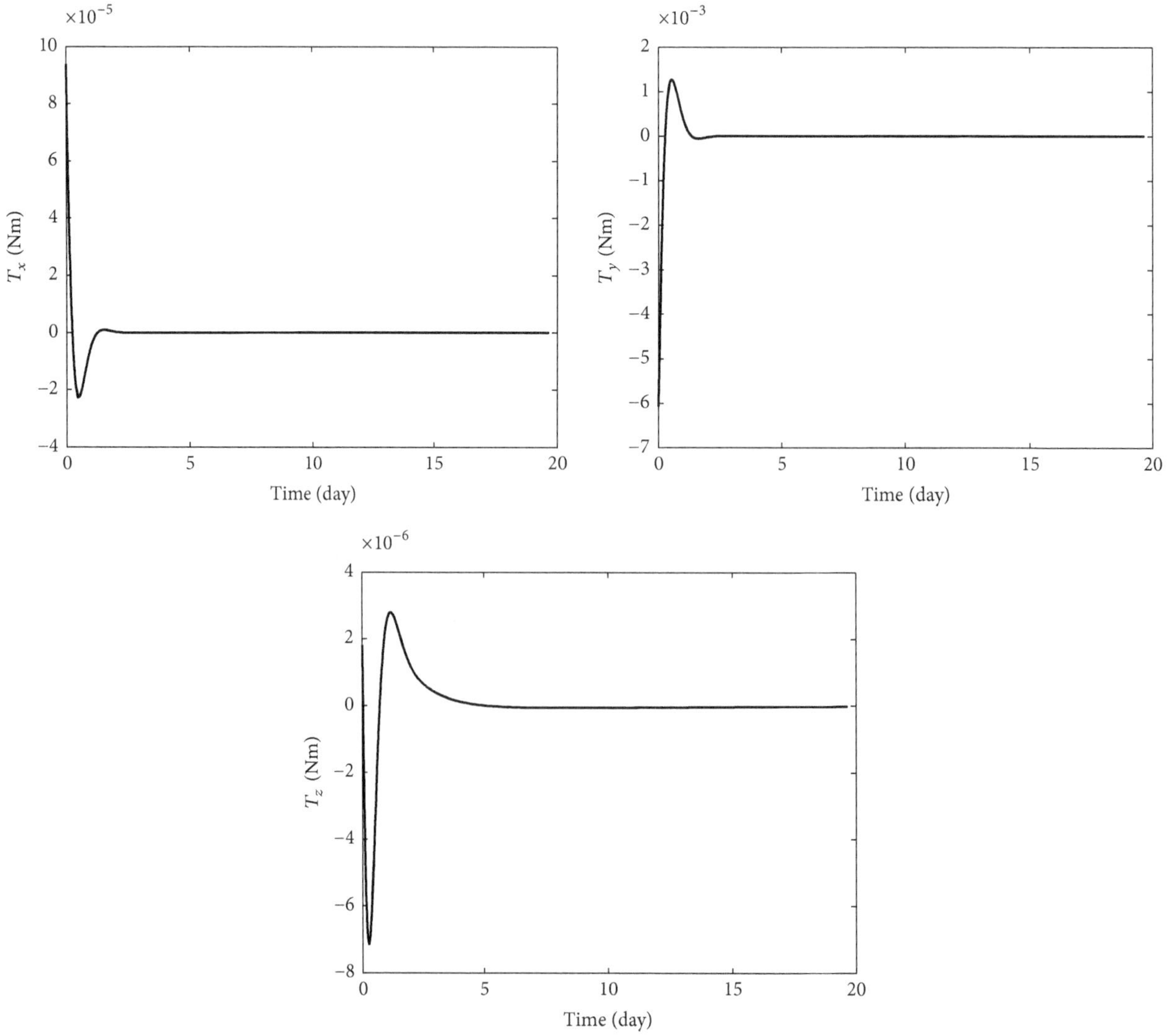

FIGURE 12: The control torques.

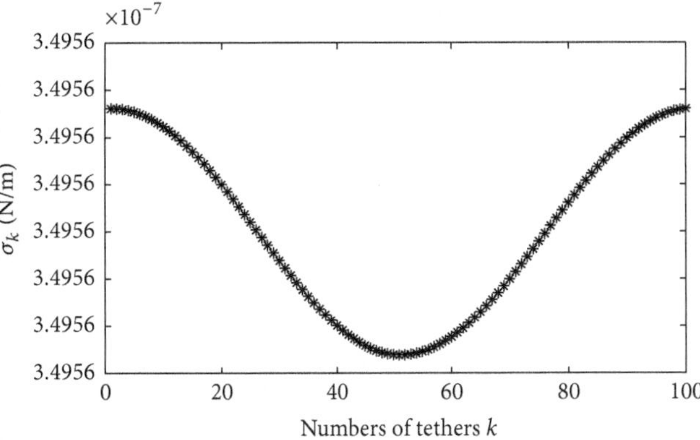

FIGURE 13: The thrust magnitudes per unit length of tethers at the beginning of the mission.

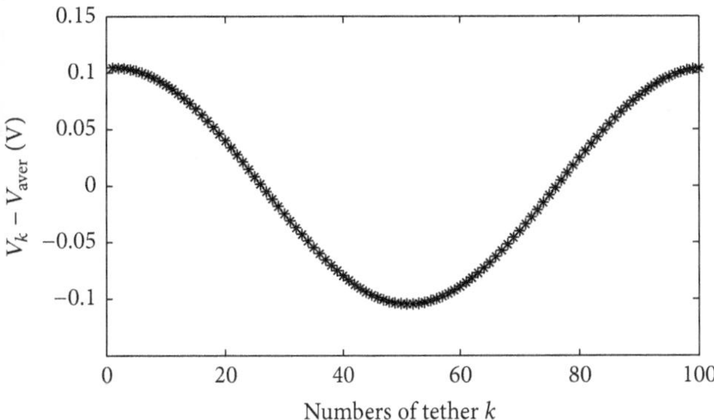

FIGURE 14: The voltage distribution of tethers at the beginning of the mission.

point of the coupled dynamical system in displaced orbit is obtained, and its stability is analyzed through a linearization. The results of stability analysis show that only some of the orbits are marginally stable. For unstable displaced orbits, linear quadratic regulator is employed to control the coupled orbit-attitude system. Numerical simulation shows that the linear quadratic regulator can control the coupled system and a small torque can stabilize both the attitude and orbit. To generate the control force and torque, the voltage distribution problem is converted into a nonlinear programming problem and solved using sequential quadratic programming in an optimal framework. The numerical results show that the control force and torque of electric sail can be realized by adjusting the voltage distribution of charged tethers.

Conflicts of Interest

The authors declare no competing financial interests.

Authors' Contributions

Mingying Huo and Yanfang Liu conceived and designed the experiments. Mingying Huo and He Liao performed the experiments and analyzed the data. Mingying Huo and Naiming Qi wrote the paper. All authors reviewed the manuscript.

Acknowledgments

This work was supported by Shanghai Academy of Spaceflight Technology (no. SAST2016039), the Open Fund of National Defense Key Discipline Laboratory of Micro-Spacecraft Technology (Grant no. HIT.KLOF.MST.201607), Heilongjiang Postdoctoral Foundation (no. LBH-Z16082), and China Postdoctoral Science Foundation (no. 2017M611372).

References

[1] P. Janhunen, P. Toivanen, J. Envall et al., "Overview of electric solar wind sail applications," *Proceedings of the Estonian Academy of Sciences*, vol. 63, no. 2S, pp. 267–278, 2014.

[2] C. R. McLnnes and J. F. L. Simmons, "Solar sail halo orbits I: heliocentric case," *Journal of Spacecraft and Rockets*, vol. 29, no. 4, pp. 466–471, 1992.

[3] C. R. McInnes, "Passive control of displaced solar sail orbits," *Journal of Guidance, Control, and Dynamics*, vol. 21, no. 6, pp. 975–982, 1998.

[4] C. R. McInnes, "Solar sail mission applications for non-keplerian orbits," *Acta Astronautica*, vol. 45, no. 4-9, pp. 567–575, 1999.

[5] S. Gong, H. Baoyin, and J. Li, "Coupled attitude-orbit dynamics and control for displaced solar orbits," *Acta Astronautica*, vol. 65, no. 5-6, pp. 730–737, 2009.

[6] S. Gong, J. Li, and H. Baoyin, "Analysis of displaced solar sail orbits with passive control," *Journal of Guidance, Control, and Dynamics*, vol. 31, no. 3, pp. 782–785, 2008.

[7] S. Gong, H. Baoyin, and J. Li, "Solar sail formation flying around displaced solar orbits," *Journal of Guidance, Control, and Dynamics*, vol. 30, no. 4, pp. 1148–1152, 2007.

[8] S. Gong, H. Baoyin, and J. Li, "Relative orbit design and control of formation around displaced solar orbits," *Aerospace Science and Technology*, vol. 12, pp. 195–201, 2008.

[9] J. Mu, S. Gong, and J. Li, "Coupled control of reflectivity modulated solar sail for GeoSail formation flying," *Journal of Guidance, Control, and Dynamics*, vol. 38, no. 4, pp. 740–751, 2015.

[10] S. Gong, J. Li, and H. Baoyin, "Formation flying solar-sail gravity tractors in displaced orbit for towing near-earth asteroids," *Celestial Mechanics & Dynamical Astronomy*, vol. 105, no. 1–3, pp. 159–177, 2009.

[11] J. Heiligers, C. R. McInnes, J. D. Biggs, and M. Ceriotti, "Displaced geostationary orbits using hybrid low-thrust propulsion," *Acta Astronautica*, vol. 71, pp. 51–67, 2012.

[12] P. Janhunen, "Electric sail for spacecraft propulsion," *Journal of Propulsion and Power*, vol. 20, no. 4, pp. 763-764, 2004.

[13] P. Janhunen, "Status report of the electric sail in 2009," *Acta Astronautica*, vol. 68, no. 5-6, pp. 567–570, 2011.

[14] G. Mengali and A. A. Quarta, "Non-Keplerian orbits for electric sails," *Celestial Mechanics and Dynamical Astronomy*, vol. 105, no. 1, pp. 179–195, 2009.

[15] N. Qi, M. Huo, and Q. Yuan, "Displaced electric sail orbits design and transition trajectory optimization," *Mathematical Problems in Engineering*, vol. 2014, Article ID 932190, 9 pages, 2014.

[16] M. Huo, J. Zhao, S. Xie, and N. Qi, "Coupled attitude-orbit dynamics and control for an electric sail in a heliocentric transfer mission," *PLoS ONE*, vol. 10, no. 5, article e0125901, 2015.

[17] P. K. Toivanen and P. Janhunen, "Spin plane control and thrust vectoring of electric solar wind sail," *Journal of Propulsion and Power*, vol. 29, no. 1, pp. 178–185, 2013.

[18] P. Janhunen, P. K. Toivanen, J. Polkko et al., "Electric solar wind sail: toward test missions," *Review of Scientific Instruments*, vol. 81, no. 11, article 111301, 2010.

14

Preliminary Capture Trajectory Design for Europa Tomography Probe

Lorenzo Federici, Alessandro Zavoli, and Guido Colasurdo

Department of Mechanical and Aerospace Engineering, Sapienza-University of Rome, Via Eudossiana 18, Rome, Italy

Correspondence should be addressed to Alessandro Zavoli; alessandro.zavoli@uniroma1.it

Academic Editor: Linda L. Vahala

The objective of this work is the preliminary design of a low-ΔV transfer from an initial elliptical orbit around Jupiter into a final circular orbit around the moon Europa. This type of trajectory represents an excellent opportunity for a low-cost mission to Europa, accomplished through a small orbiter, as in the proposed Europa Tomography Probe mission, a European contribution to NASA's Europa Multiple-Flyby Mission (or Europa Clipper). The mission strategy is based on the v_∞ leveraging concept, and the use of resonant orbits to exploit multiple gravity-assist from the moon. Possible sequences of resonant orbits are selected with the help of the Tisserand graph. Suitable trajectories are provided by an optimization code based on the parallel running of several differential evolution algorithms. Different solutions are finally compared in terms of propellant consumption and flight time.

1. Introduction

The Jovian moon Europa is a celestial body of primary interest for astrophysicists. The likely existence of a global subsurface ocean, proved by measurements carried out during Galileo mission, makes Europa one of the most promising environments in the Solar System to sustain human habitability. The presence of an ocean may also imply that Europa hosts (or, at least, hosted) life [1]. The importance of the determination of the ice-water layer characteristics is clearly stated in NASA's 2013–2022 Decadal Survey [2].

Europa Clipper is the next mission planned by NASA with the aim of exploring Europa. Because of the extremely harsh Jovian environment in the proximity of Europa, the initial concept of an orbiter was abandoned in favour of a multi-flyby strategy, the same considered for Galileo mission. The present mission profile, with more than 40 flybys of Europa, allows for a paramount investigation of Europa surface and subsurface properties, but is not very favourable to the investigation of Europa's deep interior structure.

A scientific enhancement to Europa Clipper mission was investigated in [3]. There, a small probe deployed on a polar orbit around Europa, hosting just one scientific instrument (a magnetometer) and a transponder required for the Intersatellite Link (ISL) with the mother spacecraft, is proved to be capable of providing crucial information on the interior structure of the moon, such as depth, thickness, and conductivity of the subsurface ocean. Also, ISL could support the reconstruction of the mother spacecraft orbit, hence significantly improving the accuracy of the topographic reconstruction of Europa's surface.

Standing on these arguments, a scientific and engineering team at Sapienza University of Rome, in collaboration with the Imperial College of London, carried out a feasibility study for a probe that could be hosted by the main spacecraft during the interplanetary cruise and released in the Jovian system with the aim at entering into a low-altitude circular quasi-polar orbit around Europa [4]. The result is a small spacecraft named Europa Tomography Probe (or ETP), which could fit the provisional 250 kg allowance that NASA has assigned to a secondary flight element hosted by the main spacecraft.

The feasibility study was carried out under the design philosophy of determining the minimum total mass and volume that allows for the scientific measurements considered in [3]. All subsystems have been described with some details, with the relevant exception of only two elements: (a) the transponder and (b) the trajectory which moves ETP into a polar orbit around Europa, which should interfere as less as possible with the mother spacecraft mission plan. This paper investigates the latter point, that is, the capture strategy for an Europa orbiter at the level of preliminary mission analysis.

The problem of optimizing the capture trajectory of an orbiter (or a lander) directed towards a moon of an outer planet, such as Europa [5], Enceladus [6], or Titan [7], has been the subject of many investigations. A two-body patched-conic approximation is usually assumed for interplanetary missions and transfers in a multibody planetary system [8]. This dynamical model retains the most prominent features of the real system, while keeping the numerical difficulties low. Three-dimensionality and eccentricity of the planetary bodies can be easily taken into account, and several flybys of different moons can be dealt with.

The same kind of missions have been also studied by using dynamical system techniques, which rely on circular restricted three-body problem (CRTBP). Low-energy trajectories are searched for, attempting the construction of "transfer tubes," whose boundaries are typically given by invariant manifolds originating from invariant sets (such as L_1 and L_2 Lyapunov orbits). As an example, dynamical chains formed by linking heteroclinic connections and homoclinic orbits [9] are proposed for the analysis of fast resonance transitions between exterior and interior resonant orbits (in the Sun-Jupiter system) [10] or "loose" capture trajectories [11]. Similar concepts are exploited for Halo-to-Halo [12] or libration-to-libration [13] transfers between planetary moons in the Jovian system, adopting a "patched" CRTBP model.

In the present problem, the probe comes from a high-energy condition and approaches the moon with a high hyperbolic excess velocity. The latter techniques are thus not efficient for attaining a solution, while a patched-conic approximation can be profitably adopted.

Delta velocity gravity assist (ΔV-GA) or v_∞ leveraging [14] has proved a powerful concept to improve the design of capture (or escape) trajectories. Large changes of the hyperbolic excess velocity at the encounter (v_∞) are obtained by using small deep-space maneuvers. When this strategy is used in conjunction with a series of resonant gravity assists, a significant reduction of propellant requirement can be achieved, with respect to a direct insertion maneuver [15].

In the present paper, the design of ETP capture trajectory using v_∞ leveraging is pursued by blending the patched-conic model and a modern global optimization procedure based on a differential evolution algorithm. The trajectory is modeled as a sequence of legs between two moon encounters; only one deep-space maneuver is permitted in each leg. This approach, proposed in [16] and hereafter referred as "MGA-1DSM," permits a quite general parameterization of the whole trajectory, which is not limited to ΔV-GA maneuvers. A preliminary solution (the sequence of resonant orbits and intercepted bodies) is defined by using two simple tools: the suboptimal solution of the v_∞ leveraging problem proposed by Sims and Longuski [17] and the Tisserand graph [18]. The former permits an easy design of a mission based only on v_∞ leveraging maneuvers, by suggesting a viable sequence of resonant orbits. The latter is a powerful graphical aid for the design of the same class of missions, when multiple bodies are intercepted, and some deep-space maneuvers are conveniently replaced by gravity assists of other moons in the planetary system.

The paper organization is here outlined. In Section 2, the physical problem of interest is described, and the adopted dynamical model and relevant assumptions are stated. A mathematical formulation (MGA-1DSM), which parameterizes a generic interplanetary trajectory as a sequence of legs containing a gravity assist and one deep-space maneuver, is outlined, leading to the purposeful definition of a global optimization problem. Section 3 presents the multipopulation differential evolution algorithm that has been used to solve the optimization problem. Fundamental tools for preliminary mission design, that is, v_∞ leveraging and Tisserand graph, are discussed in Section 4. A tentative solution is devised and used to prune global optimization search. Numerical results of this investigation are presented in Section 5. A conclusion section ends the paper.

2. Problem Statement and Mathematical Modeling

2.1. Problem Overview. According to the ongoing proposal [4], the probe is assumed to be released by the main spacecraft after a few Europa flybys have been completed. In particular, ETP starts its own transfer at the apocenter of a Jovian orbit of period four times the period of Europa (T_{Eu}), pericenter equal to the Europa semi-major axis, and coplanar with the Europa orbit; the orientation of the major axis is left free. A target circular quasi-polar orbit around Europa is desired, of assigned altitude $h_{\mathrm{EOI}} = 250$ km over Europa's surface. Four R-6D bipropellant thrusters form the primary propulsion system of the probe, which allows for a total thrust of 88 N with specific impulse $I_{\mathrm{sp}} = 294$ s. This propulsion system will be used for deep-space maneuvers, Europa Orbit Insertion (EOI) maneuver, and orbit maintenance during the scientific part of the mission.

A probe "net" mass $m_u = 146.6$ kg, which does not account for the propellant m_p and tank m_s masses, was estimated in [4]. Assuming a structural coefficient $\epsilon = m_s + m_p/m_s = 6$, which is a reasonable value for liquid propellant systems, a maximum value of velocity increment $\Delta V_{\max} = 1240.8$ m/s can be obtained if the spacecraft wet mass $m_0 = m_u + m_p + m_s$ is constrained at 250 kg. This value of $\Delta V_{\max}$ must cover the orbit maintenance (about 43.2 m/s for a 6-month mission) and capture cost. The goal is to reduce the ΔV required for the capture, so that a convenient safety margin is left.

2.2. Dynamical Model. A patched-conic model is assumed for the present analysis. Flybys are modeled as instantaneous changes in velocity. Subscripts "−" and "+" are used to distinguish between values immediately before or after the discontinuity, respectively. The radius of the sphere of influence (SOI) of the secondary bodies and the travel time inside these regions are assumed to be negligible. Powered flybys are neglected, as considered useless to reduce propellant consumption [19]. An impulsive-thrust model is adopted. This assumption well suits deep space maneuvers (DSMs), which are performed at a large distance from the main body and require usually a quite short time if compared to the orbital period, as chemical engines are here considered. This assumption is also used for the EOI maneuver, even though finite-thrust losses might be considered. Only one DSM is permitted between a flyby and the other. As a further assumption, Jovian moons move on Keplerian orbits (even though the proposed procedure is soon applicable to the general case that uses planetary ephemeris).

Despite its simplicity, this model allows to capture the most prominent features of the mission, while keeping the analysis simple enough. In fact, under the hypothesis of impulsive thrust, the trajectory can be computed analytically, without involving the numerical integration of the complete equations of motion.

2.3. Trajectory Parameterization. Let us assume that a sequence $M = \{M_j \mid j = 1, \ldots, N+1\}$ of $N+1$ body encounters, where $M_j \in [1..4]$ identifies the encountered body (1 = Io, 2 = Eu, 3 = Ga, and 4 = Ca), has been established. Europa is the first and last body in the series. The spacecraft trajectory can be modeled according to the multiple gravity assist-one deep space maneuver (MGA-1DSM) formulation [16]. The trajectory is broken down into a series of body-to-body legs. Each leg starts with a flyby and is made up of two ballistic arcs, joined by an impulsive maneuver.

This general formulation for a multigravity assist trajectory can be adapted to the problem at hand by adding an initial leg, which moves the probe from the assigned initial conditions to the first encounter with Europa. The mission ends with a last approach to Europa's surface, where an impulsive maneuver inserts the probe into the assigned polar orbit.

2.3.1. Departure Leg. The departure leg, which connects ETP release position to the first encounter with Europa, is modeled as a Lambert arc, where release epoch t_0, flight angle $\Delta\theta$, and flight time ΔT_0 are design parameters to optimize.

Let $\{\hat{a}_r, \hat{a}_t, \hat{a}_n\}$ be a Jovicentric radial-traversal-normal reference frame connected to a Europa position at epoch $t_1 = t_0 + \Delta T_0$, that is,

$$\hat{a}_r = \frac{r_1}{\|r_1\|}\,\hat{a}_n = \frac{r_1 \times v_{m,1}}{\|r_1 \times v_{m,1}\|}\,\hat{a}_t = \hat{a}_n \times \hat{a}_r, \quad (1)$$

where $r_1 = r_{M_{[1]}}(t_1)$ and $v_{m,1} = v_{M_{[1]}}(t_1)$ indicates position and velocity vectors of Europa ($M_{[1]} = 2$) at time t_1, respectively, which are provided by the ephemeris.

The probe departure point is located on a circle of radius r_0 which lies on the plane $\hat{\boldsymbol{a}}_r - \hat{\boldsymbol{a}}_t$; hence, it can be expressed as

$$r_0 = \tilde{r}_a^{4:1}(\cos\Delta\theta\,\hat{a}_r - \sin\Delta\theta\,\hat{a}_t), \quad (2)$$

while the velocity immediately before the release maneuver is

$$v_{0-} = \tilde{v}_a^{4:1}(\sin\Delta\theta\,\hat{a}_r + \cos\Delta\theta\,\hat{a}_t), \quad (3)$$

where the values $\tilde{r}_a^{4:1}$ and $\tilde{v}_a^{4:1}$ are, respectively, the radius and velocity magnitude at the apocenter of a Jovian orbit with pericenter equal to the Europa semi-major axis, and period $4T_{Eu}$. In this respect, the problem solution will eventually define the optimal orientation of the line of apses of the initial orbit with respect to Europa's orbit.

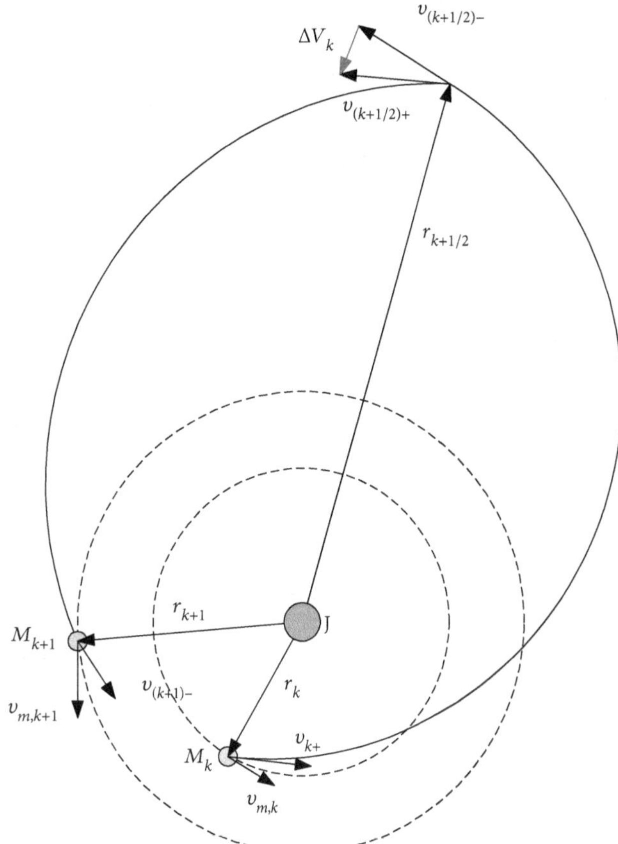

Figure 1: Trajectory sketch for an intermediate leg.

The velocity vectors after the release maneuver v_{0_+} and immediately before the first flyby v_{1_-} can now be evaluated by solving the associated Lambert problem:

$$\text{Lambert}\,(r_0, r_1, \Delta T_0) \longrightarrow [v_{0_+}, v_{1-}]. \quad (4)$$

The propulsive cost of the release maneuver is evaluated as

$$\Delta V_0 = \|v_{0_+} - v_{0_-}\|. \quad (5)$$

2.3.2. Intermediate Legs. The kth leg can be parametrized by using four parameters $r_{\pi,k}$, β_k, ΔT_k, and η_k, which represent, respectively, the flyby radius, the flyby plane orientation, the (overall) leg flight time, and the fraction of the leg flight time at which DSM occurs. The trajectory associated to a generic intermediate leg is presented in Figure 1.

Beginning from time t_1 at the first encounter, the epochs of the following encounters can be evaluated recursively as $t_{k+1} = t_k + \Delta T_k$, while the DSM epochs are $t_{k+1/2} = t_k + \eta_k \Delta T_k$. Moon position $r_k = r_{M[k]}(t_k)$ and velocity $v_{m,k} = v_{M[k]}(t_k)$ are obtained from the body ephemeris. The position of the spacecraft at the flyby must be the same as the intercepted body; hence, $r_{k-} = r_{k+} = r_k$.

Let's define a body-centered reference system $\{\hat{\mathbf{e}}_1, \hat{\mathbf{e}}_2, \hat{\mathbf{e}}_3\}$, where $\hat{\mathbf{e}}_1$ is directed along the spacecraft incoming relative velocity, $\hat{\mathbf{e}}_2$ is orthogonal to the moon orbital plane, and $\hat{\mathbf{e}}_3$ closes the right-hand side term

$$\begin{aligned} \hat{\mathbf{e}}_1 &= \frac{v_{\infty_{k-}}}{\left\| v_{\infty_{k-}} \right\|} \\ \hat{\mathbf{e}}_2 &= \frac{\hat{\mathbf{e}}_1 \times v_{m,k}}{\left\| \hat{\mathbf{e}}_1 \times v_{m,k} \right\|} \\ \hat{\mathbf{e}}_3 &= \hat{\mathbf{e}}_1 \times \hat{\mathbf{e}}_2, \end{aligned} \tag{6}$$

where $v_{\infty_k} = v_{k-} - v_{m,k}$ is the probe relative velocity before the flyby. The probe velocity v_{k+} after the flyby is evaluated as

$$\begin{aligned} v_{k+} = v_{m,k} &+ v_{\infty,k} \\ &\cdot (\cos(\delta_k)\hat{\mathbf{e}}_1 + \cos(\beta_k)\sin(\delta_k)\hat{\mathbf{e}}_2 \\ &+ \sin(\beta_k)\sin(\delta_k)\hat{\mathbf{e}}_3), \end{aligned} \tag{7}$$

where the rotation δ_k of the hyperbolic excess velocity in the flyby plane is

$$\delta_k = 2a\sin\left(\frac{\mu_{M[k]}}{\mu_{M[k]} + r_{\pi,k} v_{\infty_k}^2}\right). \tag{8}$$

Once the spacecraft state after the flyby is fully known, the position and velocity just before DSM, which are $r_{k+1/2}$ and $v_{(k+1/2)-}$, respectively, can be evaluated analytically by using propagation formulas for Keplerian orbits [20].

Position of the spacecraft is known at both DSM maneuver and next flyby; the transfer time $(1 - \eta_k)\Delta T_k$ is also known. Velocity vectors $v_{(k+1/2)+}$ immediately after the DSM and $v_{(k+1)-}$ just before the subsequent flyby can be evaluated by solving the associated Lambert problem

$$\text{Lambert}(r_{k+1/2}, r_{k+1}, \Delta T_k) \longrightarrow \left[v_{(k+1/2)+}, v_{(k+1)-}\right] \tag{9}$$

Evaluation of the DSM propulsive cost is now straightforward

$$\Delta V_k = \left\| v_{(k+1/2)+} - v_{(k+1/2)-} \right\| \tag{10}$$

2.3.3. Europa Orbit Insertion. The ETP injection into the target polar orbit around Europa is modeled as an impulsive burn applied at the hyperbola pericenter, which is purposely located on the target circular orbit.

$$\begin{aligned} v_{\text{EOI}-} &= \sqrt{v_\infty^2 + \frac{2\mu_{\text{Eu}}}{r_{\text{EOI}}}} \\ v_{\text{EOI}+} &= \sqrt{\frac{\mu_{\text{Eu}}}{r_{\text{EOI}}}}. \end{aligned} \tag{11}$$

The cost of the EOI maneuver is evaluated as

$$\Delta V_{N+1} = \left\| v_{\text{EOI}+} - v_{\text{EOI}-} \right\|. \tag{12}$$

2.4. Optimization Problem. For an assigned sequence of encountered body M_{seq}, the capture problem can be formally defined as follows:

$$P_{(M_{\text{seq}})} = \begin{cases} \min\limits_{x} & \Delta V_{\text{tot}}(\mathbf{x}) \\ \text{s.t.} & \mathbf{x}_L < \mathbf{x} < \mathbf{x}_u \end{cases} \tag{13}$$

where $\Delta V_{\text{tot}} = \sum_{k=0}^{N+1} \Delta V_k$ is the overall cost of the capture trajectory, accounting for deep-space, release, and EOI maneuvers; $\mathbf{x}$ is a vector of design variables given by

$$\mathbf{x} = [t_0, \Delta T_0, \Delta\theta] \cup \left[\left\{r_{\pi,j}, \beta_j, \Delta T_j, \eta_j\right\}_{j=1,..,N}\right], \tag{14}$$

and $\mathbf{x}_L$ and $\mathbf{x}_U$ are, respectively, lower and upper bounds on the design variables. The whole trajectory is thus parametrized by using $3 + 4N$ parameters.

3. Optimization Algorithm

The global optimization problem stated in Section 2 presents several features that make it hard to solve with local optimization approaches (e.g., Newton-like methods) as (i) multiple local optima exist, (ii) the solution may not be defined for some "unfortunate" set of optimization variables, and (iii) the gradient of the objective function is often not available or does not exist at all. In these cases, stochastic algorithms are usually preferred.

In the present application, an optimization algorithm based on differential evolution (DE) has been employed. DE is a population-based algorithm, featuring simple and efficient heuristics for global optimization problems defined over a continuous space [21]. Its good performance on several benchmarks and real-world problems drew the attention of many researchers all over the world, who further improved the effectiveness of the algorithm, by devising many variants. A review of the state of art can be found in [22]. The implementation adopted in this study collects several of these ideas.

3.1. Standard Differential Evolution. A brief description of the standard DE algorithm is here provided. Let us consider the minimization problem

$$\begin{aligned} &\min \quad x f(\mathbf{x}) \\ &\text{s.t.} \quad \mathbf{x}_l \leq \mathbf{x} \leq \mathbf{x}_U. \end{aligned} \tag{15}$$

A population of N_P candidate solutions $\text{Pop} = \{\mathbf{x}_i, i = 1 ..N_P\}$ is randomly created, and for each individual (or agent) $\mathbf{x}_i \in \mathbb{R}^{N_D}$, the corresponding fitness $f(\mathbf{x}_i)$ is evaluated. Then, a new population Pop^{new} is constructed by repeating, for each vector $\mathbf{x}_i$ belonging to Pop, a sequence of mutation/crossover/selection steps, defined as follows.

3.1.1. Mutation. A mutated vector $\mathbf{v}_i$ is created as a linear combination of a few population members. More precisely several mutation rules were proposed in order to attain either a better exploration of the search space or a faster convergence (exploitation). In the present implementation the following four strategies, among those available in literature, are adopted:

$$\begin{aligned} \mathbf{v}_i &= \mathbf{x}_{r_1} + F\left(\mathbf{x}_{r_2} - \mathbf{x}_{r_3}\right), \\ \mathbf{v}_i &= \mathbf{x}_{\text{Gbest}} + F\left(\mathbf{x}_{r_1} - \mathbf{x}_{r_2}\right), \\ \mathbf{v}_i &= \mathbf{x}_i + F(\mathbf{x}_{\text{Gbest}} - \mathbf{x}_i) + F\left(\mathbf{x}_{r_1} - \mathbf{x}_{r_2}\right), \\ \mathbf{v}_i &= \mathbf{x}_{\text{Gbest}} + F\left(\mathbf{x}_{r_1} - \mathbf{x}_{r_2}\right) + F\left(\mathbf{x}_{r_3} - \mathbf{x}_{r_4}\right), \end{aligned} \tag{16}$$

where F is a parameter controlling the mutation scale, $\mathbf{x}_{\text{Gbest}}$ is the best individual (or agent) of the current generation, and $r_1, \ldots, r_4$ represent randomly chosen, nonrepeated, indexes belonging to $\{1..N_P\}$.

Each strategy has weaknesses and strengths: Strategies based on mutation of the best individual (strategies 2 and 4) typically show a faster converge rate toward an (often local) minimum, whereas strategies based on randomly chosen individuals (strategies 1 and 3) better explore the whole search space.

3.1.2. Crossover. A trial vector $\mathbf{u}_i$ is obtained by a crossover between the target vector $\mathbf{x}_i$ and the mutated vector $\mathbf{v}_i$

$$\mathbf{u}_{i,j} = \begin{cases} \mathbf{v}_{i,j} & \text{if } p_j \leq C_R \text{ or } j = j_{\text{rand}} \\ \mathbf{x}_{i,j} & \text{else} \end{cases} \quad j = 1, \ldots, N_D, \tag{17}$$

where p_j is a random number between 0 and 1, C_R is an algorithm parameter (typical values about 0.5), and j_{rand} is a randomly chosen index in the range $\{1..N_D\}$.

3.1.3. Selection. Target and trial vectors are compared. The best one is retained and inserted in the new population

$$\mathbf{x}_i^{\text{new}} = \begin{cases} \mathbf{u}_i & \text{if } f(\mathbf{u}_i) < f(\mathbf{x}_i) \\ \mathbf{x}_i & \text{else.} \end{cases} \tag{18}$$

This process is repeated iteratively, creating at each generation a new population which replaces the previous one. The procedure ends after a fixed number of generations (N_G). The best attained solution is deemed the optimal problem solution.

3.2. Self-Adaptation of Control Parameters. A common practical issue for many stochastic algorithms concerns the selection of suitable values for the control parameters. A fine tuning is often required in order to make the algorithm suitable for complex, real-world problems. DE is privileged in this respect, as the number of its parameters is very low. Apart from the the population size N_P, the performance of the DE algorithm depends on an appropriate selection of the scale factor F, which controls the mutation phase, and crossover probability C_R, which controls the crossover phase.

In order to avoid the manual tuning of DE control parameters, a self-adaptation scheme [23] has been implemented. The values of F and C_R are encoded into the individuals, which enter the optimization procedure, and randomly initialized within the intervals $[F_{\min}, F_{\max}]$ and $[C_{R_{\min}}, C_{R_{\max}}]$, respectively. Better values of these encoded control parameters will presumably lead to better individuals who, in turn, are more prone to survive and produce offspring, thus propagating these "superior" control parameters.

In order to maintain a certain diversity of the control parameters among the population, at the end of each generation, each individual undergoes a random uniform mutation of his control parameters which happens with probability p_τ, that is,

$$\begin{aligned} F^{G+1} &= \begin{cases} U(F_{\min}, F_{\max}), & \text{if } p_1 \leq p_\tau \\ F^{G}, & \text{otherwise,} \end{cases} \\ C_R^{G+1} &= \begin{cases} U\left(C_{R_{\min}}, C_{R_{\max}}\right), & \text{if } p_2 \leq p_\tau \\ C_R^{G}, & \text{otherwise,} \end{cases} \end{aligned} \tag{19}$$

where p_1 and p_2 are random numbers in $[0, 1]$, the apexes G and G + 1 refer to the current and next generation, respectively, and $U(a, b)$ indicates to a randomly sampled number in the range $[a, b]$. In the present implementation, the following values are used: $F_{\min} = 0.1$, $F_{\max} = 1$, $C_{R_{\min}} = 0.5$, $C_{R_{\max}} = 1$, and $p_\tau = 0.1$.

3.3. Balancing Local and Global Search. A key point in designing a global optimization algorithm concerns the delicate balance between two opposite needs: "global" exploration and "local" exploitation. Exploration refers to the algorithm capability of probing wide portions of the search space, with the hope of finding promising solutions that are yet to refine, while exploitation is the ability of improving a previous solution by probing a small region of the search space around it. A proper balance is required for the success of the algorithm: favouring local search reduces computational time, at the risk of being trapped in local optima, whereas favouring the global search requires longer computation time as a wide portion of the search space has to be tested. In order to achieve a good balance between exploration and exploitation, the proposed algorithm encompasses the creation of different populations, or tribes, each located on an "island."

Each tribe evolves independently from the others and features one specific mutation strategy among the four proposed variants. As shown in Figure 2, the different islands are arranged in a radial configuration, so that tribes on inner islands feature the less exploiting (i.e., most exploring) mutation strategy, whereas tribes on the outer islands feature the most exploiting mutation strategy. The order of the strategies from the outer to the inner islands is then 2, 4, 3, and 1. Every 100 generations, a migration is performed: Each tribe passes its best three agents to the "following one" (if it does exist, according to the direction of migration), in which these agents replace the three worst agents. Outward and inward migration tides alternate at each migration event. The proposed scheme allows an easy parallelization. A minimum of 4 islands are required to make it effective, but it scales well on any architecture with $4x$ cores.

3.4. Termination Criterion. The termination criterion is mainly based on the generation number, that is, on the available computational budget. This parameter strongly depends on the complexity of the analyzed problem. The maximum

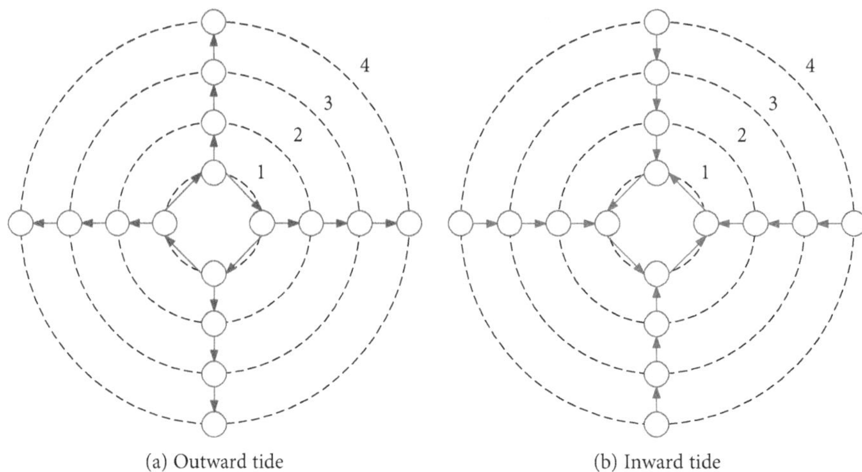

FIGURE 2: Migration scheme: forward (red) and backward (blue), for the 16-island case.

number of generation is chosen in such a way that the outcomes of independent runs of the code bring, in almost all cases, to similar results. In this respect, maintaining a certain diversity of the population in terms of its distribution on the search space is mandatory, to avoid a premature convergence on a local optimum.

A partial-restart mechanism, hereafter named "Epidemic," is adopted to handle this issue. More precisely, population diversity is evaluated for each tribe at the end of each generation, by using as metric the Euclidean distance between pairs of solutions. If the diversity score falls under a certain threshold, a large part (90%) of the population is randomly reinitialized over the entire search space. The maximum number of epidemic events that may occur in any run is fixed (2, in this application) in order not to compromise the overall efficiency of the search.

The reported solutions have been obtained by exploiting an 8-tribe optimization engine, with 512 agents per tribe and a maximum number of generations equal to 10000. In order to mitigate potential issues due to the stochastic nature of the algorithm, that is, an unfortunate premature convergence on a suboptimal solution, several runs can be performed starting from a different initial population, increasing the confidence on the attained result. In the present application, each optimization is repeated 25 times, and the best found solution is assumed as putative optimal solution.

4. Preliminary Analysis

A preliminary analysis is carried out in order to define the main features of the tentative trajectory that undergoes the DE optimization process. For the sake of simplicity, the orbits of the Galilean moons are assumed to be circular and coplanar. The Tisserand graph and the numerical suboptimal solution of the v_∞ leveraging or ΔV-Europa gravity assist (ΔV-EGA) maneuver are effective tools in defining the preliminary solution. These tools are here described with reference to the dual problem of escaping from Europa, because it is more intuitive and several papers are found in literature, dealing with v_∞ leveraging to move away from the Earth.

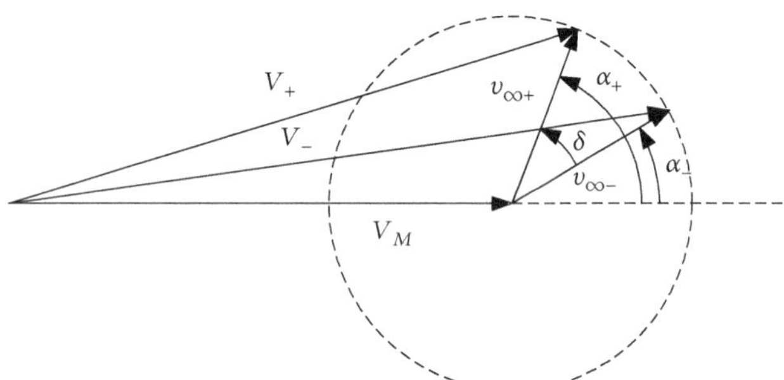

FIGURE 3: Velocity before (−) and after (+) the flyby.

4.1. Tisserand Graph. The Tisserand graph is a powerful instrument for the preliminary design of trajectories of a spacecraft that exploits multiple gravity assists in a multibody planetary system. Several versions have been discussed by different authors. The two-dimensional apoapsis-periapsis (or $r_a - r_p$) plot is adopted here.

Any point on the graph ($r_a > r_p > 0$) identifies a Keplerian elliptical orbit of the spacecraft around the central body (Jupiter). For a given secondary body M, moving on a circular orbit of radius r_M, the region $\mathcal{A}_M = \{(r_a, r_p) \mid r_p < r_M < r_a\}$ encompasses all spacecraft orbits that intersect the moon orbit. For each point (r_a, r_p) in this region, one can easily compute the values of magnitude and direction of the hyperbolic excess velocity v_∞ in case of encounter. The direction of v_∞ with respect to the moon velocity v_m is defined by the pump angle α (see Figure 3).

Curves of constant hyperbolic excess velocity are typically superimposed on this plot. Each curve collects all Jovicentric orbits approaching the moon with the same v_∞ magnitude and pump angle ranging from 0 degrees ($r_p = r_M$, rightmost point) to 180 degrees ($r_a = r_M$, leftmost point).

Lines of constant orbital period (i.e., constant energy) can also be plotted. This feature is mainly used to represent $K:L$ resonant orbits, where K and L indicate the number of revolutions completed by moon and spacecraft, respectively, between two successive encounters.

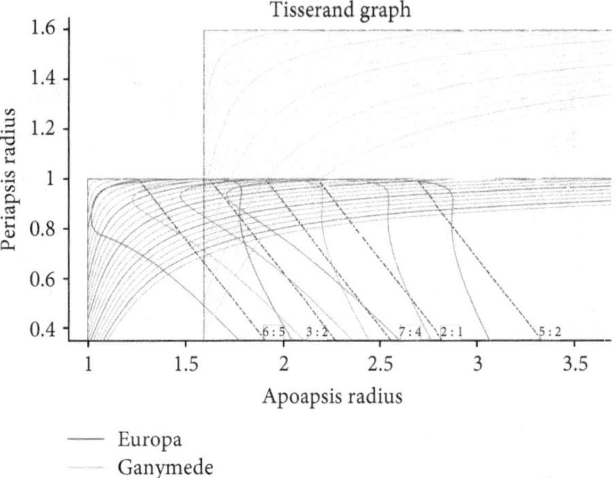

FIGURE 4: Tisserand graph for the Jovian planetary system. Contour lines of constant v_∞, at steps of 2% of V_m. Sample resonant orbits with Europa and corresponding maximum-deviation curves for a minimum flyby altitude $h_{\pi,\min} = 30$ km are also depicted.

Starting from any point on the graph, a shift to another orbit along a curve of constant v_∞ can be obtained at no cost, as result of a gravity assist. The displacement along the curve depends on the rotation of the hyperbolic excess velocity δ, which is given by 8.

Sequences of flybys can be performed, provided that the spacecraft is able to re-encounter the moon. The role of resonant orbits is now apparent, as hopping between resonant orbits guarantees the re-encounter in a known time.

For each resonant orbit, one can draw a pair of maximum deviation curves, assuming that a minimum-altitude flyby is enforced. These curves encompass all Jovicentric orbits achievable from that resonant orbit, by means of a gravity assist. Maximum deviation curves show, for any initial v_∞ value, which resonances are available for the next jump.

Points that do not lay on the same v_∞ curve can be connected only by means of some propulsive maneuver. As an example, vertical displacements on the graph can be obtained as a result of a tangential impulse at the apocenter.

Figure 4 shows a Tisserand plot for the Jovian planetary system. Contour lines of constant v_∞ value are displayed for Europa and Ganymede. Resonant orbits corresponding to resonance 6:5, 3:2, 7:4, 2:1, and 5:2 with Europa and associated maximum-deviation curves are shown as an example.

4.2. Suboptimal Solution for V Leveraging Transfers. The maneuver known as v_∞ leveraging, illustrated in Figure 5, provides an efficient way to move between points on the Tisserand graph that do not lay on the same v_∞ curve, as a large change of v_∞ is obtained by means of a small ΔV. The spacecraft leaves the secondary body (Europa) with a given hyperbolic excess velocity v_∞, entering a near-resonant orbit around the central body (Jupiter). A deep-space maneuver is performed near the apoapsis in order to modify v_∞ (increase, if one is considering an escape trajectory, or decrease, if one instead considers a capture) at the next encounter of the probe with the secondary body. Here,

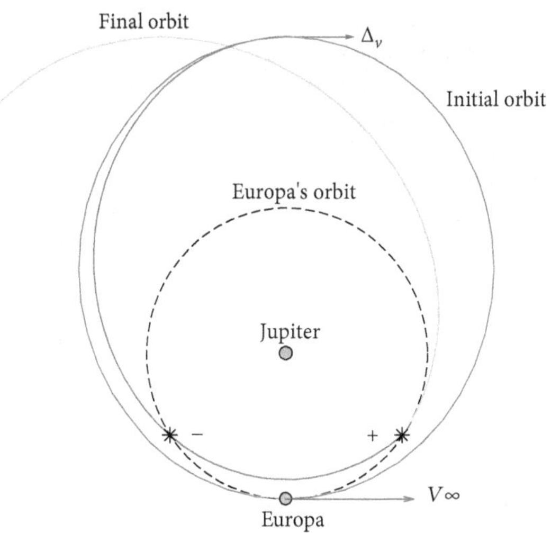

FIGURE 5: ΔV-EGA trajectory.

a flyby permits to rotate the hyperbolic excess of velocity, allowing the probe to enter a Jovicentric orbit with a larger apocenter. A generic ΔV-EGA maneuver can be labeled as $K:L(M)\pm$, where K and L indicate the number of revolutions that moon and probe complete, respectively, before the next encounter; $M \le L$ denotes the spacecraft orbit where DSM occours; the sign $\pm$ is used to distinguish between the possible encounter locations: just before (−) or after (+) traversing the line of apsides.

A suboptimal solution for ΔV-EGA maneuvers is easily achieved [14] under the assumption that (i) the spacecraft initial orbit is tangent to the moon's orbit and (ii) a tangential ΔV is applied at the apocenter of the nearly-resonant $K:L$ orbit. The maneuver is slightly improved if the aforementioned assumptions are removed [24]. Dealing with high-v_∞ flybys, the replacement of the patched-conic model with the RTBP model does not modify the mission and the numerical performance [25].

5. Numerical Results

5.1. Tentative Solution. The dual problem of escaping from Europa is considered in this section. A sequence of ΔV-EGA maneuvers with increasing $K:L$ resonances is used to progressively augment the probe apocenter, by targeting, after each flyby, a larger near-resonant orbit. Eventually, the spacecraft enters a hyperbolic trajectory and escapes Jupiter SOI [15]. In the present case, the probe trajectory ends at the apocenter of a 4:1 resonant orbit.

The tentative solution depends on the performance index that is minimized. In a minimum-propellant problem, the total ΔV is minimized by aligning vector v_∞, after each flyby, to Europa orbital velocity. This corresponds to maximizing the energy after the flyby (max-$\mathscr{E}$ strategy), assuming assigned v_∞. Low increments of v_∞ correspond to very small deep-space impulses. The optimization problem is ill-defined without any kind of time constraints, as the optimal solution would require an infinite number of ΔV-EGA maneuvers, each with infinitesimal v_∞ increment. Overall flight time is approximately equal to $T_{Eu}\sum K$. As a consequence, high-K resonances should be avoided.

Assuming $K_{max} = 16$, the suboptimal minimum-propellant solution presents $\Delta V = 1.04318$ km/s with flight time of 627.8 days. However, by skipping some resonances, while keeping max-$\mathscr{E}$ strategy, flight time can be considerably reduced with a minimum increment of propellant expenditure. Europa, due to its small mass, can provide only small rotations of the hyperbolic excess velocity, and long resonances, such as 9 : 5 and 11 : 5, must be included in order to exploit max-$\mathscr{E}$ strategy. Flight time remains incompatible with ETP mission requirements.

High-K resonances can be skipped by privileging the energy increment provided by each gravity assist, instead of the energy after the flyby. The maximum increment of energy (max-$\Delta\mathscr{E}$ strategy) implies the maximum change of the semi-major axis, that is, period. In an unconstrained gravity assist, max-$\Delta\mathscr{E}$ strategy requires the alignment of the hyperbola axis to the velocity of the secondary body ($\alpha_+ = \pi - \alpha_-$). Moreover, a v_∞ magnitude exists that maximizes $\Delta\mathscr{E}$. The present problem is, at a large extent, different. The resonant orbit after the previous gravity assist is assigned, and the pump angle α_- before the flyby is not free but a function of v_∞. Moreover, the new resonance is "a priori" selected in order to contain the trip time (low K is preferred). Therefore, $\Delta\mathscr{E}$ is assigned and the mission designer just selects a hyperbolic excess velocity in the range that permits the desired resonance jump. The best v_∞ value depends on the entire mission and will be a result of the optimization process. When the max-$\Delta\mathscr{E}$ strategy (here improperly so called) is adopted, v_∞ and pump angle after the gravity assist are large, and a further flyby, without any DSM (and leveraging), could align the spacecraft velocity to the moon orbital velocity.

The mission can be further improved by taking advantage from Europa's eccentricity and by removing constraints on position and direction of the deep-space impulse. This task is carried out by means of the proposed optimization algorithm, based on the general MGA-1DSM formulation described in Section 2. A reasonable tentative sequence of resonances is used to prune the global search. A careful exam of the Tisserand graph, in the light of the previous concepts, permits the exclusion of resonances with $K > 7$. Lower and upper bounds of the optimization variables are conveniently adjusted to improve convergence. In particular, for a generic $K:L(M)$ ΔV-EGA maneuver, one has

$$\begin{aligned} r_\pi &\in [r_{\pi,\min}, r_{\pi,\max}], \\ \beta &\in [-\pi, \pi], \\ \Delta T &\in [(K-0.1)T_{Eu}, (K+0.1)T_{Eu}], \\ \eta &\in \left[\frac{(M-1)}{L} + 1e-5, \frac{M}{L}\right]. \end{aligned} \tag{20}$$

Flyby parameters r_π and β do not depend on the prescribed resonance.

A capture trajectory is first attained under the hypothesis of a circular orbit for Europa (solution "A"). Next, this assumption is removed, and a slightly different trajectory is attained (solution "B"). Eventually, a third trajectory, which also exploits a flyby of Ganymede, is proposed (solution "C").

Table 1: Solution "A"—mission features.

	Event	Time [days]	v_∞[m/s]	Resonance	DSM [m/s]
0	Departure	0.00		4 : 1	1.931
A	Eu flyby	7.07	3683.4	7 : 2	13.09
B	Eu flyby	32.13	3519.7	3 : 1	15.16
C	Eu flyby	42.74	3351.2	5 : 2	21.00
D	Eu flyby	60.46	3150.8	2 : 1	97.19
E	Eu flyby	67.29	2353.0	7 : 4	40.55
F	Eu flyby	91.98	2016.7	3 : 2	62.49
G	Eu flyby	102.39	1560.1	4 : 3	56.11
H	Eu flyby	116.33	1176.8	6:5	60.56
I	EOI	137.32	803.2		715.18
	Totals	137.32			1083.27

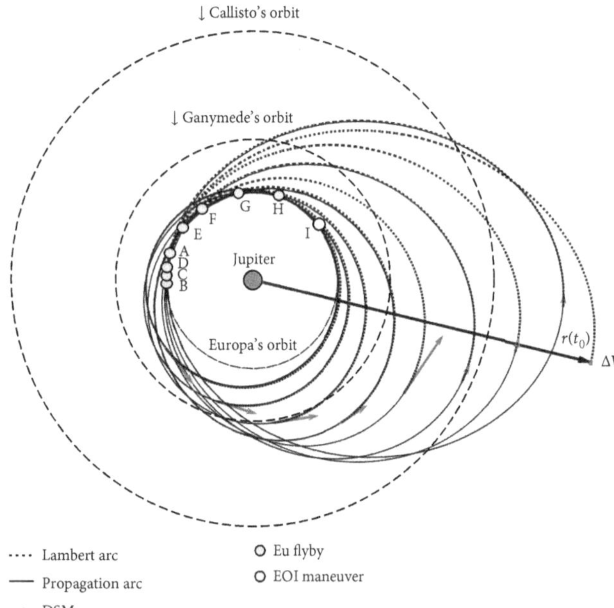

Figure 6: Jovicentric trajectory of solution A.

5.2. Solution "A"—Europa Circular Orbit. Solutions according to max-$\mathscr{E}$ strategy are impractical as the overall flight time is incompatible with ETP mission requirements. The Tisserand graph actually suggests that a reduction of the flight time can be achieved performing maneuvers based on max-$\Delta\mathscr{E}$ strategy. Solution "A" is obtained by assuming that Europa moves on a circular orbit, in order to match hypothesis and indications of the Tisserand graph. A suitable sequence of resonant orbits is assumed, and an optimal solution is provided by the DE algorithm. Features of the mission are summarized in Table 1. Each row presents initial time, v_∞ magnitude at flyby, resonance, and deep-space impulse of each leg. The last row refers to the impulsive injection into the target orbit around Europa.

Capture ΔV is reduced substantially with respect to a direct insertion (2781.9 m/s), allowing to reduce ETP total

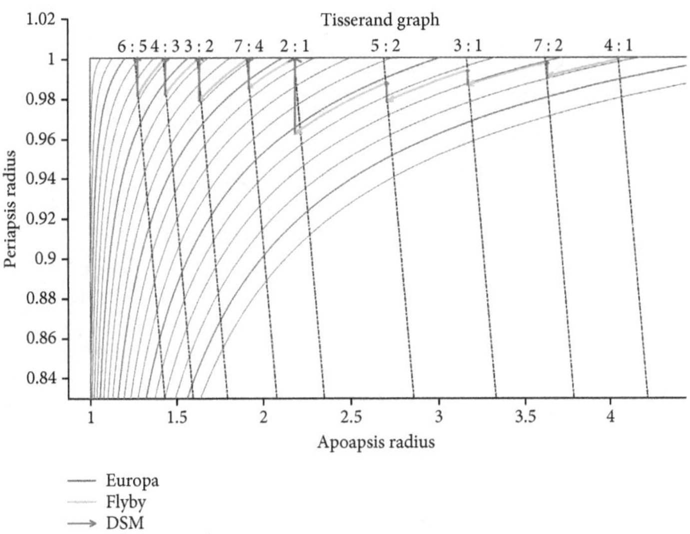

FIGURE 7: Tisserand graph of solution A. Axes are normalized with respect to Europa semi-major axis.

TABLE 2: Solution "B"—mission features.

	Event	Time [days]	v_∞[m/s]	Resonance	DSM [m/s]
0	Departure	1.40		4:1	0.01
A	Eu flyby	8.56	3702.4	7:2	14.65
B	Eu flyby	33.36	3518.4	3:1	16.32
C	Eu flyby	43.96	3335.6	5:2	21.58
D	Eu flyby	61.67	3127.5	2:1	79.93
E	Eu flyby	68.60	2478.3	7:4	57.09
F	Eu flyby	93.26	2012.6	3:2	63.14
G	Eu flyby	103.67	1548.9	4:3	55.76
H	Eu flyby	117.62	1166.5	6:5	59.85
I	EOI	137.21	797.0		712.76
	Totals	137.88			1081.10

mass (including propellant and structures) to 239.58 kg. The Jovicentric trajectory is plotted in Figure 6. Deep-space maneuvers are performed substantially at the apogee of the resonant orbits. Flyby locations on Europa orbit move clockwise, indicating that all maneuvers belong to class (−). Only the first flyby does not follow this rule, probably due to the peculiar features of the transition from the initial leg to the first intermediate leg.

The Tisserand graph is presented in Figure 7. For resonances lower than 2 : 1, after the flyby the spacecraft enters a Jovicentric orbit with perigee equal to Europa radius. This matches the suboptimal solution proposed by Longuski, where the hyperbolic excess of velocity is aligned with the moon velocity. The alignment condition is not verified for resonance higher than 2 : 1, when max-$\Delta\mathscr{E}$ maneuvers are carried out in order to minimize the number of ΔV-EGA maneuvers, skipping the most time-consuming legs.

5.3. Solution "B"—Europa Elliptic Orbit. The hypothesis on circularity of Europa's orbit is removed, and a new solution for the same mission scheme is searched for. Table 2 describes the attained solution. Figures 8 and 9 present the spacecraft trajectory and the corresponding Tisserand graph, respectively. Minor, yet interesting differences, can be observed. The transition between resonances 2 : 1 and 7 : 4 is now obtained by means of a max-$\Delta\mathscr{E}$ maneuver. The final flybys and EOI maneuver occur in close proximity of Europa pericenter, in order to benefit from a lower v_∞. The propellant consumption is coherent with solution A, and the mission takes slight advantage from the eccentricity of Europa orbit.

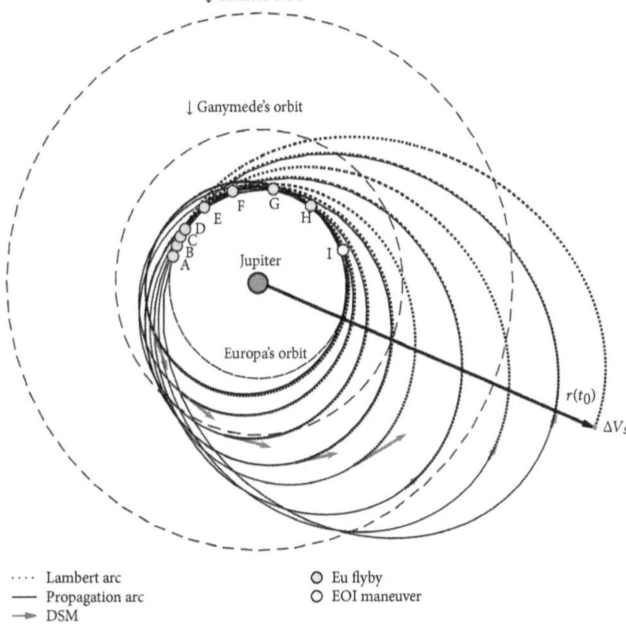

FIGURE 8: Jovicentric trajectory of solution B.

5.4. Solution "C"—Ganymede Flyby. Previous solutions show that ETP trajectory crosses Callisto and Ganymede orbits during its journey. An improved solution is searched for,

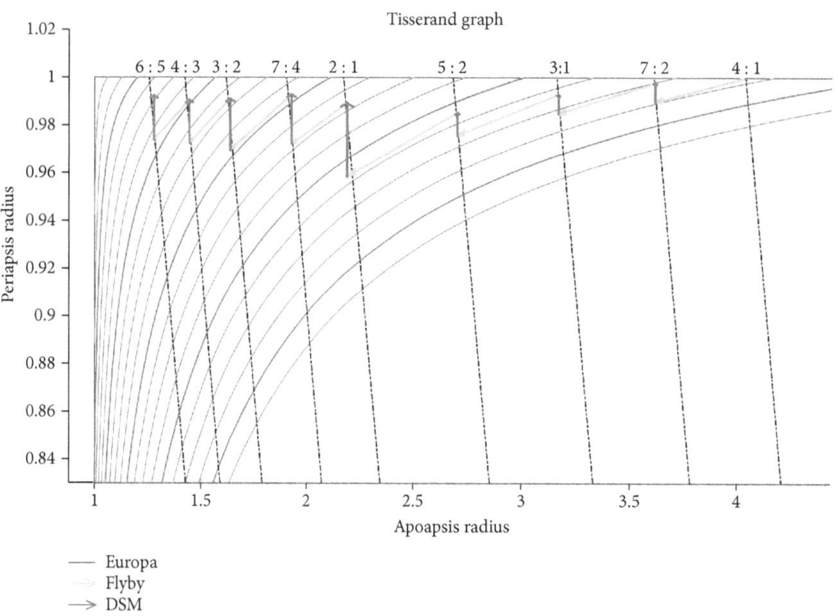

Figure 9: Tisserand graph of solution B. Axes are normalized with respect to the Europa semi-major axis.

Table 3: Solution "C"—mission features.

	Event	Time [days]	v_∞[m/s]	Resonance	DSM [m/s]
0	Departure	3.03		4 : 1	0.64
A	Eu flyby	10.05	3769.2	7 : 2	0.47
B	Eu flyby	34.92	3772.7	3 : 1	0.01
C	Eu flyby	45.58	3772.7	5 : 2	0.63
D	Eu flyby	63.34	3767.9	2 : 1	0.05
E	Eu flyby	70.45	3767.8	5:3	0.06
F	Eu flyby	88.22	3767.7		0.11
G	Ga flyby	95.65	2666.3		0.01
H	Eu flyby	99.68	1784.5	4 : 3	90.24
I	Eu flyby	113.61	1213.2	6:5	67.19
J	EOI	134.59	806.9		716.64
	Totals	131.56			882.04

aiming at replacing some deep-space maneuvers with Ganymede or Callisto flybys.

In particular, the semi-major axis of Ganymede (1.595 R_{Eu}) almost matches the apoapsis radius of the nominal 3 : 2 orbit ($1.621 R_{\text{Eu}}$). The 3 : 2(2) arc from Europa to Europa is thus substituted by two legs: the first from Europa to Ganymede and the second from Ganymede to Europa. Also, a 5 : 3 resonant orbit is used instead of 7 : 4, reducing the flight time by $2T_{\text{Eu}}$.

Results for this mission are summarized in Table 3. The Jovicentric trajectory and the Tisserand graph are presented in Figures 10 and 11, respectively. The flybys progressively rotate the initial hyperbolic excess velocity of the spacecraft, moving the spacecraft through a series of resonant orbits, until the apocenter is close to the Ganymede orbit, where a gravity assist from this moon makes the spacecraft orbit almost tangent to Europa orbit. No significant DSM is performed during the legs preceding the Ganymede flyby; thus, the Europa position at the encounters A–F is the same.

The Tisserand graph for this solution is presented in Figure 11, confirming that one Ganymede flyby is sufficient to change v_∞ in a so large amount that no deep space maneuver is necessary before two final Europa gravity assists permit the achievement of the 6 : 5 resonant orbit which ends with the EOI maneuver. A closer inspection shows small changes of v_∞ in the first part of the mission, despite no impulsive maneuver is performed. This is explained by recalling that v_∞ curves on the Tisserand plot are drawn assuming circular and coplanar orbits of the moons. In particular, Ganymede and Europa orbits are not coplanar (their inclinations are 0.2 and 0.4 degree, resp.); therefore, unlike previous solutions, this ETP trajectory is not coplanar with the Europa orbit. Spacecraft inclination and velocity vary at each flyby, resulting into a deceptive v_∞ change on the planar circular Tisserand graph.

The DE algorithm was able to find a solution with negligible deep-space impulses between flybys A and H. This kind of optimization method suffers the very low variation of the performance index in the proximity of the optimal solution and cannot be very accurate. The quality of the solution can be improved by replacing the general MGA-1DSM formulation with another one that considers ballistic legs between flybys, when required by theoretical considerations.

By reducing the capture ΔV to 882.04 m/s, ETP mass becomes 218.61 kg, saving approximately 21 kg with respect to previous solutions. The exploitation of a gravity assist by Ganymede also permits a 6-day reduction of the mission time length. A major drawback is the requirement of a suitable initial phasing between Europa and Ganymede, or a

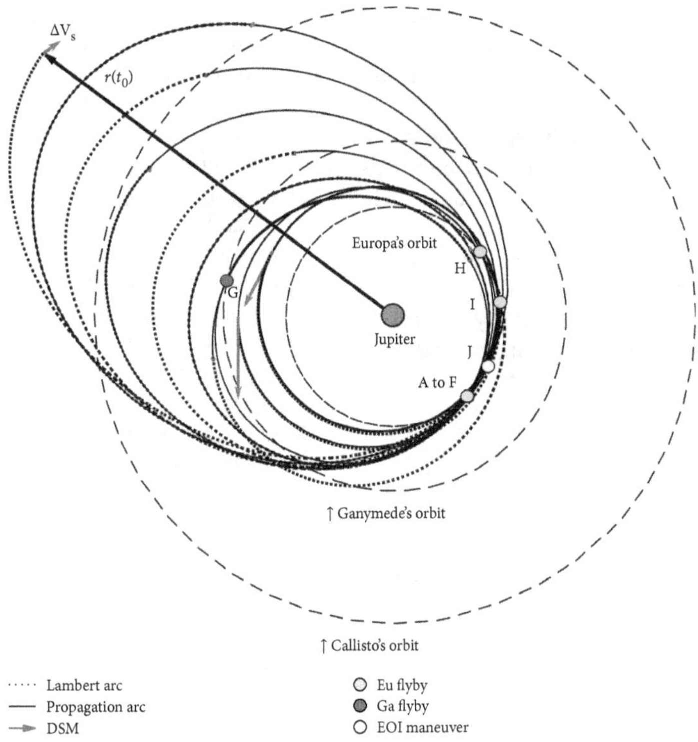

FIGURE 10: Jovicentric trajectory of solution C.

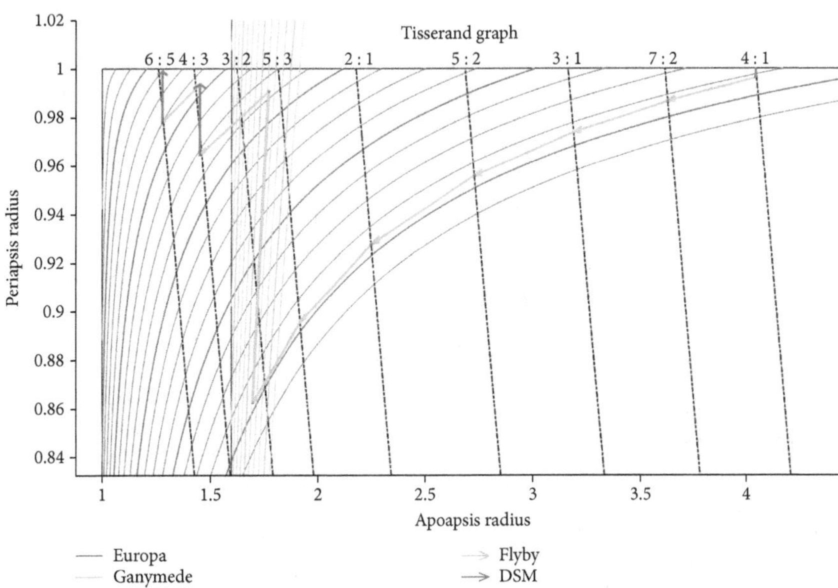

FIGURE 11: Tisserand graph of solution C. Axes are normalized with respect to the Europa semi-major axis.

precise timing for ETP release, which repeats every 7.058 days, that is, the Ganymede-Europa synodic period. Even though this issue might be thought irrelevant, as nowadays missions withstand many complex trajectory constraints, this is quite a delicate point for a probe that aims at minimal impact on the mission of the mother spacecraft. In this respect, a similar solution should be searched for, only after the baseline mission has been defined. The ETP release condition will be accordingly constrained.

6. Conclusions

In the search for a low-ΔV capture of a light-weight spacecraft into a polar orbit around Europa, v_∞ leveraging represents an excellent technique, which reduces the propellant consumption, at the cost of increased flight time. Moreover, several deep-space maneuvers along Europa-resonant orbits can be replaced by a single Ganymede gravity assist; total ΔV is further reduced with a small saving of the flight time. The Tisserand graph has proved to be an effective tool for the preliminary analysis. The same graph is also useful to introduce in the mission the gravity assist from other Galilean moons.

The search for the optimal solution is made complex by the need that the actual mission must take eccentricity and inclination of the secondary bodies into account. The exploitation of an effective heuristic optimization code allowed a wide exploration of the solution space, and an optimal solution is found in a relatively easy and fast manner. The performance of this Europa capture trajectory confirms the feasibility of a low-cost exploration mission, carried out by a small orbiter, compatible with the specifics for the secondary payload of Europa Clipper mission.

Conflicts of Interest

The authors declare that there is no conflict of interest regarding the publication of this paper.

Acknowledgments

This research was financed by University of Rome-Sapienza.

References

[1] C. F. Chyba and C. B. Phillips, "Europa as an abode of life," *Origins of Life and Evolution of the Biosphere*, vol. 32, no. 1, pp. 47–67, 2002.

[2] National Research Council, *Vision and Voyages for Planetary Science in the Decade 2013-2022*, National Academies Press, 2012.

[3] M. di Benedetto and V. Notaro, *Augmenting NASA Europa Clipper by a Small Probe: Europa Tomography Probe (ETP) Mission Concept*, IAC, 2016.

[4] V. Notaro, L. Iess, and A. Zavoli, *Europa Tomography Probe (ETP) Mission Feasibility -Spacecraft Design*, IAC, 2016.

[5] J. Johannesen and L. D'Amario, "Europa orbiter mission trajectory design," in *AAS/AIAA Astrodynamics Specialist Conference*, Girdwood, Alaska, August 1999.

[6] N. J. Strange, S. Campagnola, and R. P. Russell, "Leveraging flybys of low mass moons to enable an enceladus orbiter," in *Paper AAS 09-435, AAS/AIAA Astrodynamics Specialist Conference and Exhibit*, Pittsburgh, PA, USA, August 2009.

[7] N. Strange, T. Spilker, D. Landau, T. Lam, D. Lyons, and J. Guzman, "Mission design for the titan saturn system mission concept," vol. 135, pp. 919–934, 2010.

[8] M. H. Kaplan, *Modern Spacecraft Dynamics and Control*, Wiley, 1976.

[9] W. S. Koon, M. W. Lo, J. E. Marsden, and S. D. Ross, "Heteroclinic connections between periodic orbits and resonance transitions in celestial mechanics," *Chaos: An Interdisciplinary Journal of Nonlinear Science*, vol. 10, no. 2, pp. 427–469, 2000.

[10] W. S. Koon, M. W. Lo, J. E. Marsden, and S. D. Ross, "Resonance and capture of Jupiter comets," *Celestial Mechanics and Dynamical Astronomy*, vol. 81, no. 1/2, pp. 27–38, 2001.

[11] R. L. Anderson, "Approaching moons from resonance via invariant manifolds," *Journal of Guidance, Control, and Dynamics*, vol. 38, no. 6, pp. 1097–1109, 2015.

[12] G. Lantoine and R. P. Russell, "Near ballistic halo-to-halo transfers between planetary moons," *The Journal of the Astronautical Sciences*, vol. 58, no. 3, pp. 335–363, 2011.

[13] G. Gomez, W. S. Koon, M. W. Lo, J. E. Marsden, J. Masdemont, and S. D. Ross, "Connecting orbits and invariant manifolds in the spatial restricted three-body problem," *Nonlinearity*, vol. 17, no. 5, pp. 1571–1606, 2004.

[14] J. Sims and J. Longuski, "Analysis of V∞ Leveraging for Interplanetary Missions," in *Astrodynamics Conference, American Institute of Aeronautics and Astronautics*, Scottsdale,AZ,U.S.A 1994.

[15] G. Colasurdo, L. Casalino, and E. Fantino, "Minimum-fuel escape from two-body sun-earth system," *Journal of Guidance, Control, and Dynamics*, vol. 22, no. 5, pp. 632–636, 1999.

[16] D. Myatt, V. Becerra, S. Nasuto, J. Bishop, and D. Izzo, *Advanced Global Optimisation for Mission Analysis and Design*, Tech. rep., European Space Agency and The University of Reading, 2004.

[17] J. A. Sims, J. M. Longuski, and A. J. Staugler, "V∞ Leveraging for Interplanetary Missions: Multiple-Revolution Orbit Techniques," *Journal of Guidance, Control, and Dynamics*, vol. 20 no. 3, pp. 409–415, 1997.

[18] S. Campagnola and R. P. Russell, "Endgame problem part 1: V∞-leveraging technique and the leveraging graph," *Journal of Guidance, Control, and Dynamics*, vol. 33, no. 2, pp. 463-475, 2010.

[19] L. Casalino, G. Colasurdo, and D. Pastrone, "Simple strategy for powered swingby," *Journal of Guidance, Control, and Dynamics*, vol. 22, no. 1, pp. 156–159, 1999.

[20] R. R. Bate, D. D. Mueller, and J. E. White, *Fundamentals of Astrodynamics*, Dover Publications, 1971, http://www.worldcat.org/isbn/0486600610.

[21] R. Storn and K. Price, "Differential evolution – a simple and efficient heuristic for global optimization over continuous spaces," *Journal of Global Optimization*, vol. 11, no. 4, pp. 341–359, 1997.

[22] S. Das and P. N. Suganthan, "Differential evolution: a survey of the state-of-the-art," *IEEE Transactions on Evolutionary Computation*, vol. 15, no. 1, pp. 4–31, 2011.

[23] J. Brest, B. Bošković, S. Greiner, V. Žumer, and M. S. Maučec, "Performance comparison of self-adaptive and adaptive differential evolution algorithms," *Soft Computing*, vol. 11, no. 7, pp. 617–629, 2007.

[24] L. Casalino, G. Colasurdo, and D. Pastrone, "Optimization of ΔV earth-gravity-assist trajectories," *Journal of Guidance, Control, and Dynamics*, vol. 21, no. 6, pp. 991–995, 1998.

[25] L. Casalino and G. Colasurdo, "Optimal deltav-earth-gravity-assist trajectories in the restricted three-body problem," in *AAS/AIAA Astrodynamics Specialist Conference*, Girdwood, AK, USA, August 1999.

15

Integrated 6-DOF Orbit-Attitude Dynamical Modeling and Control Using Geometric Mechanics

Ling Jiang, Yue Wang, and Shijie Xu

School of Astronautics, Beihang University, Beijing 100191, China

Correspondence should be addressed to Yue Wang; ywang@buaa.edu.cn

Academic Editor: Enrico C. Lorenzini

The integrated 6-DOF orbit-attitude dynamical modeling and control have shown great importance in various missions, for example, formation flying and proximity operations. The integrated approach yields better performances than the separate one in terms of accuracy, efficiency, and agility. One challenge in the integrated approach is to find a unified representation for the 6-DOF motion with configuration space SE(3). Recently, exponential coordinates of SE(3) have been used in dynamics and control of the 6-DOF motion, however, only on the kinematical level. In this paper, we will improve the current method by adopting exponential coordinates on the dynamical level, by giving the relation between the second-order derivative of exponential coordinates and spacecraft's accelerations. In this way, the 6-DOF motion in terms of exponential coordinates can be written as a second-order system with a quite compact form, to which a broader range of control theories, such as higher-order sliding modes, can be applied. For a demonstration purpose, a simple asymptotic tracking control law with almost global convergence is designed. Finally, the integrated modeling and control are applied to the body-fixed hovering over an asteroid and verified by a simulation, in which absolute motions of the spacecraft and asteroid are simulated separately.

1. Introduction

The integrated 6-DOF orbit-attitude dynamical modeling and control of spacecraft have shown great importance in various space missions, such as formation flying [1–7], proximity operations [8–14], and proximity operations about minor celestial bodies [15–17].

Necessity of the integrated approach for spacecraft's 6-DOF motion lies in two aspects. Firstly, the orbit and attitude motions are actually kinematically coupled in spacecraft relative motions [1, 7] and are dynamically coupled due to effects of external forces and torques, such as the gravitational orbit-attitude coupling in close proximity of minor celestial bodies [18–21] and the orbit-attitude coupling of high area-to-mass ratio (HAMR) objects caused by the solar radiation pressure (SRP) [22, 23]. Secondly, and more importantly, proximity operations and some formation flying missions, such as proximity rendezvous, docking, landing on and sampling an asteroid, interferometric observation, and coordinated pointing, usually require the relative position and attitude of spacecraft to follow the desired trajectory simultaneously.

Unlike the traditional approach that treats orbit and attitude motions separately, such as Zhang et al. [24, 25], the integrated approach models and controls the orbit and attitude motions in a unified framework. Therefore, integrated orbit-attitude dynamical modeling and control will capture the system's dynamics better and will lead the 6-DOF motion in the phase space more accurately and effectively. That is, the integrated approach yields better performances than the separate one in terms of accuracy, efficiency, and agility.

There are two main challenges in the integrated dynamical modeling and control. The first one is to find an adequate unified mathematical representation for the 6-DOF orbit-attitude motion, which must be amenable for the application of various nonlinear control theories. The unified representation of the 6-DOF motion is also required to be applicable to a wide range of spacecraft maneuvers.

The second challenge is that the 6-DOF orbit-attitude motion is highly nonlinear and is subjected to external disturbances and parameter uncertainties. Consequently, adequate control theories are needed in the controller design, such as the sliding mode control [9], adaptive control [12, 13], adaptive sliding mode control [3, 17], adaptive terminal sliding mode control [2, 14], finite-time control [5, 16, 26], and state-dependent Riccati equation (SDRE) method [4, 7].

As for the unified mathematical representation of the 6-DOF motion, the spacecraft can be considered as a rigid body and the configuration space is the Lie group SE(3), the special Euclidean group, which is the set of position and attitude of a rigid body moving in a three-dimensional Euclidean space. The dual quaternion has been widely used as a unified representation of the 6-DOF rigid body motion, but the constraints between its elements need to be dealt with carefully [10, 11]. Sanyal et al. [8, 27] have used the matrix form of the Lie group SE(3) in the dynamical modeling of rigid body motion and achieved controllers with almost global convergence, which is the best that can be achieved for a system evolving on a noncontractible state space, like the 6-DOF rigid body motion, with continuous feedback [6]. However, due to the complexity of matrix calculations, the matrix form of SE(3) is not convenient for controller design.

Recently, exponential coordinates of Lie group SE(3) have been used as a unified representation of the 6-DOF orbit-attitude (rigid body) motion in spacecraft formation flying and asteroid hovering [6, 9, 14–17]. The exponential coordinates are in vector form with six elements, and thus, unlike the dual quaternion, have no constraint between elements. Besides, compared with the matrix form of SE(3), the set of exponential coordinates with a vector form is more convenient for applications of nonlinear control theories. In the mentioned studies, the exponential coordinates are used to describe the 6-DOF kinematics of the spacecraft, and the control scheme is designed to reduce errors of configuration and velocities from almost any given initial state, except those that differ in orientation by a π radian rotation from the desired states where exponential coordinates for the attitude are not uniquely defined [6]. Because the set of such initial states is an embedded lower-dimensional subspace of the state space, the tracking control scheme designed in terms of exponential coordinates can achieve almost global convergence, which makes it applicable in practice. Exponential coordinates of SE(3) used by Lee et al. [6, 15, 16] have provided a good unified representation for the 6-DOF motion with several advantages: having a vector form, having no constraint between elements, being convenient for controller design, and achieving almost global convergence in the controller.

In the present paper, based on the results by Bullo and Murray [28], we will improve the dynamical modeling method for the 6-DOF rigid body motion suggested by Lee et al. [6, 15, 16]. Lee et al. used exponential coordinates of SE(3) only in the kinematics of spacecraft, that is, only gave the relation between the first-order time derivative of exponential coordinates and the (angular) velocity of spacecraft. We will use exponential coordinates of SE(3) on the dynamical level further, that is, give the relation between the second-order time derivative of exponential coordinates and the (angular) acceleration of spacecraft. By using this relation on the dynamical level, the system can be written as a second-order system with a compact form. Then, the set of exponential coordinates becomes more convenient for controller design and a broader range of control theories, such as higher-order sliding modes, can be applied.

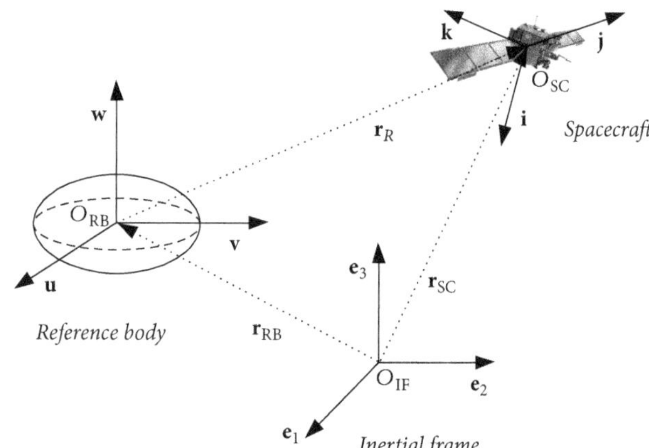

Figure 1: The spacecraft moving in proximity of the reference body.

Another improvement suggested in the current work is to introduce a desired trajectory for the 6-DOF motion, that is, the guidance on SE(3). The controller will lead the system to follow the desired trajectory until reaching the desired final state. In the earlier studies by Lee et al. [6, 15, 16], Lee [9], and Lee and Vukovich [14, 17], the controller aimed at the desired final state directly. Therefore, the true trajectory to the final state could not be specified. Compared with Lee's approach, our approach has two advantages: firstly, it is applicable to diverse spacecraft proximity operations requiring a specific 6-DOF orbit-attitude trajectory, for example, approaching along an obstacle-free trajectory; secondly, the initial tracking error is much smaller and the control saturation in the beginning can be avoided.

Based on the second-order system in terms of exponential coordinates of SE(3), a simple asymptotic tracking control law with almost global convergence is designed for a demonstration purpose. Finally, both the integrated dynamical modeling and control law are applied to body-fixed orbit-attitude hovering over an asteroid.

2. Integrated Modeling of 6-DOF Relative Dynamics

As shown in Figure 1, the 6-DOF orbit-attitude (rigid body) motion of a spacecraft in proximity of the reference body is considered. The reference body can be another spacecraft in spacecraft formation flying and proximity operations or the asteroid in minor celestial body proximity operations. The inertial frame is denoted by $S_{\mathrm{IF}} = \{\mathbf{e}_1, \mathbf{e}_2, \mathbf{e}_3\}$ with O_{IF} as its origin. The body-fixed principal-axis frames of the reference body and spacecraft are denoted by $S_{\mathrm{RB}} = \{\mathbf{u}, \mathbf{v}, \mathbf{w}\}$ and $S_{\mathrm{SC}} = \{\mathbf{i}, \mathbf{j}, \mathbf{k}\}$ with O_{RB} and O_{SC} as their origins, respectively.

2.1. Dynamics of Spacecraft. The attitude of the spacecraft with respect to the inertial frame S_{IF} is described by matrix $\mathbf{A}_{\mathrm{SC}}$:

$$\mathbf{A}_{\mathrm{SC}} = [\boldsymbol{\alpha}_{\mathrm{SC}}, \boldsymbol{\beta}_{\mathrm{SC}}, \boldsymbol{\gamma}_{\mathrm{SC}}]^T \in \mathrm{SO}(3), \quad (1)$$

where $\boldsymbol{\alpha}_{SC}$, $\boldsymbol{\beta}_{SC}$, and $\boldsymbol{\gamma}_{SC}$ are coordinates of unit vectors $\mathbf{e}_1$, $\mathbf{e}_2$, and $\mathbf{e}_3$ of the inertial frame S_{IF} expressed in the body-fixed frame of spacecraft S_{SC}, respectively. Matrix $\mathbf{A}_{SC}$ is also the coordinate transformation matrix from the body-fixed frame S_{SC} to the inertial frame S_{IF}. SO(3) is the three-dimensional special orthogonal group. The position vector of the spacecraft with respect to the origin O_{IF} expressed in the inertial frame S_{IF} is denoted by $\mathbf{r}_{SC}$.

The configuration space of 6-DOF orbit-attitude motion of the spacecraft is the Lie group:

$$Q_{SC} = \mathrm{SE}(3), \tag{2}$$

known as the special Euclidean group of three-dimensional space with elements $(\mathbf{A}_{SC}, \mathbf{r}_{SC})$, which is the semidirect product of SO(3) and $\mathbb{R}^3$, that is, $\mathrm{SO}(3) \ltimes \mathbb{R}^3$. The velocity phase space of 6-DOF motion of the spacecraft is the tangent bundle TSE(3), which can be represented by SE(3) × se(3) through the left translation, where se(3) is the Lie algebra of Lie group SE(3).

The configuration of spacecraft on the special Euclidean group SE(3) can be represented by the following 4×4 matrix:

$$\mathbf{g}_{SC} = \begin{bmatrix} \mathbf{A}_{SC} & \mathbf{r}_{SC} \\ \mathbf{0} & 1 \end{bmatrix} \in \mathrm{SE}(3). \tag{3}$$

The velocities of spacecraft, which are contained within Lie algebra se(3), can be represented by the 6×1 vector:

$$\boldsymbol{\xi}_{SC} = \begin{bmatrix} \boldsymbol{\Omega}_{SC} \\ \mathbf{V}_{SC} \end{bmatrix} \in \mathbb{R}^6, \tag{4}$$

where $\boldsymbol{\Omega}_{SC}$ and $\mathbf{V}_{SC}$ are the spacecraft's angular velocity and velocity with respect to the inertial frame, respectively, both expressed in its body-fixed frame S_{SC}.

Then, the 6-DOF orbit-attitude kinematics of the spacecraft can be given by [15]

$$\dot{\mathbf{g}}_{SC} = \mathbf{g}_{SC} (\boldsymbol{\xi}_{SC})^{\vee}, \tag{5}$$

where the map $(\cdot)^{\vee} : \mathbb{R}^6 \rightarrow \mathrm{se}(3)$ is the Lie algebra isomorphism defined by

$$(\boldsymbol{\xi}_{SC})^{\vee} = \begin{bmatrix} (\boldsymbol{\Omega}_{SC})^{\times} & \mathbf{V}_{SC} \\ \mathbf{0} & 0 \end{bmatrix} \in \mathrm{se}(3). \tag{6}$$

The map $(\cdot)^{\times} : \mathbb{R}^3 \rightarrow \mathrm{so}(3)$ is the Lie algebra isomorphism, which is also the cross-product operator for 3×1 vectors, defined by

$$(\mathbf{w})^{\times} = \begin{bmatrix} w^x \\ w^y \\ w^z \end{bmatrix}^{\times} = \begin{bmatrix} 0 & -w^z & w^y \\ w^z & 0 & -w^x \\ -w^y & w^x & 0 \end{bmatrix} \in \mathrm{so}(3), \tag{7}$$

where so(3) is the Lie algebra of Lie group SO(3).

The 6-DOF orbit-attitude dynamics of the spacecraft can be given by [15]

$$\mathbb{I}_{SC}\dot{\boldsymbol{\xi}}_{SC} = \mathrm{ad}^*_{\boldsymbol{\xi}_{SC}} (\mathbb{I}_{SC}\boldsymbol{\xi}_{SC}) + \boldsymbol{\tau}^n_{SC} + \boldsymbol{\tau}^c_{SC}, \tag{8}$$

where $\mathbb{I}_{SC}$ is the combined matrix of the mass and inertia of the spacecraft given by

$$\mathbb{I}_{SC} = \begin{bmatrix} \mathbf{I}_{SC} & \mathbf{0} \\ \mathbf{0} & m_{SC}\mathbf{I}_{3\times3} \end{bmatrix} \in \mathbb{R}^{6\times6}, \tag{9}$$

$\mathbf{I}_{SC}$ and m_{SC} are the inertia tensor and mass of the spacecraft, respectively, and $\mathbf{I}_{3\times3}$ is the 3×3 identity matrix. The adjoint operator $\mathrm{ad}_{\boldsymbol{\xi}_{SC}}$ and coadjoint operator $\mathrm{ad}^*_{\boldsymbol{\xi}_{SC}}$ on the Lie algebra se(3) can be given as matrixes:

$$\begin{aligned} \mathrm{ad}_{\boldsymbol{\xi}_{SC}} &= \begin{bmatrix} (\boldsymbol{\Omega}_{SC})^{\times} & \mathbf{0} \\ (\mathbf{V}_{SC})^{\times} & (\boldsymbol{\Omega}_{SC})^{\times} \end{bmatrix} \in \mathbb{R}^{6\times6}, \\ \mathrm{ad}^*_{\boldsymbol{\xi}_{SC}} &= \left(\mathrm{ad}_{\boldsymbol{\xi}_{SC}}\right)^T = \begin{bmatrix} -(\boldsymbol{\Omega}_{SC})^{\times} & -(\mathbf{V}_{SC})^{\times} \\ \mathbf{0} & -(\boldsymbol{\Omega}_{SC})^{\times} \end{bmatrix} \in \mathbb{R}^{6\times6}. \end{aligned} \tag{10}$$

The environmental torque $\mathbf{T}^n_{SC}$ and force $\mathbf{F}^n_{SC}$ acting on the spacecraft, both expressed in the body-fixed frame S_{SC}, are written together in vector $\boldsymbol{\tau}^n_{SC}$:

$$\boldsymbol{\tau}^n_{SC} = \begin{bmatrix} \mathbf{T}^n_{SC} \\ \mathbf{F}^n_{SC} \end{bmatrix} \in \mathbb{R}^6. \tag{11}$$

Keep in mind that, in minor celestial body proximity operations, the environmental torque and force $\boldsymbol{\tau}^n_{SC}$ include the gravitational torque and force by the minor body, while, in spacecraft formation flying and proximity operations, the gravitational torque and force between two spacecraft will be neglected.

The control torque $\mathbf{T}^c_{SC}$ and force $\mathbf{F}^c_{SC}$ acting on the spacecraft, both expressed in the body-fixed frame S_{SC}, are written together in vector $\boldsymbol{\tau}^c_{SC}$:

$$\boldsymbol{\tau}^c_{SC} = \begin{bmatrix} \mathbf{T}^c_{SC} \\ \mathbf{F}^c_{SC} \end{bmatrix} \in \mathbb{R}^6. \tag{12}$$

2.2. Dynamics of Reference Body. The 6-DOF orbit-attitude dynamics of the reference body is described similarly as that of the spacecraft. The attitude of reference body with respect to the inertial frame S_{IF} is described by matrix $\mathbf{A}_{RB}$:

$$\mathbf{A}_{RB} = \left[\boldsymbol{\alpha}_{RB}, \boldsymbol{\beta}_{RB}, \boldsymbol{\gamma}_{RB}\right]^T \in \mathrm{SO}(3), \tag{13}$$

where $\boldsymbol{\alpha}_{RB}$, $\boldsymbol{\beta}_{RB}$, and $\boldsymbol{\gamma}_{RB}$ are coordinates of unit vectors $\mathbf{e}_1$, $\mathbf{e}_2$, and $\mathbf{e}_3$ expressed in the body-fixed frame of reference body S_{RB}, respectively. The position vector of the reference body with respect to the origin O_{IF} expressed in the inertial frame S_{IF} is denoted by $\mathbf{r}_{RB}$. The configuration space of reference body, Q_{RB}, is also Lie group SE(3) with elements $(\mathbf{A}_{RB}, \mathbf{r}_{RB})$. The configuration of the reference body is denoted by

$$\mathbf{g}_{RB} = \begin{bmatrix} \mathbf{A}_{RB} & \mathbf{r}_{RB} \\ \mathbf{0} & 1 \end{bmatrix} \in \mathrm{SE}(3). \tag{14}$$

The velocities of the reference body are represented by

$$\boldsymbol{\xi}_{RB} = \begin{bmatrix} \boldsymbol{\Omega}_{RB} \\ \mathbf{V}_{RB} \end{bmatrix} \in \mathbb{R}^6, \tag{15}$$

where $\boldsymbol{\Omega}_{\text{RB}}$ and $\mathbf{V}_{\text{RB}}$ are angular velocity and velocity of the reference body with respect to the inertial frame, respectively, both expressed in the reference body's body-fixed frame S_{RB}.

The 6-DOF orbit-attitude kinematics of the reference body is given by

$$\dot{\mathbf{g}}_{\text{RB}} = \mathbf{g}_{\text{RB}} (\boldsymbol{\xi}_{\text{RB}})^{\vee}, \quad (16)$$

where

$$(\boldsymbol{\xi}_{\text{RB}})^{\vee} = \begin{bmatrix} (\boldsymbol{\Omega}_{\text{RB}})^{\times} & \mathbf{V}_{\text{RB}} \\ \mathbf{0} & 0 \end{bmatrix} \in \text{se}(3). \quad (17)$$

The 6-DOF orbit-attitude dynamics of the reference body can be given by

$$\mathbb{I}_{\text{RB}}\dot{\boldsymbol{\xi}}_{\text{RB}} = \text{ad}^*_{\boldsymbol{\xi}_{\text{RB}}} (\mathbb{I}_{\text{RB}}\boldsymbol{\xi}_{\text{RB}}) + \boldsymbol{\tau}_{\text{RB}}, \quad (18)$$

where $\mathbb{I}_{\text{RB}}$ is the combined matrix of the mass and inertia of the reference body

$$\mathbb{I}_{\text{RB}} = \begin{bmatrix} \mathbf{I}_{\text{RB}} & \mathbf{0} \\ \mathbf{0} & m_{\text{RB}}\mathbf{I}_{3\times3} \end{bmatrix} \in \mathbb{R}^{6\times6}, \quad (19)$$

$\mathbf{I}_{\text{RB}}$ and m_{RB} are the inertia tensor and mass of the reference body, respectively, and

$$\text{ad}^*_{\boldsymbol{\xi}_{\text{RB}}} = \begin{bmatrix} -(\boldsymbol{\Omega}_{\text{RB}})^{\times} & -(\mathbf{V}_{\text{RB}})^{\times} \\ \mathbf{0} & -(\boldsymbol{\Omega}_{\text{RB}})^{\times} \end{bmatrix} \in \mathbb{R}^{6\times6}. \quad (20)$$

Resultant torque $\mathbf{T}_{\text{RB}}$ and force $\mathbf{F}_{\text{RB}}$ acting on the reference body, both expressed in the body-fixed frame S_{RB}, are written together in vector $\boldsymbol{\tau}_{\text{RB}}$:

$$\boldsymbol{\tau}_{\text{RB}} = \begin{bmatrix} \mathbf{T}_{\text{RB}} \\ \mathbf{F}_{\text{RB}} \end{bmatrix} \in \mathbb{R}^{6}. \quad (21)$$

The torque and force $\boldsymbol{\tau}_{\text{RB}}$ will include control torque and force acting on the reference body, if they exist, in spacecraft formation flying and proximity operations.

2.3. Relative Dynamics of Spacecraft. In space missions, what really matters is the relative dynamics of spacecraft with respect to the reference body, rather than the absolute dynamics of spacecraft in the inertial space. The 6-DOF relative dynamics of spacecraft, which will be written in a similar form with the absolute dynamics, can be obtained based on the results in Sections 2.1 and 2.2.

The relative configuration of spacecraft with respect to the reference body, which is also on the special Euclidean group SE(3), is denoted by

$$\mathbf{g}_R = (\mathbf{g}_{\text{RB}})^{-1} \mathbf{g}_{\text{SC}} = \begin{bmatrix} \mathbf{A}_R & \mathbf{r}_R \\ \mathbf{0} & 1 \end{bmatrix} \in \text{SE}(3), \quad (22)$$

where matrix $\mathbf{A}_R$ is the relative attitude of spacecraft with respect to the reference body, which is also the coordinate transformation matrix from the body-fixed frame of spacecraft S_{SC} to the body-fixed frame of reference body S_{RB}, and vector $\mathbf{r}_R$ is the relative position of spacecraft with respect to the reference body O_{RB}, expressed in the body-fixed frame of reference body S_{RB}.

The relative attitude matrix $\mathbf{A}_R$ can be written as

$$\mathbf{A}_R = [\boldsymbol{\alpha}_R, \boldsymbol{\beta}_R, \boldsymbol{\gamma}_R]^T, \quad (23)$$

where $\boldsymbol{\alpha}_R$, $\boldsymbol{\beta}_R$, and $\boldsymbol{\gamma}_R$ are coordinates of unit vectors $\mathbf{u}$, $\mathbf{v}$, and $\mathbf{w}$ of the reference body's body-fixed frame S_{RB} expressed in the spacecraft's body-fixed frame S_{SC}, respectively.

According to (3) and (14), the relative configuration of spacecraft can be obtained as

$$\begin{aligned} \mathbf{g}_R &= (\mathbf{g}_{\text{RB}})^{-1} \mathbf{g}_{\text{SC}} = \begin{bmatrix} \mathbf{A}_{\text{RB}}^{-1} & -\mathbf{A}_{\text{RB}}^{-1}\mathbf{r}_{\text{RB}} \\ \mathbf{0} & 1 \end{bmatrix} \begin{bmatrix} \mathbf{A}_{\text{SC}} & \mathbf{r}_{\text{SC}} \\ \mathbf{0} & 1 \end{bmatrix} \\ &= \begin{bmatrix} \mathbf{A}_{\text{RB}}^{-1}\mathbf{A}_{\text{SC}} & \mathbf{A}_{\text{RB}}^{-1} (\mathbf{r}_{\text{SC}} - \mathbf{r}_{\text{RB}}) \\ \mathbf{0} & 1 \end{bmatrix}. \end{aligned} \quad (24)$$

Then, the 6-DOF relative orbit-attitude kinematics of the spacecraft with respect to the reference body can be given by

$$\begin{aligned} \dot{\mathbf{g}}_R &= (\mathbf{g}_{\text{RB}})^{-1} \dot{\mathbf{g}}_{\text{SC}} + \frac{d}{dt} \left[(\mathbf{g}_{\text{RB}})^{-1} \right] \mathbf{g}_{\text{SC}} \\ &= (\mathbf{g}_{\text{RB}})^{-1} \dot{\mathbf{g}}_{\text{SC}} - (\mathbf{g}_{\text{RB}})^{-1} \dot{\mathbf{g}}_{\text{RB}} (\mathbf{g}_{\text{RB}})^{-1} \mathbf{g}_{\text{SC}} \\ &= (\mathbf{g}_{\text{RB}})^{-1} \mathbf{g}_{\text{SC}} (\boldsymbol{\xi}_{\text{SC}})^{\vee} \\ &\quad - (\mathbf{g}_{\text{RB}})^{-1} \mathbf{g}_{\text{RB}} (\boldsymbol{\xi}_{\text{RB}})^{\vee} (\mathbf{g}_{\text{RB}})^{-1} \mathbf{g}_{\text{SC}} \\ &= \mathbf{g}_R (\boldsymbol{\xi}_{\text{SC}})^{\vee} - (\boldsymbol{\xi}_{\text{RB}})^{\vee} \mathbf{g}_R \\ &= \mathbf{g}_R \left[(\boldsymbol{\xi}_{\text{SC}})^{\vee} - (\mathbf{g}_R)^{-1} (\boldsymbol{\xi}_{\text{RB}})^{\vee} \mathbf{g}_R \right]. \end{aligned} \quad (25)$$

The relative kinematics of spacecraft (25) can be written in the same form as the absolute kinematics (5) and (16):

$$\dot{\mathbf{g}}_R = \mathbf{g}_R (\boldsymbol{\xi}_R)^{\vee}, \quad (26)$$

where the 6×1 vector $\boldsymbol{\xi}_R$ is the relative velocity of spacecraft:

$$\boldsymbol{\xi}_R = \begin{bmatrix} \boldsymbol{\Omega}_R \\ \mathbf{V}_R \end{bmatrix} \in \mathbb{R}^{6}. \quad (27)$$

Vectors $\boldsymbol{\Omega}_R$ and $\mathbf{V}_R$ are relative angular velocity and relative velocity of spacecraft with respect to the reference body, respectively, both expressed in the body-fixed frame of spacecraft S_{SC}.

According to (25) and (26), we can obtain the relative velocity of spacecraft $(\boldsymbol{\xi}_R)^{\vee}$ as follows:

$$(\boldsymbol{\xi}_R)^{\vee} = (\boldsymbol{\xi}_{\text{SC}})^{\vee} - (\mathbf{g}_R)^{-1} (\boldsymbol{\xi}_{\text{RB}})^{\vee} \mathbf{g}_R, \quad (28)$$

which can be written in a compact form

$$\boldsymbol{\xi}_R = \boldsymbol{\xi}_{\text{SC}} - \text{Ad}_{(\mathbf{g}_R)^{-1}} \boldsymbol{\xi}_{\text{RB}}. \quad (29)$$

The operator $\text{Ad}_{(\mathbf{g}_R)^{-1}}$, the adjoint action of $(\mathbf{g}_R)^{-1} \in \text{SE}(3)$ on $(\boldsymbol{\xi}_{\text{RB}})^{\vee} \in \text{se}(3)$, satisfies

$$\left(\text{Ad}_{(\mathbf{g}_R)^{-1}} \boldsymbol{\xi}_{\text{RB}} \right)^{\vee} = (\mathbf{g}_R)^{-1} (\boldsymbol{\xi}_{\text{RB}})^{\vee} \mathbf{g}_R, \quad (30)$$

and the matrix form of operator $\mathrm{Ad}_{(\mathbf{g}_R)^{-1}}$ can be given by

$$\mathrm{Ad}_{(\mathbf{g}_R)^{-1}} = \begin{bmatrix} \mathbf{A}_R^{-1} & \mathbf{0} \\ \left(-\mathbf{A}_R^{-1}\mathbf{r}_R\right)^{\times} \mathbf{A}_R^{-1} & \mathbf{A}_R^{-1} \end{bmatrix} \in \mathbb{R}^{6\times 6}. \quad (31)$$

As for the 6-DOF relative orbit-attitude dynamics of the spacecraft with respect to the reference body, the time derivative of relative velocity $\boldsymbol{\xi}_R$, that is, the relative acceleration of the spacecraft, can be obtained according to (29):

$$\dot{\boldsymbol{\xi}}_R = \dot{\boldsymbol{\xi}}_{\mathrm{SC}} - \frac{d}{dt}\left[\mathrm{Ad}_{(\mathbf{g}_R)^{-1}}\boldsymbol{\xi}_{\mathrm{RB}}\right]. \quad (32)$$

According to Lee et al. [15],

$$\frac{d}{dt}\left[\mathrm{Ad}_{(\mathbf{g}_R)^{-1}}\boldsymbol{\xi}_{\mathrm{RB}}\right] = -\mathrm{ad}_{\boldsymbol{\xi}_R}\mathrm{Ad}_{(\mathbf{g}_R)^{-1}}\boldsymbol{\xi}_{\mathrm{RB}} + \mathrm{Ad}_{(\mathbf{g}_R)^{-1}}\dot{\boldsymbol{\xi}}_{\mathrm{RB}}, \quad (33)$$

where the adjoint operator $\mathrm{ad}_{\boldsymbol{\xi}_R}$ on the Lie algebra se(3) is defined in (10).

Therefore, the relative acceleration of the spacecraft $\dot{\boldsymbol{\xi}}_R$ can be written as

$$\dot{\boldsymbol{\xi}}_R = \dot{\boldsymbol{\xi}}_{\mathrm{SC}} + \mathrm{ad}_{\boldsymbol{\xi}_R}\mathrm{Ad}_{(\mathbf{g}_R)^{-1}}\boldsymbol{\xi}_{\mathrm{RB}} - \mathrm{Ad}_{(\mathbf{g}_R)^{-1}}\dot{\boldsymbol{\xi}}_{\mathrm{RB}}. \quad (34)$$

According to (8) and (34), the 6-DOF relative dynamics of the spacecraft with respect to the reference body can be given by

$$\begin{aligned} \mathbb{I}_{\mathrm{SC}}\dot{\boldsymbol{\xi}}_R &= \mathbb{I}_{\mathrm{SC}}\dot{\boldsymbol{\xi}}_{\mathrm{SC}} + \mathbb{I}_{\mathrm{SC}}\left[\mathrm{ad}_{\boldsymbol{\xi}_R}\mathrm{Ad}_{(\mathbf{g}_R)^{-1}}\boldsymbol{\xi}_{\mathrm{RB}} - \mathrm{Ad}_{(\mathbf{g}_R)^{-1}}\dot{\boldsymbol{\xi}}_{\mathrm{RB}}\right] \\ &= \mathrm{ad}^*_{\boldsymbol{\xi}_{\mathrm{SC}}}\mathbb{I}_{\mathrm{SC}}\boldsymbol{\xi}_{\mathrm{SC}} + \boldsymbol{\tau}^n_{\mathrm{SC}} + \boldsymbol{\tau}^c_{\mathrm{SC}} \\ &\quad + \mathbb{I}_{\mathrm{SC}}\left\{\mathrm{ad}_{\boldsymbol{\xi}_R}\mathrm{Ad}_{(\mathbf{g}_R)^{-1}}\boldsymbol{\xi}_{\mathrm{RB}} - \mathrm{Ad}_{(\mathbf{g}_R)^{-1}}\dot{\boldsymbol{\xi}}_{\mathrm{RB}}\right\}. \end{aligned} \quad (35)$$

Equations (26) and (35) form complete equations for the 6-DOF relative motion.

3. Tracking Error Modeling and Control Law Design

In formation flying and proximity operations, the 6-DOF relative orbit-attitude motion of the spacecraft with respect to the reference body is usually required to track the desired trajectory. The goal of the controller is to reduce the tracking error. Here, introducing a desired trajectory for the 6-DOF orbit-attitude motion, that is, guidance on SE(3), is an improvement compared with the earlier works by Lee et al. [6, 15, 16] and Lee and Vukovich [14, 17], where the controllers aimed at the desired final state directly. In our approach with a reference trajectory, the true orbit-attitude trajectory to the final state can be specified, and the control saturation caused by the large initial tracking error can be avoided.

3.1. Desired Relative Trajectory and Tracking Error. The desired trajectory is denoted by $(\mathbf{g}_d(t), \boldsymbol{\xi}_d(t), \dot{\boldsymbol{\xi}}_d(t))$, where

$$\dot{\mathbf{g}}_d = \mathbf{g}_d\left(\boldsymbol{\xi}_d\right)^{\vee}. \quad (36)$$

The goal of the orbit-attitude motion controller is $(\mathbf{g}_R(t), \boldsymbol{\xi}_R(t)) \to (\mathbf{g}_d(t), \boldsymbol{\xi}_d(t))$.

The tracking error in configuration space of the relative motion, which is also on the special Euclidean group SE(3), is defined by

$$\mathbf{g}_e = \left(\mathbf{g}_d\right)^{-1}\mathbf{g}_R = \begin{bmatrix} \mathbf{A}_e & \mathbf{r}_e \\ \mathbf{0} & 1 \end{bmatrix} \in \mathrm{SE}(3). \quad (37)$$

By using the same method as in Section 2.3, the kinematics of tracking error of the spacecraft's relative motion is given by

$$\dot{\mathbf{g}}_e = \mathbf{g}_e\left(\boldsymbol{\xi}_e\right)^{\vee}, \quad (38)$$

where the 6×1 vector $\boldsymbol{\xi}_e$ is the tracking error in the relative velocity:

$$\boldsymbol{\xi}_e = \begin{bmatrix} \boldsymbol{\Omega}_e \\ \mathbf{V}_e \end{bmatrix} = \boldsymbol{\xi}_R - \mathrm{Ad}_{(\mathbf{g}_e)^{-1}}\boldsymbol{\xi}_d. \quad (39)$$

The time derivative of velocity tracking error $\boldsymbol{\xi}_e$ can be obtained:

$$\dot{\boldsymbol{\xi}}_e = \dot{\boldsymbol{\xi}}_R + \mathrm{ad}_{\boldsymbol{\xi}_e}\mathrm{Ad}_{(\mathbf{g}_e)^{-1}}\boldsymbol{\xi}_d - \mathrm{Ad}_{(\mathbf{g}_e)^{-1}}\dot{\boldsymbol{\xi}}_d. \quad (40)$$

According to (35), the dynamics of the 6-DOF orbit-attitude tracking error can be given by

$$\begin{aligned} \mathbb{I}_{\mathrm{SC}}\dot{\boldsymbol{\xi}}_e &= \mathrm{ad}^*_{\boldsymbol{\xi}_{\mathrm{SC}}}\mathbb{I}_{\mathrm{SC}}\boldsymbol{\xi}_{\mathrm{SC}} + \boldsymbol{\tau}^n_{\mathrm{SC}} + \boldsymbol{\tau}^c_{\mathrm{SC}} + \mathbb{I}_{\mathrm{SC}}\left\{\mathrm{ad}_{\boldsymbol{\xi}_e}\mathrm{Ad}_{(\mathbf{g}_e)^{-1}}\boldsymbol{\xi}_d\right. \\ &\quad \left. - \mathrm{Ad}_{(\mathbf{g}_e)^{-1}}\dot{\boldsymbol{\xi}}_d + \mathrm{ad}_{\boldsymbol{\xi}_R}\mathrm{Ad}_{(\mathbf{g}_R)^{-1}}\boldsymbol{\xi}_{\mathrm{RB}} - \mathrm{Ad}_{(\mathbf{g}_R)^{-1}}\dot{\boldsymbol{\xi}}_{\mathrm{RB}}\right\}, \end{aligned} \quad (41)$$

where the motion of reference body $(\mathbf{g}_{\mathrm{RB}}(t), \boldsymbol{\xi}_{\mathrm{RB}}(t), \dot{\boldsymbol{\xi}}_{\mathrm{RB}}(t))$ and the desired relative motion $(\mathbf{g}_d(t), \boldsymbol{\xi}_d(t), \dot{\boldsymbol{\xi}}_d(t))$ are known.

By using the following relations,

$$\begin{aligned} \mathbf{g}_R &= \mathbf{g}_d\mathbf{g}_e, \\ \mathbf{g}_{\mathrm{SC}} &= \mathbf{g}_{\mathrm{RB}}\mathbf{g}_R, \\ \boldsymbol{\xi}_R &= \boldsymbol{\xi}_e + \mathrm{Ad}_{(\mathbf{g}_e)^{-1}}\boldsymbol{\xi}_d, \\ \boldsymbol{\xi}_{\mathrm{SC}} &= \boldsymbol{\xi}_R + \mathrm{Ad}_{(\mathbf{g}_R)^{-1}}\boldsymbol{\xi}_{\mathrm{RB}}, \end{aligned} \quad (42)$$

(38) and (41) can form complete equations for the tracking error $(\mathbf{g}_e, \boldsymbol{\xi}_e)$ of the 6-DOF relative motion. The goal of the controller is to achieve $\mathbf{g}_e \to \mathbf{I}_{4\times 4}$ and $\boldsymbol{\xi}_e \to \mathbf{0}$, where $\mathbf{I}_{4\times 4}$ is the 4×4 identity matrix.

3.2. Tracking Error in Exponential Coordinates of SE(3). By using exponential coordinates of SE(3), the tracking error in configuration space of the orbit-attitude motion can be denoted by

$$\left(\boldsymbol{\eta}_e\right)^{\vee} = \log\left(\mathbf{g}_e\right) = \log\left(\begin{bmatrix} \mathbf{A}_e & \mathbf{r}_e \\ \mathbf{0} & 1 \end{bmatrix}\right) \in \mathrm{se}(3), \quad (43)$$

where the vector of exponential coordinates $\boldsymbol{\eta}_e$ is denoted by

$$\boldsymbol{\eta}_e = \begin{bmatrix} \boldsymbol{\Theta}_e \\ \mathbf{b}_e \end{bmatrix} \in \mathbb{R}^6, \quad (44)$$

and the corresponding element of the Lie algebra se(3) is denoted by

$$(\eta_e)^{\vee} = \begin{bmatrix} (\Theta_e)^{\times} & \mathbf{b}_e \\ \mathbf{0} & 0 \end{bmatrix} \in \mathrm{se}(3). \tag{45}$$

The specific expressions of the logarithm map log : SE(3) → se(3) are given by

$$\begin{aligned} (\Theta_e)^{\times} &= \frac{\phi}{2\sin\phi}\left[\mathbf{A}_e - \mathbf{A}_e^T\right], \\ \mathbf{b}_e &= \mathbf{A}^{-1}(\Theta_e)\,\mathbf{r}_e, \end{aligned} \tag{46}$$

where ϕ satisfies $\cos\phi = (1/2)[\mathrm{tr}(\mathbf{A}_e) - 1]$, $\phi < \pi$, and

$$\begin{aligned} \mathbf{A}^{-1}(\Theta_e) &= \mathbf{I}_{3\times3} - \frac{1}{2}(\Theta_e)^{\times} \\ &\quad + \left(\frac{1}{\phi^2} - \frac{1+\cos\phi}{2\phi\sin\phi}\right)(\Theta_e)^{\times}(\Theta_e)^{\times}. \end{aligned} \tag{47}$$

Actually, $\Theta_e = \phi\mathbf{n}$ is the principal rotation vector of attitude matrix $\mathbf{A}_e$; $\phi = |\Theta_e|$ and unit vector $\mathbf{n}$ are the principal rotation angle and axis, respectively. The logarithm map log : SE(3) → se(3) is bijective when the principal rotation angle of $\mathbf{A}_e$ is less than a π radian, that is, $\phi < \pi$, but it is not uniquely defined when ϕ is exactly a π radian [15].

The kinematics of tracking error of the spacecraft's relative motion (38) can be rewritten in terms of exponential coordinates η_e:

$$\dot{\eta}_e = \mathbf{G}(\eta_e)\,\xi_e. \tag{48}$$

The kinematical matrix $\mathbf{G}(\eta_e)$ is given by [28]

$$\begin{aligned} \mathbf{G}(\eta_e) &= \mathbf{I}_{6\times6} + \frac{1}{2}\mathrm{ad}_{\eta_e} + A(\phi)\left(\mathrm{ad}_{\eta_e}\right)^2 \\ &\quad + B(\phi)\left(\mathrm{ad}_{\eta_e}\right)^4, \end{aligned} \tag{49}$$

where $\mathbf{I}_{6\times6}$ is the 6×6 identity matrix:

$$\mathrm{ad}_{\eta_e} = \begin{bmatrix} (\Theta_e)^{\times} & \mathbf{0} \\ (\mathbf{b}_e)^{\times} & (\Theta_e)^{\times} \end{bmatrix}, \tag{50}$$

$$A(\phi) = \frac{2}{\phi^2}\left[1 - \alpha(\phi)\right] + \frac{1}{2\phi^2}\left[\alpha(\phi) - \beta(\phi)\right], \tag{51}$$

$$B(\phi) = \frac{1}{\phi^4}\left[1 - \alpha(\phi)\right] + \frac{1}{2\phi^4}\left[\alpha(\phi) - \beta(\phi)\right], \tag{52}$$

$$\alpha(\phi) = \frac{\phi}{2}\cot\left(\frac{\phi}{2}\right), \tag{53}$$

$$\beta(\phi) = \left[\frac{(\phi/2)}{\sin(\phi/2)}\right]^2. \tag{54}$$

Equations (48) and (41) can form complete equations of motion for the tracking error of spacecraft's 6-DOF relative motion in terms of exponential coordinates (η_e, ξ_e). The goal of the controller is to achieve $\eta_e \to \mathbf{0}$ and $\xi_e \to \mathbf{0}$.

3.3. Dynamics of Tracking Error as a Second-Order System. In equations of motion (48) and (41), exponential coordinates of SE(3) are used only on the kinematical level. That is, (48) has only given the relation between the first-order time derivative of exponential coordinates and the velocity. In the following, we will extend exponential coordinates to the dynamical level, by giving the relation between the second-order time derivative of exponential coordinates and the acceleration. This extension is an important improvement compared with earlier works by Lee et al. [6, 15, 16].

By taking time derivative of both sides of the kinematics of tracking error (48), we can have the second-order derivative of exponential coordinates:

$$\ddot{\eta}_e = \mathbf{G}(\eta_e)\,\dot{\xi}_e + \dot{\mathbf{G}}(\eta_e)\,\xi_e, \tag{55}$$

where, according to (49), $\dot{\mathbf{G}}(\eta_e)$ can be obtained as

$$\begin{aligned} \dot{\mathbf{G}}(\eta_e) &= \frac{1}{2}\dot{\mathrm{ad}}_{\eta_e} + \dot{A}(\phi)\left(\mathrm{ad}_{\eta_e}\right)^2 + A(\phi)\left(\dot{\mathrm{ad}}_{\eta_e}\mathrm{ad}_{\eta_e}\right. \\ &\quad \left. + \mathrm{ad}_{\eta_e}\dot{\mathrm{ad}}_{\eta_e}\right) + \dot{B}(\phi)\left(\mathrm{ad}_{\eta_e}\right)^4 + B(\phi)\left[\dot{\mathrm{ad}}_{\eta_e}\left(\mathrm{ad}_{\eta_e}\right)^3\right. \\ &\quad + \mathrm{ad}_{\eta_e}\dot{\mathrm{ad}}_{\eta_e}\left(\mathrm{ad}_{\eta_e}\right)^2 + \left(\mathrm{ad}_{\eta_e}\right)^2\dot{\mathrm{ad}}_{\eta_e}\mathrm{ad}_{\eta_e} \\ &\quad \left. + \left(\mathrm{ad}_{\eta_e}\right)^3\dot{\mathrm{ad}}_{\eta_e}\right]. \end{aligned} \tag{56}$$

According to (50), $\dot{\mathrm{ad}}_{\eta_e}$ in (56) is given by

$$\dot{\mathrm{ad}}_{\eta_e} = \begin{bmatrix} (\dot{\Theta}_e)^{\times} & \mathbf{0} \\ (\dot{\mathbf{b}}_e)^{\times} & (\dot{\Theta}_e)^{\times} \end{bmatrix}, \tag{57}$$

where $\dot{\Theta}_e$ and $\dot{\mathbf{b}}_e$ are given by (48). By using (51) and (52), $\dot{A}(\phi)$ and $\dot{B}(\phi)$ in (56) can be derived as

$$\begin{aligned} \dot{A}(\phi) &= \frac{dA}{d\phi}\frac{d\phi}{dt} \\ &= \frac{1}{\phi^3}\left[\frac{3}{2}\alpha(\phi) + \frac{3}{2}\beta(\phi) + \alpha(\phi)\beta(\phi) - 4\right]\frac{\Theta_e\cdot\dot{\Theta}_e}{\phi}, \\ \dot{B}(\phi) &= \frac{dB}{d\phi}\frac{d\phi}{dt} \\ &= \frac{1}{\phi^5}\left[\frac{3}{2}\alpha(\phi) + \frac{3}{2}\beta(\phi) + \alpha(\phi)\beta(\phi) - 4\right]\frac{\Theta_e\cdot\dot{\Theta}_e}{\phi}. \end{aligned} \tag{58}$$

Then, (55) and (41) can form a second-order equation of tracking error of the 6-DOF relative motion in terms of exponential coordinates (η_e, ξ_e):

$$\ddot{\eta}_e = \mathbf{G}(\eta_e)\,\dot{\xi}_e + \dot{\mathbf{G}}(\eta_e)\,\xi_e, \tag{59}$$

where $\dot{\xi}_e$, $\mathbf{G}(\eta_e)$, and $\dot{\mathbf{G}}(\eta_e)$ are given by (41), (49), and (56), respectively.

3.4. Integrated Tracking Control Law with Almost Global Convergence. Based on the dynamics of tracking error of the 6-DOF relative motion, appropriate control theories can be applied, and integrated tracking control laws can be designed. To show the application potential of our integrated 6-DOF dynamical modeling (59) in terms of exponential coordinates, a simple integrated tracking control law will be designed for a demonstration purpose.

Since the exponential coordinates are not uniquely defined when the rotation angle is exactly a π radian, the controllers can work with almost all initial tracking errors, except those with a rotation angle of a π radian. Because the set of such initial states is an embedded lower-dimensional subspace of the state space, the tracking control law designed in terms of exponential coordinates can achieve almost global convergence and then is applicable to space missions and operations in practice.

By using a simple feedback scheme, the continuous tracking control law $\boldsymbol{\tau}_{\text{SC}}^{c}$ for the integrated orbit-attitude motion of spacecraft can be designed as

$$\begin{aligned}\boldsymbol{\tau}_{\text{SC}}^{c} &= -\text{ad}_{\boldsymbol{\xi}_{\text{SC}}}^{*}\mathbb{I}_{\text{SC}}\boldsymbol{\xi}_{\text{SC}} - \boldsymbol{\tau}_{\text{SC}}^{n} - \mathbb{I}_{\text{SC}}\mathbf{G}^{-1}(\boldsymbol{\eta}_e)\left[\dot{\mathbf{G}}(\boldsymbol{\eta}_e)\boldsymbol{\xi}_e\right.\\ &\quad \left.+ \mathbf{K}_d\dot{\boldsymbol{\eta}}_e + \mathbf{K}_p\boldsymbol{\eta}_e\right] - \mathbb{I}_{\text{SC}}\left\{\text{ad}_{\boldsymbol{\xi}_e}\text{Ad}_{(\mathbf{g}_e)^{-1}}\boldsymbol{\xi}_d - \text{Ad}_{(\mathbf{g}_e)^{-1}}\dot{\boldsymbol{\xi}}_d\right.\\ &\quad \left.+ \text{ad}_{\boldsymbol{\xi}_R}\text{Ad}_{(\mathbf{g}_R)^{-1}}\boldsymbol{\xi}_{\text{RB}} - \text{Ad}_{(\mathbf{g}_R)^{-1}}\dot{\boldsymbol{\xi}}_{\text{RB}}\right\},\end{aligned} \tag{60}$$

where the control gains $\mathbf{K}_d$ and $\mathbf{K}_p$ can be tuned to achieve desired stability and dynamic performance of the closed-loop system.

The dynamics of closed-loop system describing tracking error of the 6-DOF relative motion is obtained by substituting the control law (60) into the second-order equation of tracking error (59):

$$\ddot{\boldsymbol{\eta}}_e + \mathbf{K}_d\dot{\boldsymbol{\eta}}_e + \mathbf{K}_p\boldsymbol{\eta}_e = \mathbf{0}, \tag{61}$$

which is a homogeneous linear differential equation with constant coefficients. By choosing appropriate control gains $\mathbf{K}_d$ and $\mathbf{K}_p$ in the control law (60), the tracking error in configuration space $\boldsymbol{\eta}_e$ and its derivative $\dot{\boldsymbol{\eta}}_e$ will both converge to $\mathbf{0}$ asymptotically, and, according to (48), the tracking error in velocity $\boldsymbol{\xi}_e$ will also converge to $\mathbf{0}$ asymptotically.

Therefore, with the feedback control law (60), the spacecraft's 6-DOF relative motion will converge to the desired trajectory asymptotically from almost all initial tracking errors, except those with a rotation angle of a π radian. Thus, the control law (60) is an integrated asymptotic tracking control law with almost global convergence.

4. Body-Fixed Orbit-Attitude Hovering over an Asteroid

In this section, we will apply integrated dynamical modeling and tracking control law for the spacecraft's 6-DOF motion obtained above to the body-fixed orbit-attitude hovering over an asteroid.

The body-fixed orbit-attitude hovering means that both the position and attitude of the spacecraft are kept to be stationary in the asteroid body-fixed frame [29]. The orbit-attitude hovering was modeled also in an integrated manner by Wang and Xu [29] by using the theory of noncanonical Hamiltonian system within the framework of gravitationally coupled orbit-attitude dynamics, in which the spacecraft was considered as a rigid body. Wang and Xu [29] have proposed a noncanonical Hamiltonian structure-based feedback control law. Here, we will use the asymptotic tracking control law designed in Section 3.4 to achieve the orbit-attitude hovering and make comparisons with the controller in Wang and Xu [29].

4.1. System Description. It is assumed that the center of mass of the asteroid, that is, the reference body in earlier parts of the paper, is stationary in the inertial space, and the asteroid is rotating with a constant angular velocity ω_T around its maximum-moment principal axis, which is assumed to be $\mathbf{w}$ axis without loss of generality. That is to say, the external torque and force acting on the asteroid, $\boldsymbol{\tau}_{\text{RB}}$, are zeros. Thus, the 6-DOF dynamics of the reference body can be simplified as

$$\begin{aligned}\mathbf{g}_{\text{RB0}} &= \mathbf{I}_{4\times 4},\\ \boldsymbol{\xi}_{\text{RB}} &\equiv [0, 0, \omega_T, 0, 0, 0]^T,\\ \dot{\boldsymbol{\xi}}_{\text{RB}} &= \mathbf{0}.\end{aligned} \tag{62}$$

To make comparisons with the controller in Wang and Xu [29], a similar mission scenario is chosen in the following. The gravity field of the asteroid is approximated by a second degree and order-gravity field with harmonics C_{20} and C_{22}. The parameters of the asteroid are chosen to be the same as in Wang and Xu [29]:

$$\begin{aligned}\mu &= 94\,\text{m}^3/\text{s}^2,\\ C_{20} &= -0.1,\\ C_{22} &= 0.04,\\ a_e &= 400\,\text{m},\\ \omega_T &= 2.9089\times 10^{-4}\,\text{s}^{-1},\end{aligned} \tag{63}$$

where $\mu = Gm_{\text{RB}}$, G is the gravitational constant, and a_e is the mean equatorial radius of the asteroid.

Since the spacecraft is in the close proximity of the asteroid and is assumed to have a low area-to-mass ratio, perturbations of solar gravity and SRP are negligible. Thus, only the gravitational torque and force by asteroid are considered in the environmental torque and force $\boldsymbol{\tau}_{\text{SC}}^{n}$. That is to say, $\boldsymbol{\tau}_{\text{SC}}^{n}$ depends only on the relative configuration of spacecraft with respect to the asteroid:

$$\boldsymbol{\tau}_{\text{SC}}^{n} = \boldsymbol{\tau}_{\text{SC}}^{n}(\mathbf{g}_R). \tag{64}$$

The explicit formulation of $\boldsymbol{\tau}_{\text{SC}}^{n}(\mathbf{g}_R)$ is given by [19]

$$\begin{aligned}\mathbf{T}_{\text{SC}}^{n} &= \frac{3\mu}{R_R^5}\mathbf{R}_R\times\mathbf{I}_{\text{SC}}\mathbf{R}_R,\\ \mathbf{F}_{\text{SC}}^{n} &= -\frac{\mu m_{\text{SC}}}{R_R^2}\overline{\mathbf{R}}_R + \frac{3\mu}{2R_R^4}\left\{\left[5\overline{\mathbf{R}}_R^T\mathbf{I}_{\text{SC}}\overline{\mathbf{R}}_R - \text{tr}(\mathbf{I}_{\text{SC}})\right.\right.\end{aligned}$$

$$+ \tau_0 m_{SC}\left(1 - 5\left(\gamma_R \cdot \overline{\mathbf{R}}_R\right)^2\right)$$
$$- 10\tau_2 m_{SC}\left(\left(\alpha_R \cdot \overline{\mathbf{R}}_R\right)^2 - \left(\beta_R \cdot \overline{\mathbf{R}}_R\right)^2\right)\Big]\overline{\mathbf{R}}_R$$
$$- 2\mathbf{I}_{SC}\overline{\mathbf{R}}_R + 2\tau_0 m_{SC}\left(\gamma_R \cdot \overline{\mathbf{R}}_R\right)\gamma_R$$
$$+ 4\tau_2 m_{SC}\left(\left(\alpha_R \cdot \overline{\mathbf{R}}_R\right)\alpha_R - \left(\beta_R \cdot \overline{\mathbf{R}}_R\right)\beta_R\right)\Big\}, \quad (65)$$

where $\mathbf{R}_R = \mathbf{A}_R^{-1}\mathbf{r}_R$ is the relative position of the spacecraft with respect to the asteroid expressed in the body-fixed frame of spacecraft S_{SC}, $\overline{\mathbf{R}}_R$ is the unit vector along $\mathbf{R}_R$, $\tau_0 = a_e^2 C_{20}$, and $\tau_2 = a_e^2 C_{22}$. In the environmental force $\mathbf{F}_{SC}^n$, the perturbation caused by the gravitational orbit-attitude coupling of the spacecraft has been included [30].

The parameters of spacecraft are chosen as

$$m_{SC} = 1 \times 10^3 \text{ kg},$$
$$\mathbf{I}_{SC} = \begin{bmatrix} 2 & 0 & 0 \\ 0 & 1 & 0 \\ 0 & 0 & 1.6 \end{bmatrix} \times 10^3 \text{ kg·m}^2. \quad (66)$$

The hovering position-attitude $\mathbf{g}_{RH}$, that is, the relative configuration of spacecraft at hovering, is chosen to be the same as in Wang and Xu [29]:

$$\mathbf{g}_{RH} = \begin{bmatrix} \mathbf{A}_{RH} & \mathbf{r}_{RH} \\ \mathbf{0} & 1 \end{bmatrix} \in SE(3), \quad (67)$$

where

$$\mathbf{A}_{RH} = \left[\alpha_{RH}, \beta_{RH}, \gamma_{RH}\right]^T,$$
$$\mathbf{r}_{RH} = \mathbf{A}_{RH}\mathbf{R}_{RH},$$
$$\alpha_{RH} = [0.9659, 0, -0.2588]^T,$$
$$\beta_{RH} = [0.067, 0.9659, 0.25]^T, \quad (68)$$
$$\gamma_{RH} = [0.25, -0.2588, 0.933]^T,$$
$$\mathbf{R}_{RH} = 500\,[0.9798, 0, 0.2]^T \text{ m}.$$

In the simulation, the desired trajectory $(\mathbf{g}_d(t), \boldsymbol{\xi}_d(t), \dot{\boldsymbol{\xi}}_d(t))$ is simply chosen as the desired hovering position-attitude:

$$\mathbf{g}_d(t) \equiv \mathbf{g}_{RH},$$
$$\boldsymbol{\xi}_d(t) \equiv \mathbf{0}, \quad (69)$$
$$\dot{\boldsymbol{\xi}}_d(t) \equiv \mathbf{0}.$$

A better reference trajectory connecting the initial position-attitude and the hovering position-attitude, which is easier to track, can be designed in the future through studies on the guidance on SE(3), which is not the main concern of this paper.

The initial relative configuration of spacecraft with respect to the asteroid $\mathbf{g}_{R0}$ is chosen as

$$\mathbf{g}_{R0} = \begin{bmatrix} \mathbf{A}_{R0} & \mathbf{r}_{R0} \\ \mathbf{0} & 1 \end{bmatrix} \in SE(3), \quad (70)$$

where

$$\mathbf{A}_{R0} = \mathbf{A}_{RH}\mathbf{L},$$
$$\mathbf{L} = \begin{bmatrix} \cos\left(\frac{\pi}{9}\right) & -\sin\left(\frac{\pi}{9}\right) & 0 \\ \sin\left(\frac{\pi}{9}\right) & \cos\left(\frac{\pi}{9}\right) & 0 \\ 0 & 0 & 1 \end{bmatrix} \begin{bmatrix} \cos\left(-\frac{\pi}{20}\right) & 0 & \sin\left(-\frac{\pi}{20}\right) \\ 0 & 1 & 0 \\ -\sin\left(-\frac{\pi}{20}\right) & 0 & \cos\left(-\frac{\pi}{20}\right) \end{bmatrix} \begin{bmatrix} 1 & 0 & 0 \\ 0 & \cos\left(\frac{\pi}{18}\right) & -\sin\left(\frac{\pi}{18}\right) \\ 0 & \sin\left(\frac{\pi}{18}\right) & \cos\left(\frac{\pi}{18}\right) \end{bmatrix}, \quad (71)$$
$$\mathbf{r}_{R0} = \mathbf{r}_{RH} + [200, 100, 250]^T \text{ m}.$$

The initial relative velocity of spacecraft with respect to the asteroid $\boldsymbol{\xi}_{R0}$ is chosen as

$$\boldsymbol{\xi}_{R0} = \begin{bmatrix} \boldsymbol{\Omega}_{R0} \\ \mathbf{V}_{R0} \end{bmatrix} = \left[-0.06\,\text{s}^{-1}, 0.05\,\text{s}^{-1}, 0.09\,\text{s}^{-1}, -1\,\text{m/s}, 2\,\text{m/s}, 1.5\,\text{m/s}\right]^T. \quad (72)$$

In simulations, absolute motions of the spacecraft and asteroid in the inertial space will be simulated separately. The relative motion of spacecraft and its tracking error that appear in the control law will be calculated by using absolute motions and the desired trajectory. With this approach, not only the control law but also the dynamical modeling method can be verified.

To initiate the simulation of spacecraft's absolute motion, the initial configuration of spacecraft in the inertial space can be calculated by

$$\mathbf{g}_{SC0} = \mathbf{g}_{RB0}\mathbf{g}_{R0}, \quad (73)$$

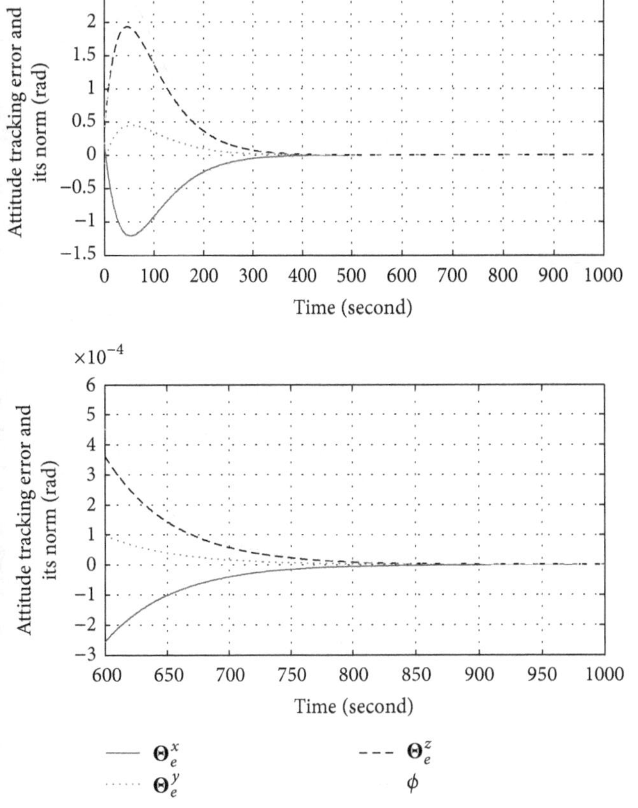

FIGURE 2: Exponential coordinates of attitude tracking error and its norm.

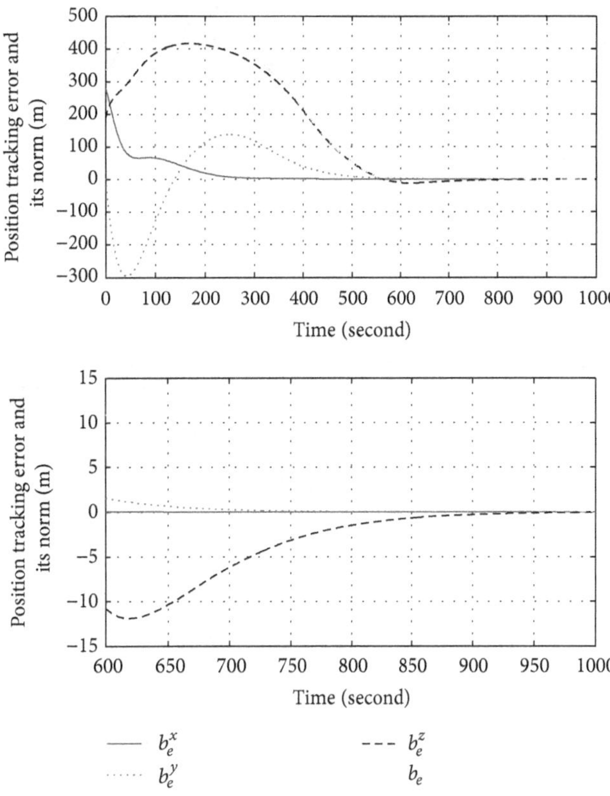

FIGURE 3: Exponential coordinates of position tracking error and its norm.

and the initial velocity of spacecraft in the inertial space can be calculated by

$$\xi_{SC0} = \xi_{R0} + \mathrm{Ad}_{(\mathbf{g}_{R0})^{-1}} \xi_{RB0}. \quad (74)$$

The control gains $\mathbf{K}_d$ and $\mathbf{K}_p$ in the control law (60) are chosen as

$$\mathbf{K}_d = 0.04\mathbf{I}_{6\times 6},$$
$$\mathbf{K}_p = 4 \times 10^{-4}\mathbf{I}_{6\times 6}. \quad (75)$$

The maximum control torque and force that can be provided by the spacecraft's control system are set to be 10 N·m and 10 N for each axis, respectively.

4.2. Numerical Simulation Results. The exponential coordinates of the position-attitude tracking error $\boldsymbol{\eta}_e = [\boldsymbol{\Theta}_e^T, \mathbf{b}_e^T]^T$ and their norms are shown in Figures 2 and 3, respectively, and the tracking errors of the spacecraft's angular velocity and velocity $\boldsymbol{\xi}_e = [\boldsymbol{\Omega}_e^T, \mathbf{V}_e^T]^T$ are shown in Figures 4 and 5, respectively. It can be seen that the motion of spacecraft converges to the hovering position-attitude and keeps motionless relative to the asteroid after about 900 seconds. The attitude motion of spacecraft has an error of about 0.026 degrees and 5 $\times 10^{-4}$ degree/s within 600 seconds and an error of about 1.5 $\times 10^{-5}$ degrees and 3×10^{-7} degree/s within 1000 seconds, while the orbital motion has an error of about 10 m and 0.15 m/s

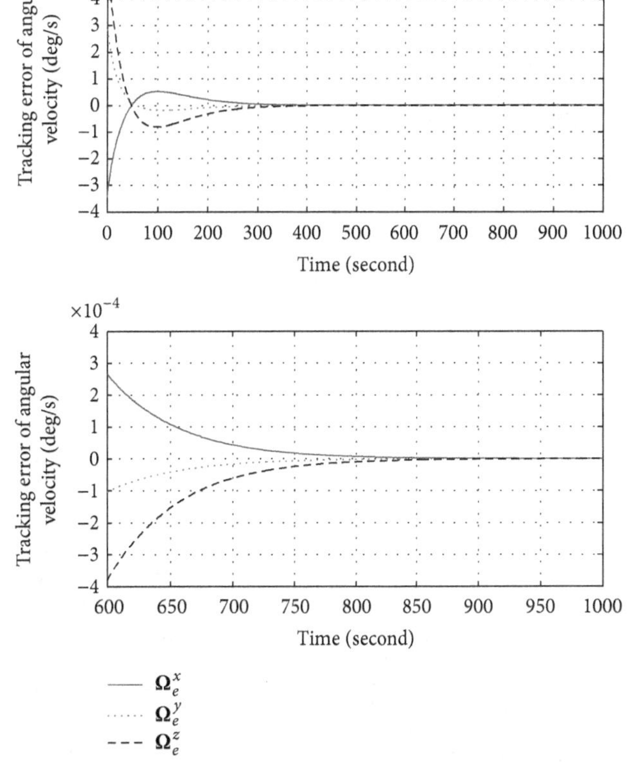

FIGURE 4: Tracking error of the angular velocity of spacecraft.

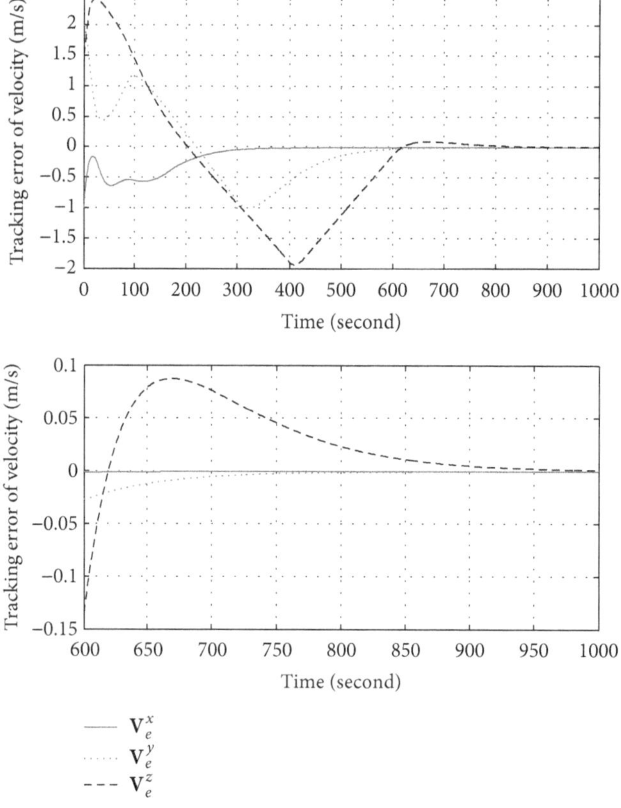

FIGURE 5: Tracking error of the velocity of spacecraft.

within 600 seconds and an error of about 0.05 m and 1×10^{-3} m/s within 1000 seconds. The attitude motion has a faster converge than the orbital motion.

The trajectory of spacecraft in the body-fixed frame of asteroid in Figure 6 shows that it does not approach the hovering position directly due to the initial velocity error. In the figure, the hovering position is denoted by the star ($*$), the initial position of spacecraft is denoted by the circle (○), and the final position of spacecraft is denoted by the pentagram (☆). Notice that the star ($*$) is overlapped with the pentagram (☆) and cannot be distinguished.

The components of control torque and force $\mathbf{T}_{\text{SC}}^c$ and $\mathbf{F}_{\text{SC}}^c$, both expressed in the spacecraft body-fixed frame S_{SC}, are shown in Figures 7 and 8, respectively. The control torque is within the spacecraft's control capacity during the whole trajectory, whereas the thrust saturation of the control force occurs on all three axes before 600 seconds. Although thrust saturation occurs for the control force, the controller is still effective.

As the spacecraft is approaching to the hovering configuration, the control torque and force are converging to values that are needed to balance the gravitational and centrifugal torque and force at hovering position-attitude. Since the gravity gradient torque $\mathbf{T}_{\text{SC}}^n$ is quite small, the control torque $\mathbf{T}_{\text{SC}}^c$ converges to nearly zero.

Compared with the hovering controller in Wang and Xu [29], the controller proposed in this paper has not utilized the Hamiltonian structure of the system. On one hand, the controller in this paper (60) is much more complicated than that in Wang and Xu [29], which is consisted of two potential shapings and one energy dissipation. However, on the other hand, the closed-loop system in this paper is a linear system and converges much faster than that in Wang and Xu [29], which is a noncanonical Hamiltonian system with energy dissipation, oscillating around the hovering position-attitude for a long duration before the final convergence.

5. Conclusions

The integrated 6-DOF orbit-attitude dynamical modeling and controller design for the relative motion of spacecraft with respect to a reference body have been studied in the framework of geometric mechanics in the present paper. The configuration and velocity of the relative motion have been represented by the Lie group SE(3) and also its exponential coordinates. By using exponential coordinates of SE(3) on the dynamical level, the dynamics of tracking error have been formulated as a second-order system with a quite compact form. A simple asymptotic tracking control law with almost global convergence was designed for a demonstration purpose. Both the integrated dynamical modeling and control law were verified in an asteroid hovering mission scenario. To verify the dynamical modeling method, individual motions of the spacecraft and asteroid in the inertial space have been simulated, rather than the equations of motion describing the relative dynamics.

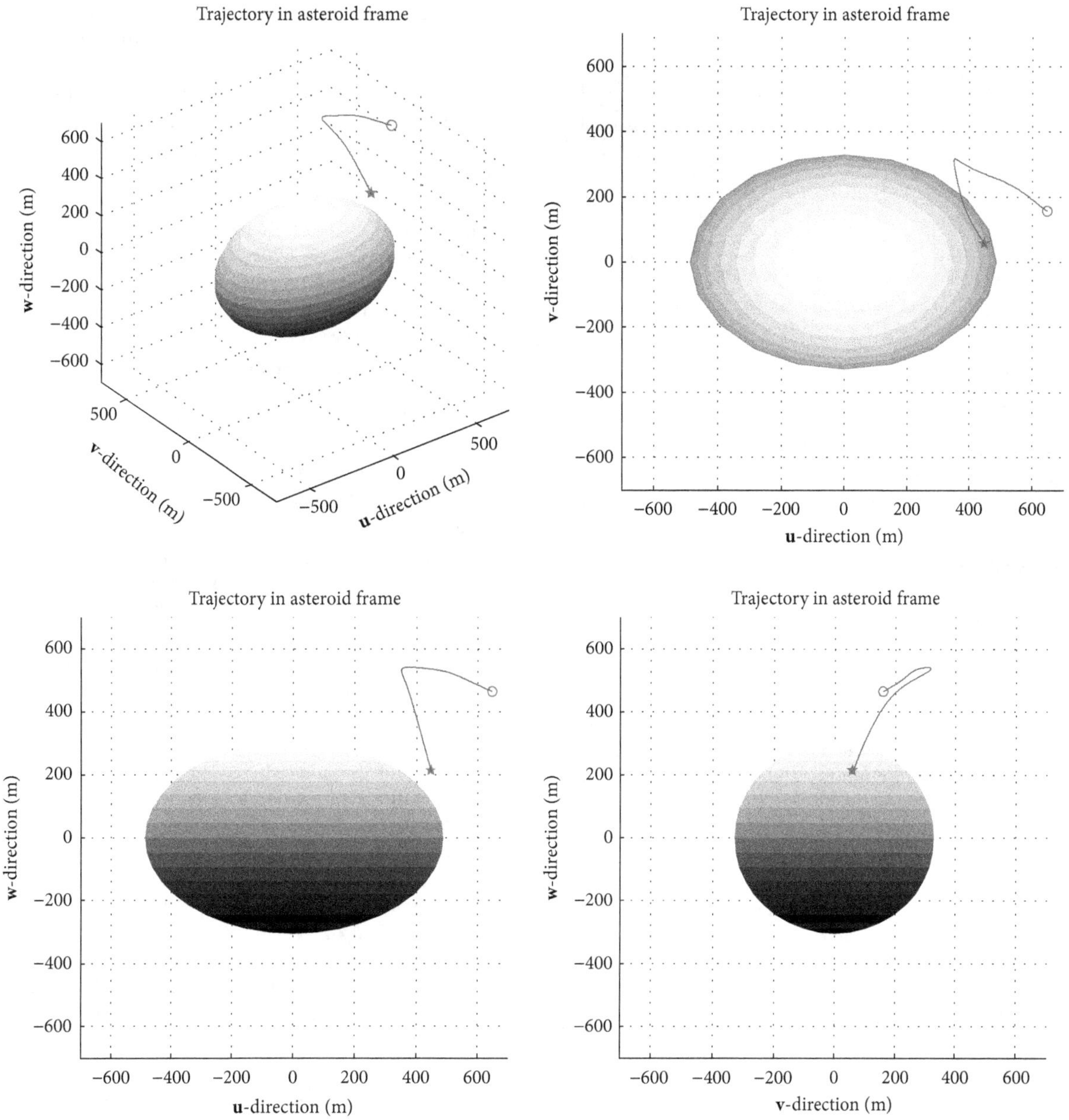

FIGURE 6: Trajectory of the spacecraft in the asteroid body-fixed frame.

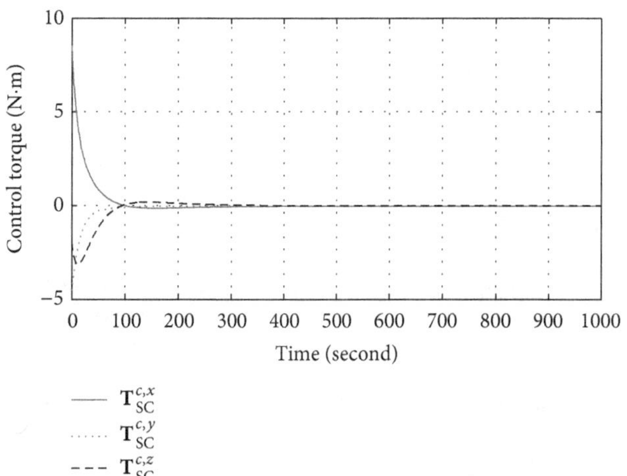

FIGURE 7: Components of control torque on the three axes of the spacecraft.

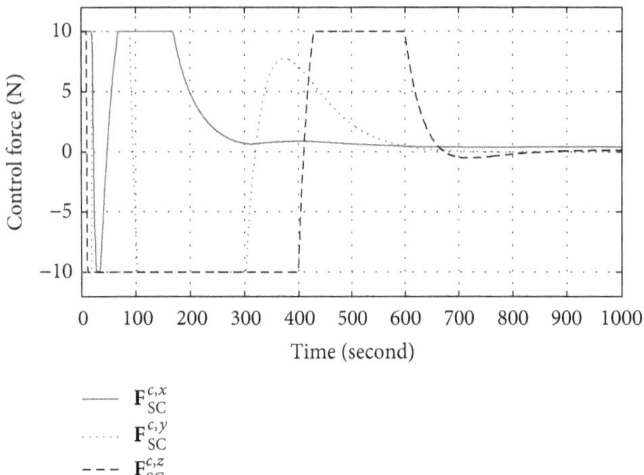

FIGURE 8: Components of control force on the three axes of the spacecraft.

Conflicts of Interest

The authors declare that they have no conflicts of interest.

Acknowledgments

This work has been supported by the National Natural Science Foundation of China under Grants 11432001 and 11602009 and the Fundamental Research Funds for the Central Universities.

References

[1] S. Segal and P. Gurfil, "Effect of kinematic rotation-translation coupling on relative spacecraft translational dynamics," *Journal of Guidance, Control, and Dynamics*, vol. 32, no. 3, pp. 1045–1050, 2009.

[2] J. Wang and Z. Sun, "6-DOF robust adaptive terminal sliding mode control for spacecraft formation flying," *Acta Astronautica*, vol. 73, pp. 76–87, 2012.

[3] J. Wu, K. Liu, and D. Han, "Adaptive sliding mode control for six-DOF relative motion of spacecraft with input constraint," *Acta Astronautica*, vol. 87, pp. 64–76, 2013.

[4] M. Massari and M. Zamaro, "Application of SDRE technique to orbital and attitude control of spacecraft formation flying," *Acta Astronautica*, vol. 94, no. 1, pp. 409–420, 2014.

[5] Q. Lan, J. Yang, S. Li, and H. Sun, "Finite-time control for 6DOF spacecraft formation flying systems," *Journal of Aerospace Engineering*, vol. 28, no. 5, Article ID 04014137, 2015.

[6] D. Lee, A. K. Sanyal, and E. A. Butcher, "Asymptotic tracking control for spacecraft formation flying with decentralized collision avoidance," *Journal of Guidance, Control, and Dynamics*, vol. 38, no. 4, pp. 587–600, 2015.

[7] D. Lee, H. Bang, E. A. Butcher, and A. K. Sanyal, "Kinematically coupled relative spacecraft motion control using the state-dependent riccati equation method," *Journal of Aerospace Engineering*, vol. 28, no. 4, Article ID 4014099, 2015.

[8] A. Sanyal, L. Holguin, and S. P. Viswanathan, "Guidance and control for spacecraft autonomous chasing and close proximity maneuvers," in *Proceedings of the 7th IFAC Symposium on Robust Control Design (ROCOND '12)*, pp. 753–758, 2012.

[9] D. Lee, "Spacecraft coupled tracking maneuver using sliding mode control with input saturation," *Journal of Aerospace Engineering*, vol. 28, no. 5, Article ID 04014136, 2015.

[10] H. Gui and G. Vukovich, "Dual-quaternion-based adaptive motion tracking of spacecraft with reduced control effort," *Nonlinear Dynamics*, vol. 83, no. 1-2, pp. 1–18, 2015.

[11] N. Filipe and P. Tsiotras, "Adaptive position and attitude-tracking controller for satellite proximity operations using dual quaternions," *Journal of Guidance, Control, and Dynamics*, vol. 38, no. 4, pp. 566–577, 2015.

[12] L. Sun and W. Huo, "Robust adaptive control of spacecraft proximity maneuvers under dynamic coupling and uncertainty," *Advances in Space Research*, vol. 56, no. 10, pp. 2206–2217, 2015.

[13] L. Sun and W. Huo, "Robust adaptive relative position tracking and attitude synchronization for spacecraft rendezvous," *Aerospace Science and Technology*, vol. 41, pp. 28–35, 2015.

[14] D. Lee and G. Vukovich, "Robust adaptive terminal sliding mode control on SE(3) for autonomous spacecraft rendezvous and docking," *Nonlinear Dynamics*, vol. 83, no. 4, pp. 2263–2279, 2016.

[15] D. Lee, A. K. Sanyal, E. A. Butcher, and D. J. Scheeres, "Almost global asymptotic tracking control for spacecraft body-fixed hovering over an asteroid," *Aerospace Science and Technology*, vol. 38, pp. 105–115, 2014.

[16] D. Lee, A. Sanyal, E. Butcher, and D. Scheeres, "Finite-time control for spacecraft body-fixed hovering over an asteroid," *IEEE Transactions on Aerospace and Electronic Systems*, vol. 51, no. 1, pp. 506–520, 2015.

[17] D. Lee and G. Vukovich, "Adaptive sliding mode control for spacecraft body-fixed hovering in the proximity of an asteroid," *Aerospace Science and Technology*, vol. 46, pp. 471–483, 2015.

[18] Y. Wang and S. Xu, "Gravitational orbit-rotation coupling of a rigid satellite around a spheroid planet," *Journal of Aerospace Engineering*, vol. 27, no. 1, pp. 140–150, 2014.

[19] Y. Wang and S. Xu, "Relative equilibria of full dynamics of a rigid body with gravitational orbit-attitude coupling in a uniformly rotating second degree and order gravity field," *Astrophysics and Space Science*, vol. 354, no. 2, pp. 339–353, 2014.

[20] Y. Wang, S. Xu, and L. Tang, "On the existence of the relative equilibria of a rigid body in the J2 problem," *Astrophysics and Space Science*, vol. 353, no. 2, pp. 425–440, 2014.

[21] Y. Wang, S. Xu, and M. Zhu, "Stability of relative equilibria of the full spacecraft dynamics around an asteroid with orbit-attitude coupling," *Advances in Space Research*, vol. 53, no. 7, pp. 1092–1107, 2014.

[22] C. Früh, T. M. Kelecy, and M. K. Jah, "Coupled orbit-attitude dynamics of high area-to-mass ratio (HAMR) objects: influence of solar radiation pressure, earth's shadow and the visibility in light curves," *Celestial Mechanics and Dynamical Astronomy* vol. 117, no. 4, pp. 385–404, 2013.

[23] C. Früh and M. K. Jah, "Coupled orbit-attitude motion of high area-to-mass ratio (HAMR) objects including efficient self-shadowing," *Acta Astronautica*, vol. 95, no. 1, pp. 227–241, 2014.

[24] J. Zhang, S. Zhao, and Y. Yang, "Characteristic analysis for elliptical orbit hovering based on relative dynamics," *IEEE Transactions on Aerospace and Electronic Systems*, vol. 49, no. 4, pp. 2742–2750, 2013.

[25] J. Zhang, S. Zhao, and Y. Zhang, "Autonomous guidance for rendezvous phasing based on special-point-based maneuvers," *Journal of Guidance, Control, and Dynamics*, vol. 38, no. 4, pp. 578–586, 2015.

[26] H. Gui and G. Vukovich, "Finite-time output-feedback position and attitude tracking of a rigid body," *Automatica*, vol. 74, pp. 270–278, 2016.

[27] A. Sanyal, N. Nordkvist, and M. Chyba, "An almost global tracking control scheme for maneuverable autonomous vehicles and its discretization," *IEEE Transactions on Automatic Control*, vol. 56, no. 2, pp. 457–462, 2011.

[28] F. Bullo and R. M. Murray, "Proportional derivative (PD) control on the Euclidean group," in *Proceedings of the European Control Conference*, vol. 2, pp. 1091–1097, European Control Association, Zurich, Switzerland, 1995.

[29] Y. Wang and S. Xu, "Body-fixed orbit-attitude hovering control over an asteroid using non-canonical Hamiltonian structure," *Acta Astronautica*, vol. 117, pp. 450–468, 2015.

[30] Y. Wang and S. Xu, "Orbital dynamics and equilibrium points around an asteroid with gravitational orbit-attitude coupling perturbation," *Celestial Mechanics & Dynamical Astronomy*, vol. 125, no. 3, pp. 265–285, 2016.

16

Autonomous Orbit Determination for Lagrangian Navigation Satellite Based on Neural Network Based State Observer

Youtao Gao,[1] Tanran Zhao,[1] Bingyu Jin,[1] Junkang Chen,[1] and Bo Xu[2]

[1] *College of Astronautics, Nanjing University of Aeronautics and Astronautics, Nanjing, China*
[2] *College of Astronomy and Space Science, Nanjing University, Nanjing, China*

Correspondence should be addressed to Youtao Gao; ytgao@nuaa.edu.cn

Academic Editor: Paolo Tortora

In order to improve the accuracy of the dynamical model used in the orbit determination of the Lagrangian navigation satellites, the nonlinear perturbations acting on Lagrangian navigation satellites are estimated by a neural network. A neural network based state observer is applied to autonomously determine the orbits of Lagrangian navigation satellites using only satellite-to-satellite range. This autonomous orbit determination method does not require linearizing the dynamical mode. There is no need to calculate the transition matrix. It is proved that three satellite-to-satellite ranges are needed using this method; therefore, the navigation constellation should include four Lagrangian navigation satellites at least. Four satellites orbiting on the collinear libration orbits are chosen to construct a constellation which is used to demonstrate the utility of this method. Simulation results illustrate that the stable error of autonomous orbit determination is about 10 m. The perturbation can be estimated by the neural network.

1. Introduction

The navigation of deep space probes is one of the main problems that restrict the deep space exploration. Generally, navigation support for deep space probes is primarily provided by the NASA's Deep Space Network (DSN). In order to improve the navigation performance by efficiently determining angular position of interplanetary spacecraft, the delta differential one-way tanging technique is also employed by some missions. Besides that, some autonomous navigation strategies are also proposed to support future deep space exploration missions. Autonomous navigation is important for deep space probes to deal with communication delay as well as reducing the dependency on ground stations. Several autonomous navigation methods have been proposed for deep space probes. As early as 1968, the sextant had been used for autonomous navigation in Apollo program [1]. In 1999, Deep Space 1 achieved autonomous orbit determination by tracking small celestial bodies with an optical sensor [2]. The comet probe "Deep Impact" carried out its navigation automatically based on an optical navigation system [3]. The rotation period of X-ray pulsar is extremely stable; therefore, time and the location of spacecraft have been proposed to be determined by tracking several X-ray pulsars in [4, 5]. The Global Positioning System (GPS) can support the navigation of deep space probes when they are orbiting in low Earth orbits and medium Earth orbits. For deep space transfer orbits and deep space target orbits, due to limited visibility, extremely low signal-to-noise ratio, and poor relative geometry among sources and users, GPS is not adequate. In 2005, Hill suggested placing navigation constellation on the periodic orbits in the vicinity of libration points of the Earth-Moon system to support deep space navigation [6]. Similar to GPS, a high-precision satellite navigation constellation which consists of libration point satellites in the Earth-Moon system is introduced to provide navigation information for deep space probes, which can be called, accordingly, the Lagrangian point satellite navigation system. The satellites which construct the Lagrangian point satellite navigation system are called Lagrangian navigation satellites. Zhang and Xu analyzed the architecture and navigation performance of the Lagrangian point satellite navigation system [7–9]. The Lagrangian navigation constellation is introduced to navigate the deep space probes autonomously. Hence the navigation constellation itself should have the ability of autonomous orbit determination (AOD). In [10], the feasibility of AOD for satellites in quasiperiodic orbits about the Earth-Moon libration point was verified. Based on circular restricted three-body problem (CRTBP), Du et al. researched the autonomous orbit determination method of satellites in halo orbits, and only satellite-to-satellite range was used as observation [11]. For the Earth navigation satellite constellation, there is a rank deficiency problem when only satellite-to-satellite range is used to determine the orbit. However, the rank deficiency problem does not exist for the Lagrangian navigation satellites because of the special dynamics near the libration points [6]. Thus, the Lagrangian navigation satellites in the navigation constellation can autonomously determine their orbits using only satellite-to-satellite range. In [12], Gao et al. discussed the algorithm of autonomous orbit determination using only the satellite-to-satellite range measurement for Lagrangian navigation constellation. The current studies about AOD of Lagrangian navigation satellites are under the CRTBP model. However, the motion of the Moon around the Earth is eccentric. Therefore, the elliptic restricted three-body problem (ERTBP) is more accurate to describe the motion of the Lagrangian navigation satellite [13]. The ERTBP has been discussed in detail in [14–16]. Our purpose here is to extend the applicability of ERTBP to the study of the AOD of Lagrangian navigation satellites. In this study, we consider ERTBP with perturbation. The perturbation is estimated using a neural network. Meanwhile an observer is designed to determine the orbit of Lagrangian satellite. We reference the design of a reduced-order modified state observer which is introduced in [17]. However, in [17], the authors assume that the position can be measured. In our study, we improve the observer which can estimate the state of Lagrangian satellite with only satellite-to-satellite range.

First the dynamical model of ERTBP with perturbation is introduced. Then we design a neural network based state observer to determine the orbit of Lagrangian satellites. Afterwards the stability of the observer is proved. Finally four Lagrangian satellites are chosen to validate the effectiveness of this AOD method.

2. Elliptic Restricted Three-Body Problem with Perturbation

As shown in Figure 1, P_1 and P_2 are the primaries in the three-body system. P_1 and P_2 are in elliptical orbits. P is the third body which is vanishingly small compared to the two primaries. Similar to the CRTBP, we study the motion of P in a rotating coordinates.

Let C-XYZ represent the barycentric synodic coordinate depicted in Figure 1; the x-axis of this frame is along the radius vector, which connects the primaries, positive in the direction pointing from P_1 to P_2. The y-axis of this frame is perpendicular to the x-axis, positive in the direction of the motion of P_2. The XY frame rotates with an angular velocity equal to the instant motion of P_2 with respect to P_1. The z-axis is perpendicular to the orbital plane of the primaries.

The equations of motion in ERTBP are dimensionless. The dimensionless units are as follows:

$$[M] = m_1 + m_2$$

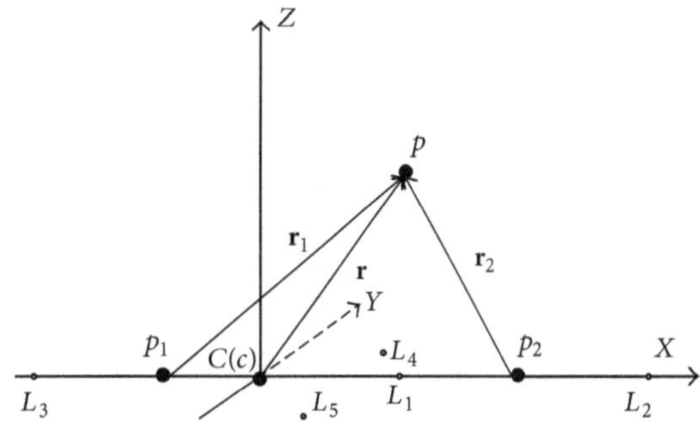

FIGURE 1: The barycenter synodic coordinates C-XYZ and the libration points.

$$[L] = \frac{a\left(1-e^2\right)}{1+e\cos f}$$

$$[T] = \left[\frac{L^3}{G\left(m_1+m_2\right)}\right]^{1/2} = \frac{\sqrt{1+e\cos f}}{\dot{f}}, \tag{1}$$

where m_1 and m_2 are the masses of the two primaries. a, e, respectively, refer to the semimajor axis and eccentricity of the two primaries' elliptical relative revolving orbit. f is the true anomaly of the secondary on the elliptic orbit. Then the motion of the Lagrangian satellite in the barycentric synodic coordinate system is governed by the following [18]:

$$\begin{aligned} X'' - 2Y' &= \frac{1}{1+e\cos f}\frac{\partial\Omega}{\partial X} \\ Y'' + 2X' &= \frac{1}{1+e\cos f}\frac{\partial\Omega}{\partial Y} \\ Z'' &= \frac{1}{1+e\cos f}\frac{\partial\Omega}{\partial Z}. \end{aligned} \tag{2}$$

f is taken as the time-like independent variable. The first and second derivatives of the coordinate with respect to f are calculated as

$$\begin{aligned} X' &= \frac{dX}{df}, \\ X'' &= \frac{d^2X}{df^2}. \end{aligned} \tag{3}$$

Ω is the pseudo-potential function of the three-body problem; it is described as follows:

$$\Omega = \frac{1}{2}\left[X^2+Y^2+Z^2+\mu\left(1-\mu\right)\right] + \frac{1-\mu}{r_1} + \frac{\mu}{r_2}, \tag{4}$$

where $\mu = m_2/(m_1+m_2)$, $r_1 = \sqrt{(X+\mu)^2+Y^2+Z^2}$, $r_2 = \sqrt{(X-1+\mu)^2+Y^2+Z^2}$.

In this study, only the motions around the collinear libration point L_1 or L_2 are investigated. Therefore the origin

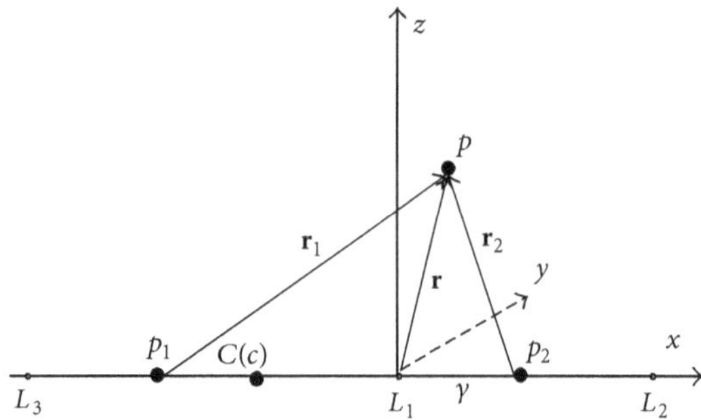

FIGURE 2: The L_1-centered synodic reference frame.

of the coordinate can be moved from the barycenter of system to the interested libration point for convenient analysis. As shown in Figure 2, the instantaneous distance between the libration point and its closest primary which is denoted as γ is chosen as the new length unit. The new reference frame is defined as the L_1- or L_2-centered synodic reference frame. The directions of the axes of this new coordinate system are coincident with the barycentric synodic reference frame.

Let (x, y, z, x', y', z') represent the state variables in L_1- or L_2-centered synodic reference frame. The transformation of coordinates between the barycentric synodic frame and the L_1- or L_2-centered synodic system is represented as follows [19]:

$$\begin{aligned} X &= \gamma (x \mp 1) + 1 - \mu \\ Y &= \gamma y \\ Z &= \gamma z, \end{aligned} \tag{5}$$

where the upper sign refers to the L_1 case and the lower one refers to the L_2 case. The linearized equations of motion in the L_1- or L_2-centered synodic system can be formulated as follows [20]:

$$\begin{aligned} x'' - 2y' - (1 + 2c_2)\, x &= \sum_{i\geq 1} \left[(-e)^i \cos^i f \,(1 + 2c_2)\, x \right] \\ y'' + 2x' - (1 - c_2)\, y &= \sum_{i\geq 1} \left[(-e)^i \cos^i f \,(1 - c_2)\, y \right] \\ z'' + c_2 z &= \sum_{i\geq 1} \left[(-e)^i \cos^i f \,(1 - c_2)\, z \right], \end{aligned} \tag{6}$$

where $c_2(\mu) = (1/\gamma_j^3)(\mu + (1-\mu)\gamma_j^2/(1 \mp \gamma_j)^3)$ and γ_j $(j = 1, 2)$ is the instantaneous distance between L_j and its closest primary.

We define a new variable $\overline{X} = [x, y, z, x', y', z']^T$; then (6) can be written as

$$\overline{X}' = A\overline{X}, \tag{7}$$

where

$$A = \begin{bmatrix} 0 & 0 & 0 & 1 & 0 & 0 \\ 0 & 0 & 0 & 0 & 1 & 0 \\ 0 & 0 & 0 & 0 & 0 & 1 \\ (1+2c_2)\left(1 + \sum_{i\geq 1}\left[(-e)^i \cos^i f\right]\right) & 0 & 0 & 0 & 2 & 0 \\ 0 & (1-c_2)\left(\sum_{i\geq 1}\left[(-e)^i \cos^i f\right] + 1\right) & 0 & -2 & 0 & 0 \\ 0 & 0 & 0 & 0 & 0 & -c_2 + \sum_{i\geq 1}\left[(-e)^i \cos^i f\,(1-c_2)\right] \end{bmatrix} \tag{8}$$

In addition to the gravitational force from the two primaries, several perturbations can affect the motion of Lagrangian satellite. When these perturbations are considered, the motion equation of the Lagrangian satellite can be described as

$$\overline{X}' = A\overline{X} + Bg\left(\overline{X}\right), \tag{9}$$

where

$$g\left(\overline{X}\right) = \begin{bmatrix} g_x \\ g_y \\ g_z \end{bmatrix} \tag{10}$$

is the nonlinear perturbing force on per unit mass. And

$$B = \begin{bmatrix} 0 & 0 & 0 \\ 0 & 0 & 0 \\ 0 & 0 & 0 \\ 1 & 0 & 0 \\ 0 & 1 & 0 \\ 0 & 0 & 1 \end{bmatrix}. \tag{11}$$

3. The Observation

In this paper, satellite-to-satellite range is the observation. The relationship between the observation and the state is

$$\overline{Y} = \rho\left(\overline{X},t\right) = \sqrt{\left(x-x_2\right)^2+\left(y-y_2\right)^2+\left(z-z_2\right)^2}, \quad (12)$$

where $[x\ y\ z]^T$ and $[x_2\ y_2\ z_2]^T$ are the position variable of Lagrangian satellite 1 and Lagrangian satellite 2, respectively, in L_1- or L_2-centered synodic reference frame.

Expanding (12) in a Taylor series with respect to estimated state yields

$$\begin{aligned}\overline{Y} &= \sqrt{\left(x-x_2\right)^2+\left(y-y_2\right)^2+\left(z-z_2\right)^2}\\ &= \hat{\overline{Y}} + \left.\frac{\partial \overline{Y}}{\partial x}\right|_{x=\hat{x}}(x-\hat{x}) + \left.\frac{\partial \overline{Y}}{\partial y}\right|_{y=\hat{y}}(y-\hat{y})\\ &\quad + \left.\frac{\partial \overline{Y}}{\partial z}\right|_{z=\hat{z}}(z-\hat{z}) + \cdots .\end{aligned} \quad (13)$$

We get the linearized relation between observation and states by neglecting the higher order terms.

$$\begin{aligned}\overline{Y} &= \hat{\overline{Y}} + \left.\frac{\partial \overline{Y}}{\partial x}\right|_{x=\hat{x}}(x-\hat{x}) + \left.\frac{\partial \overline{Y}}{\partial y_1}\right|_{y=\hat{y}}(y-\hat{y})\\ &\quad + \left.\frac{\partial \overline{Y}}{\partial z}\right|_{z=\hat{z}}(z-\hat{z})\\ &= \hat{\overline{Y}} + \frac{\hat{x}-x_2}{\rho}(x-\hat{x}) + \frac{\hat{y}-y_2}{\rho}(y-\hat{y})\\ &\quad + \frac{\hat{z}-z_2}{\rho}(z-\hat{z}).\end{aligned} \quad (14)$$

Now define the state estimate error as

$$\widetilde{X} = \overline{X} - \hat{\overline{X}}. \quad (15)$$

Further, the residual error of observation is defined as

$$\widetilde{Y} = \overline{Y} - \hat{\overline{Y}}. \quad (16)$$

Then we get

$$\widetilde{Y} = C\widetilde{X}, \quad (17)$$

where

$$C = \left[\frac{\hat{x}-x_2}{\rho}\ \ \frac{\hat{y}-y_2}{\rho}\ \ \frac{\hat{z}-z_2}{\rho}\ \ 0\ \ 0\ \ 0\right]. \quad (18)$$

Usually we only can get the estimated value of the position of Lagrangian satellite 2; thus the matrix C can be calculated by using the following:

$$C = \left[\frac{\hat{x}-\hat{x}_2}{\rho}\ \ \frac{\hat{y}-\hat{y}_2}{\rho}\ \ \frac{\hat{z}-\hat{z}_2}{\rho}\ \ 0\ \ 0\ \ 0\right]. \quad (19)$$

4. Design of the Adaptive Observer Based on Neural Network

The perturbation $g(\overline{X})$ in (9) can be estimated using a neural network over a compact set $D_{\overline{X}}$ as follows [17]:

$$g\left(\overline{X}\right) = W^T\phi\left(\overline{X}\right) + \varepsilon\left(\overline{X}\right) \quad \|\varepsilon\| \le \varepsilon^*,\ \overline{X} \in D_{\overline{X}}, \quad (20)$$

where W is a matrix of unknown neural network weights, $\phi(\overline{X})$ is a known vector of bounded basis functions, and ε^* is a uniform bound on the approximation error.

In order to estimate the state of Lagrangian satellite using the satellite-to-satellite range, an observer is designed as

$$\hat{\overline{X}}' = A\hat{\overline{X}} + B\left[\hat{g}\left(\hat{\overline{X}}\right) - v(f)\right] + K\widetilde{Y}, \quad (21)$$

where $v(f)$ is a robust term that will be determined later. f is the time-like independent variable. K is a user-specified gain matrix. $\hat{g}(\hat{\overline{X}})$ is the estimated uncertainty vector which is given by

$$\hat{g}\left(\hat{\overline{X}}\right) = \widehat{W}^T\phi\left(\hat{\overline{X}}\right). \quad (22)$$

A weight estimate update $\dot{\widehat{W}}$ is designed as

$$\dot{\widehat{W}} = F\left(\phi\left(\hat{\overline{X}}\right)\widetilde{Y}^T - \sigma\widehat{W}\right), \quad (23)$$

where F is a positive-definite symmetric matrices that will be determined later. And

$$v(f) = -D\frac{\widetilde{Y}}{\left\|\widetilde{Y}\right\|} - \varepsilon_{\max}\widetilde{Y}. \quad (24)$$

$\phi(\hat{\overline{X}})$ is calculated as follows [17]:

$$\phi\left(\hat{\overline{X}}\right) = \left[1\ \ \frac{\hat{x}}{R_\oplus}\ \ \frac{\hat{y}}{R_\oplus}\ \ \frac{\hat{z}}{R_\oplus}\ \ \frac{\hat{x}\hat{y}}{R_\oplus^2}\ \ \frac{\hat{x}\hat{z}}{R_\oplus^2}\ \ \frac{\hat{y}\hat{z}}{R_\oplus^2}\right]. \quad (25)$$

Now we have completed the design of the observer. However when this method is used to determine the orbits of the Lagrangian satellite in a navigation constellation, the number of the Lagrangian navigation satellites must satisfy a constraint condition. In the following, we will derive the minimum number of the Lagrangian navigation satellites which is needed in this method.

In order to prove that the designed observer in (21) can estimate the state of Lagrangian satellites accurately, we define the errors in the neural network weights and the basis vector ϕ as follows:

$$\begin{aligned}\widetilde{W} &= W - \widehat{W}\\ \widetilde{\phi} &= \phi\left(\overline{X}\right) - \phi\left(\hat{\overline{X}}\right).\end{aligned} \quad (26)$$

And a candidate Lyapunov functions is chosen as [17]

$$V = \frac{1}{2}\widetilde{X}^T P\widetilde{X} + \frac{1}{2}\mathrm{tr}\left[\widetilde{W}^T F^{-1}\widetilde{W}\right], \quad (27)$$

where P and F are positive-definite symmetric matrices. By differentiating (27), the Lyapunov function's derivative is illustrated as

$$\begin{aligned}\dot{V} &= \widetilde{X}^T P \dot{\widetilde{X}} + \operatorname{tr}\left[\widetilde{W}^T F^{-1} \dot{\widetilde{W}}\right] = \widetilde{X}^T P \left\{(A - KC)\widetilde{X}\right. \\ &\quad + B\left[W^T \phi(\overline{X}) - \widehat{W}^T \phi(\widehat{\overline{X}}) + v(f)\right] + B\varepsilon(\overline{X})\Big\} \\ &\quad + \operatorname{tr}\left[\widetilde{W}^T F^{-1} \dot{\widetilde{W}}\right].\end{aligned} \tag{28}$$

We define

$$A_{cl} = A - KC. \tag{29}$$

Then it yields

$$\begin{aligned}\dot{V} &= \widetilde{X}^T P \left\{A_{cl}\widetilde{X} + B\left[W^T \phi(\overline{X}) - W^T \phi(\widehat{\overline{X}})\right.\right. \\ &\quad + W^T \phi(\widehat{\overline{X}}) - \widehat{W}^T \phi(\widehat{\overline{X}}) + v(f)\Big] + B\varepsilon(\overline{X})\Big\} \\ &\quad + \operatorname{tr}\left[\widetilde{W}^T F^{-1} \dot{\widetilde{W}}\right] = \widetilde{X}^T P \left\{A_{cl}\widetilde{X} + B\left[W^T \widetilde{\phi}\right.\right. \\ &\quad + \widetilde{W}^T \phi(\widehat{\overline{X}}) + v(f) + \varepsilon(\overline{X})\Big]\Big\} + \operatorname{tr}\left[\widetilde{W}^T F^{-1} \dot{\widetilde{W}}\right] \\ &= \frac{1}{2}\widetilde{X}^T \left[PA_{cl} + A_{cl}^T P\right]\widetilde{X} + \widetilde{X}^T PB\left[W^T \widetilde{\phi}\right. \\ &\quad + \widetilde{W}^T \phi(\widehat{\overline{X}}) + v(f) + \varepsilon(\overline{X})\Big] + \operatorname{tr}\left[\widetilde{W}^T F^{-1} \dot{\widetilde{W}}\right].\end{aligned} \tag{30}$$

If $PB = C^T$, it will yield

$$\begin{aligned}\dot{V} &= \frac{1}{2}\widetilde{X}^T \left[PA_{cl} + A_{cl}^T P\right]\widetilde{X} + (C\widetilde{X})^T \left[W^T \widetilde{\phi}\right. \\ &\quad + \widetilde{W}^T \phi(\widehat{\overline{X}}) + v(f) + \varepsilon(\overline{X})\Big] + \operatorname{tr}\left[\widetilde{W}^T F^{-1} \dot{\widetilde{W}}\right] \\ &= \frac{1}{2}\widetilde{X}^T \left[PA_{cl} + A_{cl}^T P\right]\widetilde{X} \\ &\quad + \operatorname{tr}\left\{\left[W^T \widetilde{\phi} + \widetilde{W}^T \phi(\widehat{\overline{X}}) + v(f) + \varepsilon(\overline{X})\right]\widetilde{Y}^T\right. \\ &\quad + \left[\widetilde{W}^T F^{-1} \dot{\widetilde{W}}\right]\Big\}.\end{aligned} \tag{31}$$

Here the Meyer-Kalman-Yakubovich (MKY) lemma [17, 21] is used to derive $PB = C^T$. Noting that $P \sim 6 \times 6$, $B \sim 6 \times 3$, if $PB = C^T$, C should satisfy $C \sim 6 \times 3$. It means that three satellite-to-satellite ranges should be provided. In other words, the navigation constellation should include four Lagrangian satellites which can generate three satellite-to-satellite ranges.

If four Lagrangian satellites construct the navigation constellation, C can be derived as

$$C = \begin{bmatrix} \dfrac{\widehat{x} - \widehat{x}_2}{\rho_2} & \dfrac{\widehat{y} - \widehat{y}_2}{\rho_2} & \dfrac{\widehat{z} - \widehat{z}_2}{\rho_2} & 0 & 0 & 0 \\ \dfrac{\widehat{x} - \widehat{x}_3}{\rho_3} & \dfrac{\widehat{y} - \widehat{y}_3}{\rho_3} & \dfrac{\widehat{z} - \widehat{z}_3}{\rho_3} & 0 & 0 & 0 \\ \dfrac{\widehat{x} - \widehat{x}_4}{\rho_4} & \dfrac{\widehat{y} - \widehat{y}_4}{\rho_4} & \dfrac{\widehat{z} - \widehat{z}_4}{\rho_4} & 0 & 0 & 0 \end{bmatrix}, \tag{32}$$

where $[\widehat{x}_3\ \ \widehat{y}_3\ \ \widehat{z}_3]^T$ and $[\widehat{x}_4\ \ \widehat{y}_4\ \ \widehat{z}_4]^T$ are the estimated position variable of Lagrangian satellite 3 and Lagrangian satellite 4, respectively.

Define matrix Q as

$$-Q = PA_{cl} + A_{cl}^T P, \quad (Q > 0). \tag{33}$$

The Lyapunov function's derivative becomes

$$\begin{aligned}\dot{V} &= -\frac{1}{2}\widetilde{X}^T Q \widetilde{X} \\ &\quad + \operatorname{tr}\left\{\left[W^T \widetilde{\phi} + \widetilde{W}^T \phi(\widehat{\overline{X}}) + v(f) + \varepsilon(\overline{X})\right]\widetilde{Y}^T\right. \\ &\quad + \widetilde{W}^T F^{-1} \dot{\widetilde{W}}\Big\}.\end{aligned} \tag{34}$$

Since $\dot{\widetilde{W}} = -\dot{\widehat{W}}$ and $\dot{\widehat{W}} = F(\phi(\widehat{\overline{X}})\widetilde{Y}^T - \sigma\widehat{W})$,

$$\begin{aligned}\dot{V} &= -\frac{1}{2}\widetilde{X}^T Q \widetilde{X} + \operatorname{tr}\left\{\left[W^T \widetilde{\phi} + v(f) + \varepsilon(\overline{X})\right]\widetilde{Y}^T\right. \\ &\quad + \widetilde{W}^T \left[\phi(\widehat{\overline{X}})\widetilde{Y}^T - F^{-1}\left(F\left(\phi(\widehat{\overline{X}})\widetilde{Y}^T - \sigma\widehat{W}\right)\right)\right]\Big\} \\ &= -\frac{1}{2}\widetilde{X}^T Q \widetilde{X} + \operatorname{tr}\left(W^T \widetilde{\phi}\widetilde{Y}^T\right) - \operatorname{tr}\left(D\frac{\widetilde{Y}\widetilde{Y}^T}{\|\widetilde{Y}\|}\right) \\ &\quad - \operatorname{tr}\left(\varepsilon_{\max}\widetilde{Y}\widetilde{Y}^T\right) + \operatorname{tr}\left[\varepsilon(\overline{X})\widetilde{Y}^T\right] + \operatorname{tr}\left(\sigma\widetilde{W}^T\widehat{W}\right) \\ &= -\frac{1}{2}\widetilde{X}^T Q \widetilde{X} + \operatorname{tr}\left(W^T \widetilde{\phi}\widetilde{Y}^T\right) - \operatorname{tr}\left(D\|\widetilde{Y}\|\right) \\ &\quad - \operatorname{tr}\left(\varepsilon_{\max}\widetilde{Y}\widetilde{Y}^T\right) + \operatorname{tr}\left[\varepsilon(\overline{X})\widetilde{Y}^T\right] + \operatorname{tr}\left(\sigma\widetilde{W}^T\widehat{W}\right).\end{aligned} \tag{35}$$

From [17], it is noted that

$$\operatorname{tr}\left[W^T \widetilde{\phi}\widetilde{Y}^T\right] \le \alpha\|\widetilde{Y}\|. \tag{36}$$

We also can get

$$\begin{aligned}-\frac{1}{2}\widetilde{X}^T Q \widetilde{X} &\le -\lambda_{\min}(Q)\|\widetilde{X}\|^2 \le -\lambda_{\min}(Q)\|\widetilde{Y}\|^2 \\ \operatorname{tr}\left(\varepsilon_{\max}\widetilde{Y}\widetilde{Y}^T\right) &= \varepsilon_{\max}\|\widetilde{Y}\|^2 \\ \operatorname{tr}\left(\sigma\widetilde{W}^T\widehat{W}\right) &\le W_{\max}\|\widehat{W}\|_F - \|\widehat{W}\|_F^2.\end{aligned} \tag{37}$$

$\|\cdot\|_F$ is calculated as the Frobenius norm.

Substituting (36)-(37) into (35), we can get

$$\begin{aligned}\dot{V} &\le -(\lambda_{\min}(Q) + \varepsilon_{\max})\|\widetilde{Y}\|^2 + (\alpha - D + \varepsilon^*)\|\widetilde{Y}\| \\ &\quad + \sigma W_{\max}\|\widehat{W}\|_F - \sigma\|\widehat{W}\|_F^2 \\ &\le -(\lambda_{\min}(Q) + \varepsilon_{\max})\|\widetilde{Y}\|^2 + (\alpha - D + \varepsilon^*)\|\widetilde{Y}\| \\ &\quad - \sigma\left(\|\widehat{W}\|_F - \frac{W_{\max}}{2}\right)^2 + \sigma\frac{W_{\max}^2}{4}.\end{aligned} \tag{38}$$

Table 1: Initial state of the Lagrangian satellites.

Satellites	S1	S2	S3	S4
$x\,(L)$	−0.009158653890369	−0.008426021835493	−0.011852455058632	−0.012153563498421
$y\,(L)$	0.057117380302627	0.063116578748419	0.018916778218648	−0.000027358900568
$z\,(L)$	−0.005441690073741	−0.009538268525430	0.000050795432543	−0.001861700547480
$\dot{x}\,(L/T)$	0.071242540915778	0.077724037910168	0.043377272215699	0.030872891241568
$\dot{y}\,(L/T)$	0.025237253422237	0.044432703977482	0.002938265403217	−0.000128391119819
$\dot{z}\,(L/T)$	0.024768142966689	0.092125880718452	−0.000042323140915	0.011290787920073

Thus, from (38), $\dot{V} < 0$ when

$$\|\tilde{Y}\| \geq \frac{(\alpha - D + \varepsilon^*) + \sqrt{(\alpha - D + \varepsilon^*)^2 + (\lambda_{\min}(Q) + \varepsilon_{\max})\,\sigma W_{\max}^2}}{2(\lambda_{\min}(Q) + \varepsilon_{\max})}, \quad (39)$$

or

$$\|\widehat{W}\|_F \geq \frac{W_{\max} + \sqrt{W_{\max}^2 + (\alpha - D + \varepsilon^*)^2 / \sigma(\lambda_{\min}(Q) + \varepsilon_{\max})}}{2}. \quad (40)$$

Note that parameters D, $\varepsilon_{\max}$, σ are user selected so they can be appropriately chosen to minimize the estimation error.

5. Simulation and Analysis

In order to illustrate the effectiveness of the method introduced in this paper, we design a constellation around the Earth-Moon L1 libration point which includes four Lagrangian satellites. The initial states of the four satellites are listed in Table 1. The data are normalized. Quasiperiodic orbits are sensitive to initial states. Thus the initial states must be accurate enough. This is why the data in Table 1 retain many valid digits after the dot. We calculate quasiperiodic orbits with different initial states. The conditions in Table 1 may be different from the actual task situation. Therefore the initial conditions and measurement noise here are assumed conditions based on the characteristics of CRTBP.

Four quasiperiodic orbits are shown in Figure 3. The ephemeris of the Moon and Earth is obtained by DE405. In order to illustrate the AOD error, the “true trajectory” of Lagrangian satellites should be provided. In this paper, the “true trajectory” is calculated by using the precise dynamical equations as follows:

$$\ddot{\mathbf{R}} = -\frac{\mu_3}{R^3}\mathbf{R} - \sum_{i=1}^{n}\mu_i\left(\frac{\mathbf{\Delta}_i}{\Delta_i^3} + \frac{\mathbf{R}_i}{R_i^3}\right), \quad (41)$$

where $\mathbf{R}$ is the position vector of the Lagrangian satellite in the J2000 coordinate. $\mathbf{\Delta}_i$ is the position vector of the ith celestial with respect to the Moon. $\mathbf{R}_i$ is the position vector of the ith celestial with respect to the Earth. μ_i is the gravitational parameter of ith celestial. The quasiperiodic orbits are calculated by using multiple shooting method. The initial targets are obtained by using approximate analytic solutions. The initial targets are expressed in the barycentric synodic coordinate. Therefore, we should translate the initial targets into J2000 coordinate. Then the initial target is modified according to the task constraint, and the ideal quasiperiodic trajectory is obtained. Since the Lagrangian satellites in the Earth-Moon system is far away from the Earth, the perturbation caused by atmospheric resistance, temporal change in gravity, oblateness, and mass distribution is not considered. The main perturbations come from the gravity of the Sun and other planets. Therefore, the above equation is appropriate to describe the precise motion of the Lagrangian satellites.

Table 2: The simulation conditions.

	Initial state error (m)	Observation error (m)
Scenario 1	1	1
Scenario 2	10	1
Scenario 3	100	1

The various tuning constants used in this scenario were chosen as follows:

$$K = \begin{bmatrix} 1.5\times10^{-3} & 0 & 0 \\ 0 & 1.5\times10^{-3} & 0 \\ 0 & 0 & 1.5\times10^{-3} \\ 6.5\times10^{-4} & 0 & 0 \\ 0 & 6.5\times10^{-4} & 0 \\ 0 & 0 & 6.5\times10^{-4} \end{bmatrix} \quad (42)$$

$$F = \operatorname{diag}\left(1.5\times10^{-5}, 3.2\times10^{-5}, 1.5\times10^{-5}, 3.2\times10^{-5}, 2.25\times10^{-5}, 1.5\times10^{-5}, 1.5\times10^{-5}\right).$$

The parameters are designed as $\varepsilon_{\max} = 1\times10^{-12}$, $D = 1\times10^{-9}$. The step of AOD is 1 hour. Three scenarios with different initial state errors and observation errors as shown in Table 2 are analyzed. The gravitational attraction from the Sun is the only perturbation considered in the simulation.

For scenario 1, we give the simulation results of two Lagrangian satellites S1 and S2. The AOD errors in 30 days are presented in Figures 4(a) and 4(b). It can be seen that the AOD error of the both Lagrangian satellites is less than 10 m. The RMS of the AOD error of S1 in x-, y-, and z-axis are 2.3 m, 2.6 m, and 2.1 m, respectively. From Figures 5(a) and 5(b), we can see that this method can estimate the perturbation with

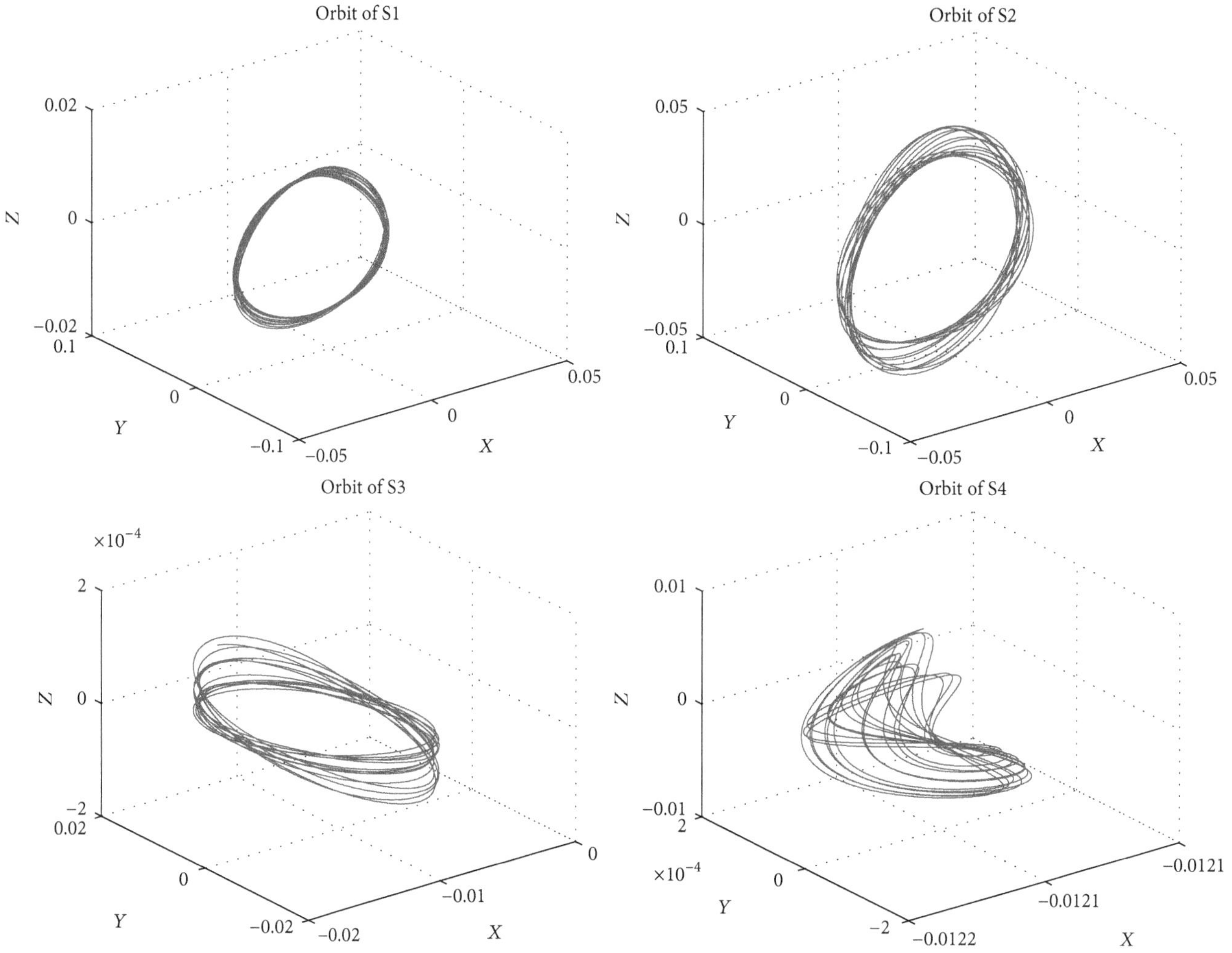

FIGURE 3: The quasiperiodic orbits around the L1 libration.

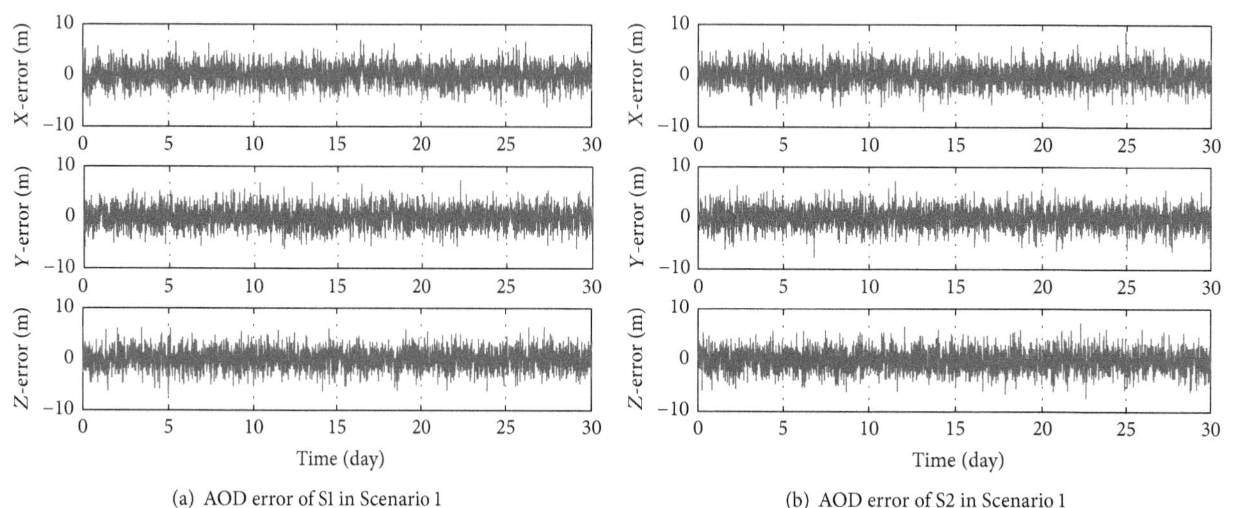

(a) AOD error of S1 in Scenario 1

(b) AOD error of S2 in Scenario 1

FIGURE 4

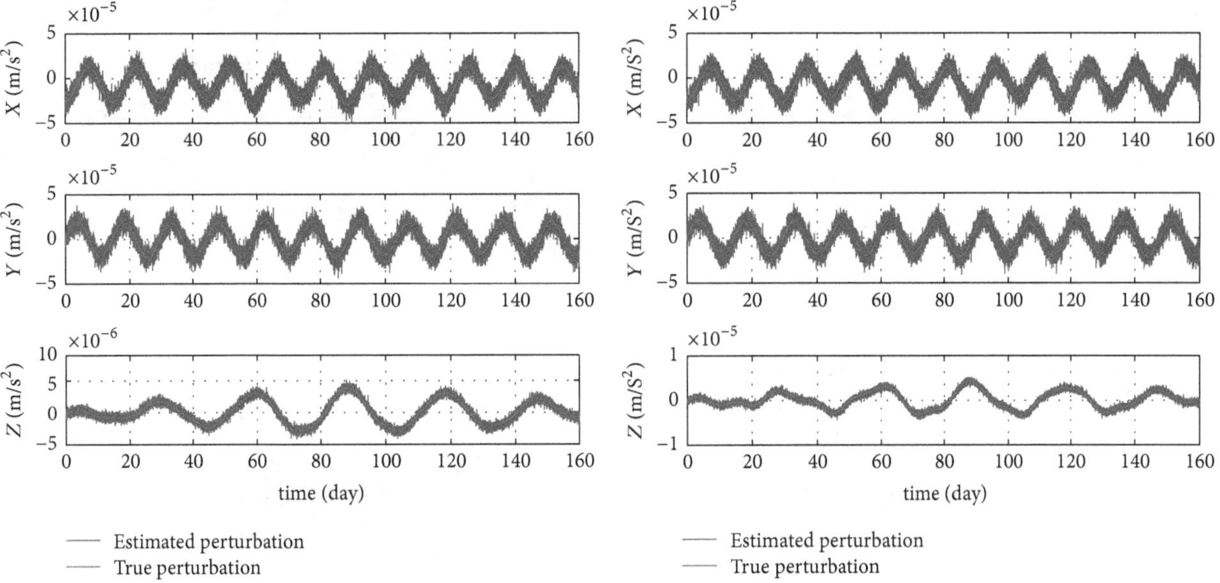

(a) True perturbation and the estimated perturbation of S1 in Scenario 1 (b) True perturbation and the estimated perturbation of S2 in Scenario 1

FIGURE 5

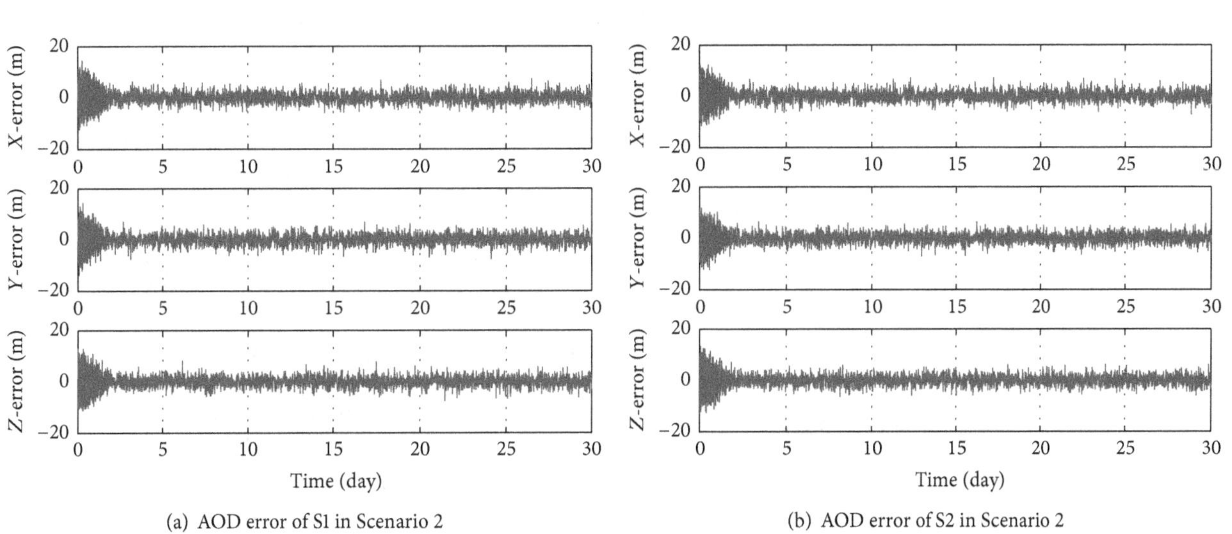

(a) AOD error of S1 in Scenario 2 (b) AOD error of S2 in Scenario 2

FIGURE 6

small errors. The red curves are the theoretical calculating value of the Sun gravitational perturbation. The blue curves are the estimation of the Sun's gravity. The RMS of the estimate error of perturbation of S1 in x-, y-, and z-axis are 6.1×10^{-6} m/s^2, 5.9×10^{-6} m/s^2, and 4.3×10^{-7} m/s^2.

In Figures 6(a), 6(b), 7(a), and 7(b), we give the results for Scenario 2 and Scenario 3. As we can see, the maximum AOD errors increase with the initial state errors. All the stable AOD errors are less than 10 m. As shown in Figure 7(a), the AOD errors can converge to less than 10 m in 5 days, even the initial errors 100 m.

6. Conclusions

We analyzed the AOD for Lagrangian navigation constellation based on accurate dynamical mode. A neural network based method is applied to estimate the solar gravitational perturbation and the state of the Lagrangian satellites. The update rate of neural network weights is obtained by convergence of Lyapunov function. Therefore, the stability of the method introduced in this paper is proven. We derived that the constellation must include more than four satellites when this method is used. The simulation results show that this method can achieve a stable AOD error about 10 m only using satellite-to-satellite range as observation. Since the CRTBP is an approximate model for the Earth-Moon system, the AOD based on perturbed ERTBP is much useful for the future Lagrangian navigation constellation. We only test a Lagrangian navigation constellation with four satellites around L1 libration in this paper. Future work will be focused on a more complex constellation with satellites around different librations.

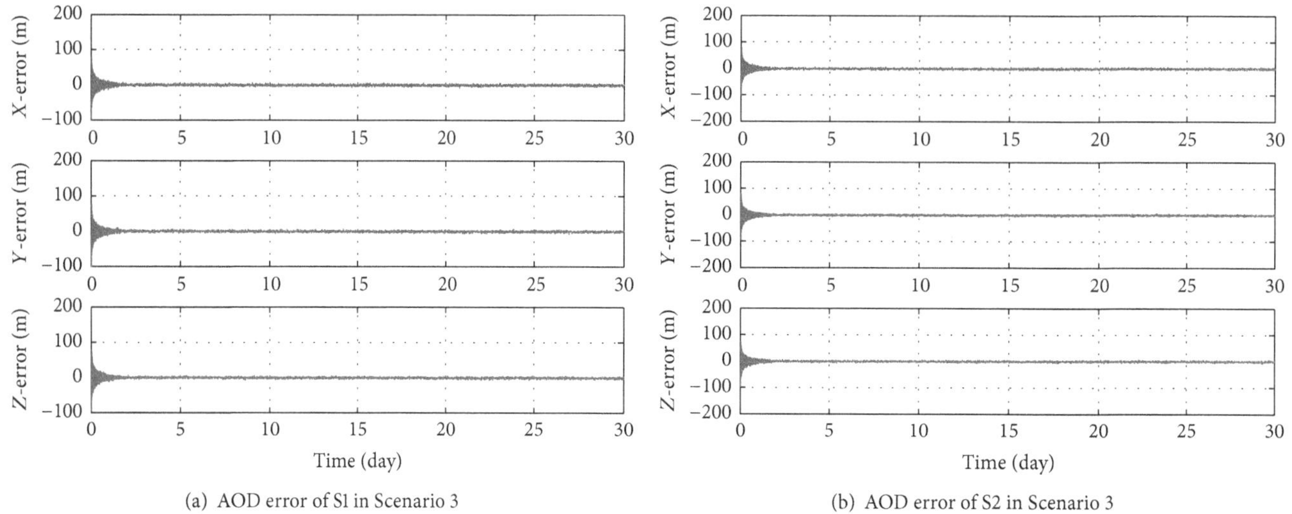

(a) AOD error of S1 in Scenario 3

(b) AOD error of S2 in Scenario 3

FIGURE 7

Conflicts of Interest

The authors declare that there are no conflicts of interest regarding the publication of this paper.

Acknowledgments

This study was supported by the Basic Scientific Research Fund of National Defense (no. 2016110C019), the Natural Science Foundation of Jiangsu Province (no. BK20160811), and Shanghai Deep Space Detection Technology Key Laboratory Open Funding (no. DS2016-01).

References

[1] R. H. Battin, *An Introduction to the Mathematics and Methods of Astrodynamics*, American Institute of Aeronautics and Astronautics, Reston, VA, USA, 1999.

[2] S. Bhaskaran, J. E. Riedel, S. P. Synnott, and T. C. Wang, "The deep space 1 autonomous navigation system: a post-flight analysis," in *Proceedings of the AIAA/AAS Astrodynamics Specialist Conference*, AIAA-2000-3935, Denver, CO, USA, 2000.

[3] N. Mastrodemos, D. G. Kubitschek, and S. P. Synnott, "Autonomous navigation for the deep impact mission encounter with comet tempel 1," *Space Science Reviews*, vol. 117, no. 1, pp. 95–121, 2005.

[4] G. S. Downs, "Interplanetary Navigation Using Pulsating Radio Sources," Tech. Rep., NASA Technical Reports, 1974, N74-34150/4.

[5] S. I. Sheikh, D. J. Pines, P. S. Ray, K. S. Wood, M. N. Lovellette, and M. T. Wolff, "Spacecraft navigation using X-ray pulsars," *Journal of Guidance, Control, and Dynamics*, vol. 29, no. 1, pp. 49–63, 2006.

[6] K. A. Hill, *Autonomous Navigation in Libration Point Orbits [Ph.D. thesis]*, University of Colorado, Colorado, USA, 2007, Ph.D. dissertation.

[7] L. Zhang and B. Xu, "A universe light house—candidate architectures of the libration point satellite navigation system," *Journal of Navigation*, vol. 67, no. 5, pp. 737–752, 2014.

[8] L. Zhang and B. Xu, "Navigation performance of the libration point satellite navigation system in cislunar space," *Journal of Navigation*, vol. 68, no. 2, pp. 367–382, 2015.

[9] L. Zhang and B. Xu, "Navigation performance of the libration point satellite navigation system for future Mars exploration," *Journal of Navigation*, vol. 69, no. 1, pp. 41–56, 2016.

[10] Y.-J. Qian, W.-X. Jing, C.-S. Gao, and W.-S. Wei, "Autonomous orbit determination for quasi-periodic orbit about the translunar libration point," *Journal of Astronautics*, vol. 34, no. 5, pp. 625–633, 2013.

[11] L. Du, Z. K. Zhang, L. Yu, and S. J. Chen, "SST orbit determination of halo-lmo constellation in CRTBP," *Acta Geodaetica et Cartographica Sinica*, pp. 184–190, 2013.

[12] Y. T. Gao, B. Xu, and L. Zhang, "Feasibility study of autonomous orbit determination using only the crosslink range measurement for a combined navigation constellation," *Chinese Journal of Aeronautics*, vol. 27, no. 5, pp. 1199–1210, 2014.

[13] L. Liu and X. Y. Hou, *Deep Space Probe Orbit Mechanics*, Publishing House of Electronics Industry, Beijing, China, 2012.

[14] M. Ollé and J. R. Pacha, "The 3D elliptic restricted three-body problem: periodic orbits which bifurcate from limiting restricted problems," *Astronomy and Astrophysics*, vol. 351, pp. 1149–1164, 1999.

[15] J. M. Cors, C. Pinyol, and J. Soler, "Periodic solutions in the spatial elliptic restricted three-body problem," *Physica D: Nonlinear Phenomena*, vol. 154, no. 3-4, pp. 195–206, 2001.

[16] J. Singh and A. Umar, "Motion in the photogravitational elliptic restricted three-body problem under an oblate primary," *The Astronomical Journal*, vol. 143, no. 5, article 109, 22 pages, 2012.

[17] N. Harl, K. Rajagopal, and S. N. Balakrishnan, "Neural network based modified state observer for orbit uncertainty estimation," *Journal of Guidance, Control, and Dynamics*, vol. 36, no. 4, pp. 1194–1209, 2013.

[18] V. Szebehely, *Theory of Orbits*, Academic Press, New York, USA, 1967.

[19] H. Lei, B. Xu, X. Hou, and Y. Sun, "High-order solutions of invariant manifolds associated with libration point orbits in the elliptic restricted three-body system," *Celestial Mechanics and Dynamical Astronomy*, vol. 117, no. 4, pp. 349–384, 2013.

[20] X. Y. Hou and L. Liu, "On motions around the collinear libration points in the elliptic restricted three-body problem," *Monthly Notices of the Royal Astronomical Society*, vol. 415, no. 4, pp. 3552–3560, 2011.

[21] P. Ioannou and J. Sun, *Robust Adaptive Control*, Prentice-Hall, Englewood Cliffs, NJ, USA, 1996.

Permissions

All chapters in this book were first published in IJAE, by Hindawi Publishing Corporation; hereby published with permission under the Creative Commons Attribution License or equivalent. Every chapter published in this book has been scrutinized by our experts. Their significance has been extensively debated. The topics covered herein carry significant findings which will fuel the growth of the discipline. They may even be implemented as practical applications or may be referred to as a beginning point for another development.

The contributors of this book come from diverse backgrounds, making this book a truly international effort. This book will bring forth new frontiers with its revolutionizing research information and detailed analysis of the nascent developments around the world.

We would like to thank all the contributing authors for lending their expertise to make the book truly unique. They have played a crucial role in the development of this book. Without their invaluable contributions this book wouldn't have been possible. They have made vital efforts to compile up to date information on the varied aspects of this subject to make this book a valuable addition to the collection of many professionals and students.

This book was conceptualized with the vision of imparting up-to-date information and advanced data in this field. To ensure the same, a matchless editorial board was set up. Every individual on the board went through rigorous rounds of assessment to prove their worth. After which they invested a large part of their time researching and compiling the most relevant data for our readers.

The editorial board has been involved in producing this book since its inception. They have spent rigorous hours researching and exploring the diverse topics which have resulted in the successful publishing of this book. They have passed on their knowledge of decades through this book. To expedite this challenging task, the publisher supported the team at every step. A small team of assistant editors was also appointed to further simplify the editing procedure and attain best results for the readers.

Apart from the editorial board, the designing team has also invested a significant amount of their time in understanding the subject and creating the most relevant covers. They scrutinized every image to scout for the most suitable representation of the subject and create an appropriate cover for the book.

The publishing team has been an ardent support to the editorial, designing and production team. Their endless efforts to recruit the best for this project, has resulted in the accomplishment of this book. They are a veteran in the field of academics and their pool of knowledge is as vast as their experience in printing. Their expertise and guidance has proved useful at every step. Their uncompromising quality standards have made this book an exceptional effort. Their encouragement from time to time has been an inspiration for everyone.

The publisher and the editorial board hope that this book will prove to be a valuable piece of knowledge for researchers, students, practitioners and scholars across the globe.

List of Contributors

Liang Sun, Guowei Zhao, Hai Huang and Ming Chen
Beihang University, Beijing 100191, China

Guangming Dai, Xiaoyu Chen, Mingcheng Zuo, Lei Peng, Maocai Wang and Zhiming Song
School of Computer Science, Hubei Key Laboratory of Intelligent Geo-Information Processing, China University of Geosciences, Lumo Road, Wuhan 430074, China

Dateng Yu, Hua Wang and Shuai Guo
College of Aerospace Science and Engineering, National University of Defense Technology, Changsha, China

Hongyu Wang
Aerospace System Engineering Shanghai, Shanghai 200000, China

Seong-Cheol Kwon and Hyun-Ung Oh
Space Technology Synthesis Laboratory, Department of Aerospace Engineering, Chosun University, 375 Seosuk-dong, Dong-gu, Gwangju 501-759, Republic of Korea

Mun-Shin Jo
Mechanical Design Team, Hanwha Systems, 304 Cheoin-gu, Yongin 449-886, Republic of Korea

Deok-Jin Lee
School of Mechanical & Automotive Engineering, Kunsan National University, Gunsan 54150, Republic of Korea

Eun Jung Choi, Sungki Cho and Jung-Hyun Jo
Center for Space Situational Awareness, Korea Astronomy and Space Science Institute, Daejeon 305-348, Republic of Korea

Tae Soo No
Department of Aerospace Engineering, Chonbuk National University, Jeonju 54896, Republic of Korea

Davide Viganò and Filippo Maggi
Department of Aerospace Science and Technology, SPLab, Politecnico di Milano, 20156 Milan, Italy

Adriano Annovazzi
Space Propulsion Design Department, AVIO S.p.A., 00034 Colleferro, Italy

Haojie Liu, Yonghui Zhao and Haiyan Hu
State Key Laboratory of Mechanics and Control of Mechanical Structures, Nanjing University of Aeronautics and Astronautics, Nanjing 210016, China

Yunfeng Dong, Xiaona Wei and Fengrui Liu
School of Astronautics, Beihang University, Beijing 100191, China

Lu Tian
Department of Satellite Application System, North China Institute of Computing Technology, Beijing 100191, China

Guangde Xu
Institute of Manned Space System Engineering, China Academy of Space Technology, Beijing 100094, China

Qun Fang, Xuefeng Wang, Chong Sun and Jianping Yuan
National Key Laboratory of Aerospace Flight Dynamics, Xi'an, Shaanxi 710072, China
School of Astronautics, Northwestern Polytechnical University, Xi'an, Shaanxi 710072, China

Bingshan Hu and Hongliu Yu
University of Shanghai for Science and Technology, Shanghai 200093, China
Shanghai Engineering Research Center for Assistive Devices, Shanghai 200093, China

Feng Chen, Liangliang Han and Huanlong Chen
Shanghai Institute of Aerospace Systems Engineering, Shanghai 201109, China

Jin Choi and Jung Hyun Jo
Korea Astronomy and Space Science Institute, Daejeon 34055, Republic of Korea
University of Science and Technology, Daejeon 34113, Republic of Korea

Kyung-Min Roh, Hong-Suh Yim, Eun-Jung Choi and Sungki Cho
Korea Astronomy and Space Science Institute, Daejeon 34055, Republic of Korea
University of Science and Technology, Daejeon 34113, Republic of Korea

Burak Yuksek and N. Kemal Ure
Aerospace Research Center, Department of Aeronautical Engineering, Istanbul Technical University, 34469 Istanbul, Turkey

Muzi Li
School of Astronomy and Space Science, Nanjing University, Nanjing 210023, China

Bo Xu
School of Aeronautics and Astronautics, Sun Yat-Sen University, Guangzhou 510006, China

Jun Sun
Shanghai Aerospace Control Technology Institute, Shanghai 201199, China

Mingying Huo, Yanfang Liu and Naiming Qi
Department of Aerospace Engineering, Harbin Institute of Technology, Harbin 150001, China

He Liao
Shanghai Satellite Engineering Research Institute, Shanghai 200240, China

Lorenzo Federici, Alessandro Zavoli and Guido Colasurdo
Department of Mechanical and Aerospace Engineering, Sapienza-University of Rome, Via Eudossiana 18, Rome, Italy

Ling Jiang, Yue Wang and Shijie Xu
School of Astronautics, Beihang University, Beijing 100191, China

Youtao Gao, Tanran Zhao, Bingyu Jin and Junkang Chen
College of Astronautics, Nanjing University of Aeronautics and Astronautics, Nanjing, China

Bo Xu
College of Astronomy and Space Science, Nanjing University, Nanjing, China

Index

CPSIA information can be obtained
at www.ICGtesting.com
Printed in the USA
BVHW010145270322
632559BV00003B/72